T0331125

APPLIED THERMODYNAMICS FOR METEOROLOGISTS

This textbook on atmospheric thermodynamics is for students of meteorology or atmospheric science. It also serves as a reference text for working professionals in meteorology and weather forecasting. It is unique because it provides complete, calculus-based derivations of basic physics from first principles and connects mathematical relationships to real-world, practical weather forecasting applications. Worked examples and practice problems are included throughout.

Sam Miller began his career in meteorology as a weather observer in the U.S. Air Force (USAF) in 1982. In 1984 he attended the weather forecaster school at Chanute Air Force Base, Illinois, and graduated with honors. He worked as a USAF weather observer in Maine, and as a forecaster in northern California, upstate New York, and Adana, Turkiye. He eventually earned the rank of Technical Sergeant and left the USAF in 1989 after more than ten years on active duty. Miller then attended the University of New Hampshire and earned a Bachelor of Science in Physics (1996), a Master of Science in Earth Sciences: Oceanography (1999), and a PhD in Earth Sciences (2003), while also working as a weather observer in Portsmouth, New Hampshire, a research scientist at the university, and a teacher in New Hampshire's public university system. From 2003 to 2005, Miller was a weather forecaster with the U.S. National Weather Service in Anchorage, Alaska. Since 2005 he has served as a professor of meteorology at Plymouth State University, New Hampshire, where he teaches courses in basic meteorological analysis, atmospheric thermodynamics, instrumentation, weather forecasting, satellite meteorology, and radar meteorology. Miller is a member of Phi Beta Kappa, the American Meteorological Society, the Royal Meteorological Society, and many other professional organizations. He has published research papers on the sea breeze and has worked extensively as a meteorological consultant in legal matters.

APPLIED THERMODYNAMICS FOR METEOROLOGISTS

SAM MILLER

Plymouth State University

CAMBRIDGE
UNIVERSITY PRESS

Shaftesbury Road, Cambridge CB2 8EA, United Kingdom

One Liberty Plaza, 20th Floor, New York, NY 10006, USA

477 Williamstown Road, Port Melbourne, VIC 3207, Australia

314–321, 3rd Floor, Plot 3, Splendor Forum, Jasola District Centre, New Delhi – 110025, India

103 Penang Road, #05–06/07, Visioncrest Commercial, Singapore 238467

Cambridge University Press is part of Cambridge University Press & Assessment, a department of the University of Cambridge.

We share the University's mission to contribute to society through the pursuit of education, learning and research at the highest international levels of excellence.

www.cambridge.org
Information on this title: www.cambridge.org/9781107100718

First published 2015

A catalogue record for this publication is available from the British Library

Library of Congress Cataloging-in-Publication data
Miller, Sam, 1961– author.
Applied thermodynamics for meteorologists / Sam Miller, Plymouth State University.
pages   cm
Includes bibliographical references and index.
ISBN 978-1-107-10071-8 (hardback)
1. Atmospheric thermodynamics.   I . Title.
QC880.4.T5M55   2015
536′.7–dc23   2015002801

ISBN   978-1-107-10071-8   Hardback

This book is dedicated to weather forecasters everywhere.

Contents

Acknowledgments

I want to sincerely thank:

- My wife Virginia and daughter Julia, for patiently indulging me while I invested more than two years of my life in this book, and my parents Catherine Carter-Hancock and Edward J. Miller, who believed in me, even when the determination to do so was justifiably viewed with suspicion.
- The many fine instructors who taught me at the U.S. Air Force's (USAF's) Weather Observer and Weather Forecaster Schools at Chanute Air Force Base, Illinois, in 1982 and 1984, such as Dr. Glenn Van Knowe (he was a technical sergeant back then), *and* the professors and staff at the University of New Hampshire, who helped me thrash through degrees in physics, oceanography, and Earth sciences, between 1989 and 2003.
- Master Sergeant Brian Hammond, who was my boss at two USAF weather stations, and Lieutenant Colonel R. Bruce Telfeyan, who was my boss at a third weather station. Both of these gentlemen provided outstanding examples of careful, conscientious operational meteorologists who took time to mentor youngsters.
- Ms. Özlem Bilgin, Mr. Zeki Çelikbaş, Dr. Ahmet Öztopal, and, *most of all*, Dr. Mikdat Kadıoğlu of Istanbul Technical University (ITU), who helped me immensely during my sabbatical leave in Turkiye in the spring of 2012, by providing me with an office and technical support in the ITU Meteorological Engineering Department, where I wrote the first draft of this book.
- ITU's Dr. Aydın Mısırlıoğlu, Dr. Sevinç Sırdaş, Dr. Ali Deniz, Dr. Selahattin İncelik, Dr. Yurdanur Ünal, Dr. Melike Nikbay, Dr. Barış Önol, Dr. Zerefşan Kaymaz, Dr. Levent Şaylan, and Mr. Mehmet Ünal for their kindness and help.
- My colleagues in Plymouth State University's meteorology program, Dr. Jim Koermer (who also served on my PhD committee; retired now), Dr. Eric Hoffman, Dr. Lourdes Aviles, Dr. Eric Kelsey, Dr. Jason Cordeira, Dr. Joe Zabransky (retired), Dr. Lisa Doner, Mr. Anthony (Toby) Fusco, and Mr. Brendon Hoch. The majority of the content of this book began as the notes I use in the thermodynamics course they allow me to teach our undergraduates.
- Ms. Marsi Wisniewski, who helped keep me on the linguistic straight and narrow. Dr. Tom Boucher of Plymouth State University's Mathematics Department checked my math in a few places. Dr. Dennis Machnik checked some of my physics in the first few chapters.

- Four anonymous reviewers, who read six chapters of this text and made several very valuable suggestions, and my editor at Cambridge University Press, who helped me through the publication process.

And most of all, my students at Plymouth State University, who – through their enthusiasm, intelligence, and sense of humor – remind me of why teachers do this kind of work.

1

Basic Concepts and Terminology

1.1. What Is Thermodynamics?

Thermodynamics is the study of energy and its transformations. Several books on this subject begin with that line, or something very much like it, and I can't think of a better way to start this one. In my experiences as a student, teacher, and user of thermodynamics, I've read several of these books, each one directed to a slightly different audience. The purpose of *this* book is to provide undergraduate meteorology students with a solid theoretical (physical and mathematical) basis for understanding "energy and its transformations" in the Earth's atmosphere, and an appreciation for both the *limitations* and the *practical usefulness* of the thermodynamic models we use to describe the atmosphere. If, at the end of this book, you know where these ideas came from, what their weaknesses are, and how to apply them to your job as a working meteorologist, then you have learned what I hoped you would learn.

It's worth saying a little more about the word *model*. Meteorologists are physical scientists, and as such we like to use equations (models) to describe what's happening in the Earth system. This is pragmatic, because it makes it possible to do quantitative research and make weather forecasts. But some scientists forget about the differences between their models and the real objects they're studying, and this is a mistake. A model (i.e., the physics we use) gives us a glimpse at an underlying reality, but the real atmosphere is much more complex. For example, there are no real "isolated parcels," that is, there are no little packets of air that do not exchange mass or heat with their surrounding environment. This idea is just an approximation that makes it possible for us to solve the equations and make weather forecasts. Over some short period of time, it isn't a *bad* approximation, but it's still only a simplified picture of the real atmosphere. Remember that when you're using these equations.

Thermodynamics evolved into a well-integrated science from several separate threads over the last three centuries. Scholars from many different parts of the world, and highly varied scientific disciplines, have contributed to its evolution. In addition to this text, I'd like to recommend that you read von Baeyer's fun and interesting *Warmth Disperses and Time Passes*,[1] which describes this history very well. One of the points that von Baeyer mentions early in his book is the mistaken belief, once widely held, that heat is some kind of fluid that "flows" through matter. This fluid – called *caloric* – was thought to come from the spaces between the atoms of a substance, and could be "squeezed out" under the right circumstances. We now know this is wrong, but you'll find that many of the ideas we use to describe heat are the same ideas we use to describe fluid flow. We still use the idea of heat as a fluid, although we know it's only an analogy, because it works very well under *some* circumstances.

1.2. Systems

The fundamental model underlying practical atmospheric thermodynamics breaks the atmosphere up into components. The first component, called a *system*, is an object, such as a quantity of matter consisting of many particles, which take up a volume of space (see Figure 1.1). It is separated from its *surroundings* by a *boundary*, which is an imaginary barrier that may or may not allow mixing between the system and the surroundings. In meteorology, we use the word *parcel* as a substitute for "system," and *environment* or *ambient* (or *background*) *air* as a substitute for "surroundings." This model assumes that changes can occur to a system completely independent of its surroundings, or as a result of some forcing *by* the surroundings, but that these internal changes have no important effect *on* the surroundings.

The parcel may be *open*, *closed*, or *isolated*.

- In an open parcel, the boundary permits the exchange of both energy and matter between the parcel and its environment.[2]
- In a closed parcel, the boundary permits the exchange of energy but prevents the exchange of matter.
- In an isolated parcel, the boundary prevents the exchange of both matter and energy with the environment.

Open parcels (exchanging both matter and energy with the environment) can often be treated as closed parcels (exchanging energy only) by considering the surface-to-volume ratios of small and large spheres. Recall that the surface area (A) of a sphere is given by:

$$A = 4\pi r^2 \tag{1.1}$$

where r is the radius of the sphere. The volume (V) of a sphere is given by:

$$V = \frac{4}{3}\pi r^3 \tag{1.2}$$

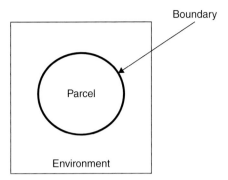

Figure 1.1. Parcel, environment, and boundary.

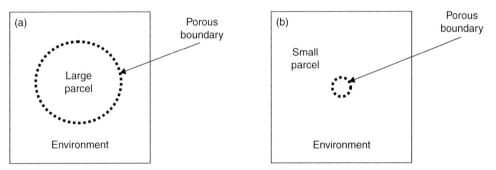

Figure 1.2. (a) Large and (b) small parcels with porous boundaries. Changes to the internal volume of the large parcel occur slowly, because of the relatively small surface area enclosing it. Changes to the internal volume of the small parcel are assumed to have already occurred.

The surface-to-volume ratio is therefore:

$$\frac{A}{V} = \frac{4\pi r^2}{\frac{4}{3}\pi r^3} = \frac{3}{r} \tag{1.3}$$

which gets smaller as the radius of the sphere increases. As the volume increases (scaled to r^3), the surface area (scaled to r^2) enclosing it doesn't increase as quickly. Because exchanges between a parcel and its environment occur through the boundary, (1.3) implies that there is a diminishing effect on the internal volume of a parcel as its radius increases.

We can treat open parcels as closed parcels by making the following simplifying assumptions (see Figure 1.2):

- For *small* parcels, the surface-to-volume ratio is *large*, and we can assume that any changes to the internal volume of the parcel that are going to occur *have already done so*.

- For *large* parcels, the surface-to-volume ratio is *small*, and we can assume that the rate of change in the internal volume of the parcel is so small that it can be ignored for short periods of time.

Two questions ought to immediately pop into your mind. The first is "what about middle-sized parcels?" The answer to this one is not very satisfying, but we'll use it anyway: there are no "middle-size" parcels in our model atmosphere. The second question is "who cares?" *Why is this important?* To answer that question, we have to talk about energy and equilibrium.

1.3. Energy and Equilibrium

Energy is an abstract idea for something that appears in nature in many different forms. In practical terms, it is often described as the ability to do *work*, and so energy and work share the same physical units (*Joules*, in the *Système International d'Unités* or SI system). Energy can be divided into categories such as *external* and *internal*. In thermodynamic terms, "external energy" is the energy some object has in relation to its surroundings, such as the energy of position (known as *potential energy*), or the energy of motion (*kinetic energy*). "Internal energy" is determined using information about the internal *state* of the system. The "state" is the collection of physical properties describing the system, such as its temperature.

The internal state of a system may undergo changes, such as phase changes between solids, liquids, or gases, but, if left undisturbed, eventually all of these changes will cease, and the system reaches *internal equilibrium*. A system in internal equilibrium has a fixed set of properties that can be measured with a high degree of precision. Any small perturbations to a system in internal equilibrium won't result in any large changes to its state; once the perturbation ends, the system will return to equilibrium. An open or closed system can also be in *external equilibrium*, which means that any exchanges between the system and the environment are exactly balanced by exchanges in the opposite direction.

One example of external equilibrium is the classic bowling ball at the bottom of a valley (Figure 1.3). A small nudge to the bowling ball to either the left or right may get it moving back and forth a little, but eventually it will settle back at the lowest point in the valley. This is equivalent to stating that the bowling ball is in a minimum potential energy state. As long as the kinetic energy of the nudge is less than the energy necessary to push the bowling ball all the way up the hill, it will never escape,[3] and over time, will return to its equilibrium position at the bottom of the valley. Another example of external equilibrium involves temperature: When you bring two objects together that begin at different temperatures, over time the temperature difference between them will disappear, and reach an equilibrium point that's somewhere between the two starting temperatures. We'll return to this example again in Chapter 3.

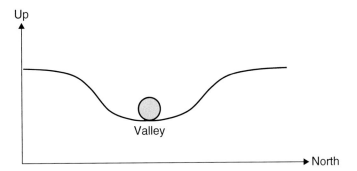

Figure 1.3. A state of external equilibrium: Bowling ball at the bottom of a valley. In classical physics, the only way to get the bowling ball out of this potential energy well is to impart enough kinetic energy to push it up the hill on either side.

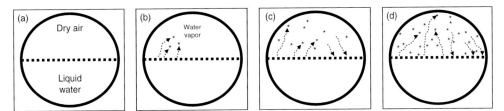

Figure 1.4. Saturation – a compound system in equilibrium. Initially (a) the system consists of dry air above and liquid water below. In (b) evaporation begins, adding water vapor to the air. The vapor molecules float around in the air above the liquid water. In (c), condensation begins, and some of the vapor molecules stick back onto the surface of the liquid water. At first the evaporation rate is much greater than the condensation rate, but as the amount of water vapor in the air increases, the condensation rate increases too. Eventually (d) the evaporation and condensation rates are the same – that is, reach an equilibrium state – and the air is saturated.

An example of internal equilibrium is a compound system consisting of a subsystem of water vapor plus dry air, and another subsystem of liquid water (Figure 1.4). The two subsystems are open to each other (but isolated from the external environment), meaning that both mass and energy can move back and forth between them (but nothing can leave or enter from outside). Water vapor molecules will escape from the surface of the liquid water (evaporate), causing it to lose mass. At the same time, vapor molecules will also stick back to the surface of the liquid water (condense), causing the liquid to gain mass. If the two phase changes are occurring at the same rate, the system is in equilibrium. Evaporation and condensation are still occurring, but the two exactly cancel each other. Once equilibrium is reached, there will be no noticeable change in the masses of the liquid water or the water vapor over time. This kind of equilibrium is called *saturation*.

1.4. Extensive and Intensive Properties; State Variables

Extensive properties are dependent on the mass of the parcel. One example of an extensive property is the *total heat capacity* (C), which can be used to describe the transfer of heat when two objects with different temperatures are brought into direct contact. In Table 1.4, volume (V) is an extensive property. *Intensive properties* are independent of mass, that is, the mass of the parcel is divided out of the property. These are also referred to *specific properties*. In Table 1.4, specific volume (α) is an intensive property, and has units of m^3/kg. As we begin to discuss work and heat and the *Ideal Gas Law*, specific volume will be used quite a bit.

A parcel is called *homogeneous* if its intensive properties are the same at every point. If a parcel is composed of an *ideal gas*,[4] and is homogeneous and in equilibrium, the Ideal Gas Law can be used to describe the relationship between its state variables. For this reason, the Ideal Gas Law belongs to a class of mathematical models known as *equations of state*. There are additional equations of state that can be invoked if a gas is not "ideal."

For a parcel composed of an ideal gas, the Ideal Gas Law describes the relationship between its *state variables* – that is, its pressure (p), volume (V), mass (m), and temperature (T). If mass is held constant, or the behavior of the parcel is described in terms of intensive (specific) properties, mass can be eliminated, leaving a total of only three state variables in the relationship. Knowledge of any two state variables provides information about the third, through an equation of state. There are additional state variables, such as chemical composition, internal energy (u), and enthalpy (h). The latter two will be discussed later.

Now, to answer the question posed in the preceding text: we want a "large" or "small" parcel (but not a "medium-size" parcel), so that we can, in turn, assume it reaches equilibrium, making an equation of state (such as the Ideal Gas Law) a valid way to describe the relationship between its state variables. Without this, we can't write an equation relating pressure, volume, and temperature to each other, which, you'll see in later chapters, would leave us unable to model most of the atmosphere.

1.5. Fundamental Quantities and Units

In classical physics (i.e., the physics before relativity and quantum mechanics intruded in the twentieth century), there were seven *fundamental* quantities. By fundamental, we mean quantities that are not dependent on one another and cannot be reduced to functions of other quantities. These are:

• Length,
• Mass,
• Time,
• Temperature,

Table 1.1. *Fundamental quantities and SI units.*

Quantity	Quantity abbreviation	SI Unit	Unit abbreviation
Length	l	Meter	m
Time	t	Second	s
Mass	m	Kilogram	kg
Temperature	T	Kelvin	K
Electric charge	q	Coulomb	C
Luminous intensity	I	Candela	cd
Amount of substance	n	Mole	mol

- Electric charge,
- Luminous intensity, and
- Amount of substance.

Each of these is associated with a distinct class of unit that is totally unlike those associated with the other fundamental quantities. Over time, several different systems of units accumulated, resulting in more than one way, for example, to express *length*. One example is the *English* system; another is the *British Imperial* system. This necessitated the creation of conversion factors to convert a length expressed in one system to the same length expressed in another.

In Earth sciences there are currently two dominant systems, both of which are based on the metric system: CGS and MKS. CGS stands for centimeter-gram-second, and is still the system preferred by some scientists. MKS stands for meter-kilogram-second, and is the system used by most physical scientists. In 1960, the MKS system was expanded into the *Système International d'Unités* (SI). The SI units for the quantities listed in the preceding text are shown in Table 1.1.

In this text, we'll stick with SI-MKS as much as possible. This simplifies the calculations and, through careful unit analysis, also provides a convenient method to check the results of the calculations.

The quantity of length is a *scalar* quantity, meaning that it has a magnitude, but no direction. The equivalent *vector* quantity, *position*, has both a magnitude and a direction. In this text, scalars will be shown in plain text (e.g., L), and vectors will be shown with an arrow (e.g., $\vec{L}$). Units will usually be specified using square brackets (e.g., [m]).

The SI temperature scale is known as the *absolute* scale, which has units of *Kelvins* (not "*degrees Kelvin*"). *Absolute zero* is defined as the temperature at which all random molecular motion stops, and is therefore the coldest temperature possible, but it doesn't actually occur anywhere in nature. There is no parallel maximum possible temperature in classical physics.

Table 1.2. *Temperature scale conversions.*

To compute	From	Use
Absolute	Celsius	$T_K = T_C + 273.16$
Absolute	Fahrenheit	$T_K = \dfrac{5}{9}T_F - 255.38$
Celsius	Absolute	$T_C = T_K - 273.16$
Celsius	Fahrenheit	$T_C = \dfrac{5}{9}(T_F - 32)$
Fahrenheit	Absolute	$T_F = \dfrac{9}{5}T_K - 459.67$
Fahrenheit	Celsius	$T_F = \dfrac{9}{5}T_C + 32$

Two additional temperature scales, probably more familiar to the reader, are the *Celsius*[5] (or "*centigrade*") scale, and the older *Fahrenheit* scale.[6] The former is still widely used in science, but must be converted to the absolute scale before performing any physical calculations. There are one or two uses of Celsius temperatures in this text. Until the 1960s, the Fahrenheit scale was the primary temperature scale used in English-speaking countries, but by the 2010s, it had almost passed completed out of use. Only the United States, the U.S. Virgin Islands, Puerto Rico (a U.S. territory), and a few smaller countries still use this scale. The rest of the world now uses the Celsius scale for everyday purposes, such as weather forecasting. The units of both scales are called *degrees*, but one degree of temperature change on the Celsius scale is larger than a degree on the Fahrenheit scale. In fact, one Celsius degree is exactly 1.8 Fahrenheit degrees. One Celsius degree *does* represent, however, the same change in temperature as one Kelvin on the absolute scale. Conversions between the absolute, Celsius, and Fahrenheit scales are shown in Table 1.2. Some common temperature reference points are shown on all three scales in Table 1.3.

1.6. Derived Quantities and Units

All physical quantities not listed in Table 1.1 are derived by combining one or more fundamental quantities. One example of this is *pressure*, which is *force* per unit *area*. Force, in turn, is reducible to mass (a fundamental quantity) and *acceleration*; acceleration to the time rate of change of *velocity*; and velocity to the time-rate-of-change of *position* (length). Area is reducible to length. Table 1.4 summarizes the derived quantities used in this text, and their associated units.

Table 1.3. *Common temperature reference points.*

All values are rounded to the nearest whole unit.

Description	Celsius scale [°C]	Fahrenheit scale [°F]	Absolute scale [K]
Water boils at sea level	100	212	373
Highest temperature ever recorded at a surface station (Libya)[i]	58	136	331
	50	122	323
	40	104	313
	30	86	303
	20	68	293
	10	50	283
Water freezes/melts at sea level	0	32	273
	−10	14	263
	−20	−4	253
	−30	−22	243
Coldest temperature at which supercooled water droplets can exist	−40	−40	233
Lowest temperature ever recorded at a surface station (Antarctica)[ii]	−89	−129	184
Absolute zero	−273	−460	0

[i,ii] Wikipedia (2012).

The unit of force is obviously named for Isaac Newton, who defined it as the product of mass and acceleration. The unit of pressure is named for Blaise Pascal, a French mathematician and physicist, and the unit of energy, the *Joule* (J), is named after James Prescott Joule, an English brewer (and scientist) who took pride in his ability to make remarkably precise measurements of temperature with an ordinary thermometer. In addition to the definition of the J shown in Table 1.4, one J is also equal to the energy required to lift one pound of water[7] nine inches, increase its speed by seven ft/s (walking speed), or heat it up by 0.001 °F.[8]

There are several other pressure units used in meteorology. The *Pascal* (*Pa*) must be used when performing most physical calculations, but it is too small a unit for practical use in weather forecasting. For this reason the *hecto-Pascal* (100 Pa, or 1 *hPa*) is now in common use. In addition to this, there are several older units in the scientific literature, such as *inches of mercury* (*in.Hg*), which are still used in some applications, such as aviation. The older unit *millibar* (one 1000th of a bar) is

Table 1.4. *Derived quantities and SI units.*

Quantity	Quantity abbreviation	Definition	SI Unit
Velocity, speed	$\vec{v}, v$	$\dfrac{d\vec{L}}{dt}$	$\dfrac{m}{s}$
Acceleration	$\vec{a}$	$\dfrac{d\vec{v}}{dt}$	$\dfrac{m}{s^2}$
Area	A	$L \times L$	m^2
Volume	V	$L \times L \times L$	m^3
Specific volume	α	$\dfrac{V}{m}$	$\dfrac{m^3}{kg}$
Density	ρ	$\dfrac{m}{V} = \dfrac{1}{\alpha}$	$\dfrac{kg}{m^3}$
Force	$\vec{F}$	$m\vec{a}$	$Newton(N) \equiv kg\dfrac{m}{s^2}$
Pressure	p	$\dfrac{F}{A}$	$Pascal(Pa) \equiv \dfrac{N}{m^2} = \dfrac{kg}{ms^2}$
Energy, work	E, W	mv^2	$Joule(J) \equiv kg\dfrac{m^2}{s^2}$
Power	P	$\dfrac{E}{t}$	$Watt(W) \equiv \dfrac{J}{s} = kg\dfrac{m^2}{s^3}$

equivalent to one hPa, so these two units are frequently used interchangeably. The hPa is preferred in recent literature. The equations in the following text summarize some factors for converting between the various pressure units.

$$1 \text{ bar} = 10^3 \text{ millibar (mb)} \tag{1.4a}$$

$$1 \text{ mb} = 10^2 \text{ Pa} \tag{1.4b}$$

$$1 \text{ hPa} = 10^2 \text{ Pa} \tag{1.4c}$$

$$1 \text{ atm} = 1.01325 \text{ bar} \tag{1.4d}$$

$$1 \text{ atm} = 1013.25 \text{ hPa} \tag{1.4e}$$

$$1 \text{ in.Hg} = 33.86 \text{ hPa} \tag{1.4f}$$

The *molar mass* is a derived quantity than can be described using other fundamental quantities. It is defined (in the MKS system) as the mass of the substance divided by the amount of substance. For example, for diatomic oxygen gas:

$$[O_2] = 32 \frac{\text{kg}}{\text{kmol}}$$

where the square brackets on the left-hand side (LHS) of the equation indicate the molar mass of O_2 molecules. This number is approximately equal to the *atomic weight* of the substance shown on periodic tables, which is usually expressed in multiples of the mass of a proton. A single oxygen atom has an atomic weight of sixteen (eight protons and eight neutrons), thus diatomic oxygen has an atomic weight of thirty-two. The quantity in the denominator of the RHS, *kmol*, refers to a thousand moles of substance. One mole refers to the standard number of particles of substance, defined by *Avogadro's Number* (N_A), where:

$N_A = 6.02 \times 10^{23}$ particles

Thus, a kmol is 1,000 times this, or 6.02×10^{26} particles.

The volume of space (V) occupied by one mole is given by the molar mass (in the case of O_2, this is 32 kg/kmol), divided by the density (ρ), that is, the inversion of the simple definition:

$$\rho = \frac{m}{V} \tag{1.5a}$$

$$V = \frac{m}{\rho} \tag{1.5b}$$

which turns out to be a constant (at a given temperature and pressure) for an ideal gas. The *molar volume* for any ideal gas, regardless of the chemical elements involved, is approximately 22.4 liters (one liter is 10^{-3} m^3) at a pressure of one atmosphere and a temperature of 0 °C. Therefore, for a kmol, the molar volume is 22.4×10^3 liters (or 22.4 m^3). It increases by about 1 percent as the pressure decreases from 1,000 to 100 hPa, and by about 10 percent as the temperature increases from 0 to 25 °C.

1.7. The Importance of Unit Analysis

College students have many places to go for help with their homework. If they're not sure they've computed some quantity correctly, they can ask their professors or teaching assistants to look at it with them, or they can check a solutions manual. Once they graduate, however, they have to start developing creative and independent ways of checking the results of their computations. This is pretty important for professions like weather forecasting: an incorrect calculation, if left uncorrected, could result in property damage, injury to people, or worse.

Unit analysis is a pretty easy way to check your results. Performing a unit analysis may not tell you if you're off by a factor of five, but it does provide a basic sanity check: If you're trying to compute a pressure, and your answer comes out in *Joules*, you have a problem. Here is an example.

According to Newton, gravity is a force described by:

$$\vec{F}_g = -\left(G\frac{m_e}{r^2} \right)m_i \hat{r} \tag{1.6}$$

where G is the Universal Gravitational Constant (6.673×10^{-11} Nm^2/kg^2), m_e is the mass of the Earth (5.988×10^{24} kg), m_i is the test mass [kg], r is the distance between the two centers of mass [m], and $\hat{r}$ is a unitless vector (length equal to 1) pointing from the center of mass of the Earth to the center of mass of the test mass. (This unit vector can be written explicitly as $\hat{r} = \vec{r} / |r|$, that is, it is a vector divided by its own magnitude.) The negative sign on the RHS of (1.6) means that the gravitational force points "down," toward the center of the Earth.

Near the Earth's surface, we can also write:

$$\vec{F}_g = \vec{g}m_i \tag{1.7}$$

where "little g" is a vector pointing down, with a magnitude of about 9.81 m/s^2; that is, it has the units of acceleration (see Table 1.4). Combining these two gives the result:

$$\boxed{\vec{g} = -\left(G\frac{m_e}{r^2} \right)\hat{r}} \tag{1.8}$$

which implies that the quantity on the RHS of (1.8) must also have the units of acceleration. If it does *not*, then there must be a mistake somewhere.

We begin the unit analysis by substituting the units for each term into (1.8):

$$\left[\frac{m}{s^2} \right] = \left[\left(\frac{N\,m^2}{kg^2} \right)\left(\frac{kg}{m^2} \right) \right] \tag{1.9}$$

where the units in the first parenthesis on the RHS represent G, and the remaining units represent m_e/r^2. The unit vector $\hat{r}$ has no units. A *Newton* is not a fundamental unit, because it is reducible to kg m/s^2, as shown in Table 1.4. Making this substitution results in:

$$\left[\frac{m}{s^2}\right] = \left[\left(\frac{kg\,m}{s^2}\right)\left(\frac{m^2}{kg^2}\right)\left(\frac{kg}{m^2}\right)\right] \tag{1.10}$$

After canceling terms on the RHS, we find that:

$$\left[\frac{m}{s^2}\right] = \left[\frac{m}{s^2}\right] \tag{1.11}$$

and confirm that that RHS of (1.8) does indeed have the units of acceleration.

1.8. Functions of More Than One Variable and Partial Derivatives

Let's say z is a function that describes the vertical height of a hill above the surrounding countryside, and that it varies only with the northward distance from some reference point. We'll call the reference point the *origin* (familiar to anyone who has studied coordinate systems, such as the Cartesian system), and we'll call the dimension stretching from the origin toward the north the y dimension. Using this convention, locations to the north of the origin would correspond to positive y values, and locations south of the origin would correspond to negative values of y. Locations above the origin would correspond to positive values of z, and locations below the origin would correspond to negative values of z (i.e., they would be below ground). In this case, you can describe the *slope* of the hill by taking the first derivative of $z(y)$ with respect to y:

$$\frac{dz}{dy} = z' \tag{1.12}$$

and the hill might look something like Figure 1.5.

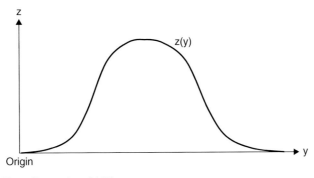

Figure 1.5. Two-dimensional hill.

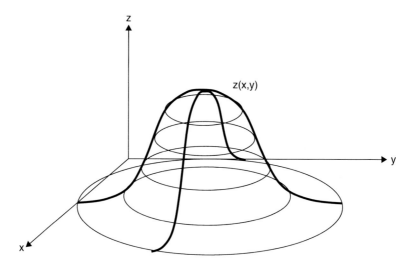

Figure 1.6. Three-dimensional hill.

A *positive* slope implies that the height of the hill is increasing as you go north, or decreasing as you go south. A *negative* slope implies that the height of the hill is decreasing as you go north, or increasing as you go south.

Very few hills are as flat as this page, so the height z of a real hill probably also varies along a horizontal dimension perpendicular to both the y and z dimensions, which we will call the x dimension. This makes z a function of two variables $z(x,y)$. Positive values of x correspond to points east of the origin, and negative values of x imply a location west of the origin. This somewhat more realistic hill might look something like Figure 1.6.

This more realistic picture of the hill makes it clear that the slope of the hillside may be different in the x dimension than it is in the y dimension; that is, it may be steeper along a line of latitude than it is along a line of longitude, or the reverse. This means that the first derivatives of $z(x,y)$, with respect to the two variables x and y may be (and usually are) different. Formally, they are called *partial derivatives* and are expressed as:

$$\frac{dz}{dx} = \left(\frac{\partial z}{\partial x}\right)_y \tag{1.13a}$$

$$\frac{dz}{dy} = \left(\frac{\partial z}{\partial y}\right)_x \tag{1.13b}$$

In (1.13a), the expression $\partial z/\partial x$ means the partial derivative of z with respect to x, with the subscript y indicating that the operation is performed with y treated as a constant. This would be the slope of the hillside in the east-west dimension, at some fixed location along the north-south dimension. Similar arguments apply to the expression shown in (1.13b). The variation of *z* with respect to *both* variables is found by multiplying (1.13a) through by *dx*, and multiplying (1.13b) through by *dy*:

$$dz = \left(\frac{\partial z}{\partial x}\right)_y dx \tag{1.14a}$$

$$dz = \left(\frac{\partial z}{\partial y}\right)_x dy \tag{1.14b}$$

and then adding the two results together as a simple linear sum:

$$\boxed{dz = \left(\frac{\partial z}{\partial x}\right)_y dx + \left(\frac{\partial z}{\partial y}\right)_x dy} \tag{1.15}$$

which is then defined as the *total* or *exact differential* of *z*.[9]

   In this example, *z* was defined as a height (a vertical distance), and the two independent variables *x* and *y* were defined as horizontal distances, but this was only for the purposes of the demonstration. The function *z(x,y)*, as well as the independent variables *x* and *y*, could have been defined as pressures, temperatures, densities, volumes, or any other physical variables that we have reason to believe are related and interdependent, such as in the Ideal Gas Law, or another equation of state. This is an important and useful idea in thermodynamics.

1.9. Exact Differentials and Line Integrals[10]

Let's say we have some small, finite quantity δ*z* that can be written as:

$$\delta z = M\,dx + N\,dy \tag{1.16}$$

where *M* and *N* are functions of the independent variables *x* and *y*. The right-hand side (RHS) of (1.16) can't be integrated unless some relationship *f(x,y) = 0* is chosen. Returning to the convenient (but not essential) notion that *x* and *y* are spatial dimensions, integrating (1.16) would define a path (or line) taken through the *(x,y)* plane. Because computing an integral is equivalent to computing an area, it follows that the result of such a line integral would depend on the path taken through the *(x,y)* plane.

   Adopting the conventions:

$$M = \left(\frac{\partial z}{\partial x}\right)_y \tag{1.17a}$$

$$N = \left(\frac{\partial z}{\partial y}\right)_x \qquad\qquad (1.17b)$$

Equation (1.16) becomes:

$$\delta z = \left(\frac{\partial z}{\partial x}\right)_y dx + \left(\frac{\partial z}{\partial y}\right)_x dy \qquad\qquad (1.18)$$

which you can compare to (1.15). The RHS of (1.18) is, therefore, both the total differential and the exact differential of *z*. In other words, in this case:

$$\delta z = dz \qquad\qquad (1.19)$$

To integrate (1.18) and obtain a closed value of *z*, we can integrate both sides of (1.19) from some initial state *i* to some final state *f* by:

$$\int_i^f \delta z = \int_i^f dz = z\left(x_f, y_f\right) - z\left(x_i, y_i\right) \qquad\qquad (1.20)$$

so that the *net change* along the path from (x_i, y_i) to (x_f, y_f) depends only on these two states. Put another way, *the final state is independent of the path taken*. If (x_i, y_i) and (x_f, y_f) are in the same location, the result of this *closed line integral* (describing a *cyclic process*) is:

$$\oint \delta z = z\left(x, y\right) = 0 \qquad\qquad (1.21)$$

and the net change is zero.

1.10. Thermodynamic Diagrams

Sometimes it's easier to understand thermodynamic processes if you can see a picture. One way you can do this is to plot the progress of a thermodynamic system from one state to another on a diagram, where the axes coincide with the state variables. Generically, these are called *thermodynamic diagrams*. One example is the *pressure-Volume (p-V) diagram* that you probably studied in your undergraduate physics courses (Figure 1.7). A related example extends the diagram to all three state variables, creating a *p-V-T diagram* (Figure 1.8). Another example is the *Skew-T Log P* (or just *Skew-T*) used in the United States and elsewhere (Figure 1.9).

In the third example, the isopleths (lines) on the diagram represent equations or constants, and are, therefore, another way to visualize thermodynamic processes. Because energy is a key idea in thermodynamics, an ideal thermodynamic diagram has isopleths drawn such that the *area on the diagram is proportional to energy*.[11] It's also desirable to define the axes so that as many of the equations appear as straight

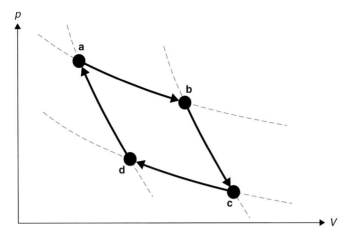

Figure 1.7. p-V diagram with cyclic process. A parcel of air begins in state *a* (pressure = p_a, Volume = V_a), then proceeds around a closed loop to states *b, c* and *d*, and back to *a*. Drawing the lines between the different states in the cyclic process is equivalent to performing an integral. In this case, the *net change* is zero (see Equation 1.21), but the *area enclosed* is nonzero (see description following Equation 1.16).

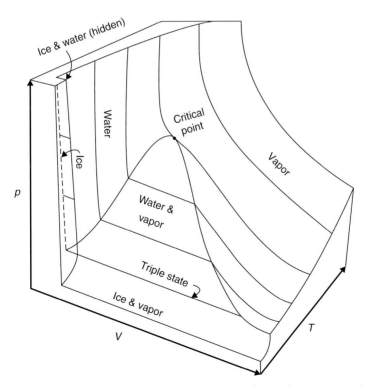

Figure 1.8. p-V-T diagram. The thermodynamic surface of water seen in three dimensions. Details of interpreting this diagram are discussed in later chapters in the text. Figure from Hess (1959), with axes relabeled p, V, T from original e, α, T.

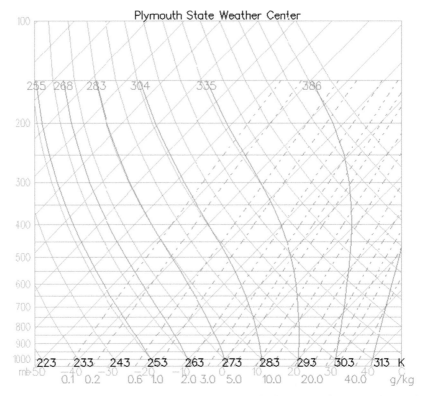

Figure 1.9. Skew-T Log P diagram. The Skew-T is just one of several thermodynamic diagrams used in operational meteorology around the world. Horizontal axes are temperature (although isotherms slant diagonally from lower left to upper right, i.e., are skewed), mixing ratio, dry adiabats, and moist adiabats. Vertical scale is pressure, and is logarithmic rather than linear (i.e., log p). See Figures 1.10–1.14 for more details. (Image courtesy of Plymouth State University meteorology program, 2012.)

(or nearly straight) lines as possible, and, so that the intersections of these lines are as close to perpendicular as possible.

Diagrams such as the Skew-T have been in operational use since the middle of the twentieth century. None of them are ideal, in that not all equations appear as straight lines, and the lines don't meet at right angles. They were originally invented at a time before compact, powerful computers were available, so the diagrams were a way to apply the equations of thermodynamics to a particular atmosphere (recorded by a balloon-mounted radiosonde) and obtain answers graphically. Put simply, thermodynamic diagrams are graphical calculators, operated by plotting observed data on the diagram, and drawing additional lines representing thermodynamic equations.

Details of the Skew-T Log p. In this text, we'll utilize the Skew-T extensively, to both illustrate the ideas we derive through pure mathematics, and to show how the concepts in thermodynamics are used in an operational meteorology setting. (Nowadays, an ordinary desktop computer can easily ingest radiosonde data and compute the

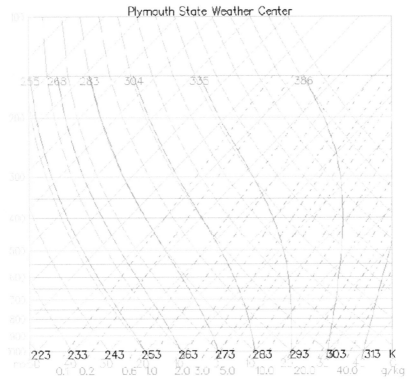

Figure 1.10. Isobars are horizontal lines; some are highlighted in yellow. Pressure scale is at left and is in units of millibars [mb], which are the same as hecto-Pascals [hPa].

parameters that an analyst formerly did manually with a Skew-T, but the results of the computer's automated analysis are still often plotted on a Skew-T.) As the text proceeds from one chapter to the next, more and more of the physics captured in the isopleths of the Skew-T will be explained.

At this point, all we need to do is label and define each of the different isopleths. Figures 1.10 through 1.14 highlight and briefly explain each of them. The isopleths shown are *isobars* (lines of equal pressure), *isotherms* (lines of equal temperature), *mixing ratio lines* (lines of equal water vapor content, discussed in Chapter 7), *dry adiabats* (the vertical cooling rate of unsaturated parcels, as well as lines of equal potential temperature and entropy, discussed in Chapters 5 and 6), and finally, *moist adiabats* (the vertical cooling rate of saturated parcels, discussed in Chapter 8 and Appendix 7). Some Skew-Ts include additional isopleths, but we won't go into that in this text.

One quantity not shown here is the vertical elevation scale. Such a scale would be linear and stretch from a few hundred meters below sea level near the bottom of the diagram to about 16 kilometers above sea level at the top of the diagram. The reason

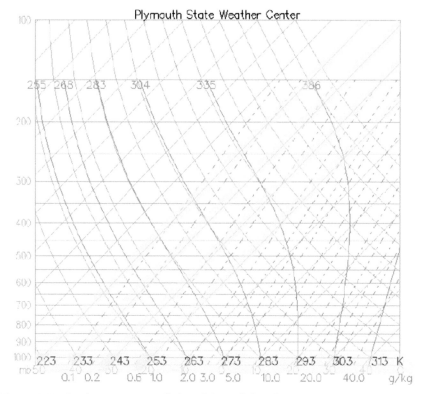

Figure 1.11. Isotherms are parallel, solid, straight, diagonal lines stretching from lower left to upper right; some are highlighted in yellow. Temperature scales ([K]; [°C]) are on the horizontal axis (also highlighted).

it isn't shown on the Skew-Ts here is that the vertical elevation associated with a given pressure (such as 850 hPa) varies from time to time and place to place, as do the vertical distances between adjacent pressure levels (a quantity called *thickness*). In some cases, printed Skew-Ts actually *do* show a height scale on the right side of the diagram, along with advice not to use it too precisely. Another way elevation is often dealt with is to plot the *in situ* elevation associated with the mandatory pressure levels of a radiosonde observation (RAOB) as a simple number (usually in meters), just below the label associated with that pressure. We'll show you some examples later in the book.

Practice Problems

1. Thermodynamics is the study of _____ and its _____.
2. Define *open*, *closed*, and *isolated* systems.
3. What assumptions are made about small and large *open* systems in order to treat them as *closed* systems without much loss of accuracy?
4. What name do we use for a system in the atmosphere?

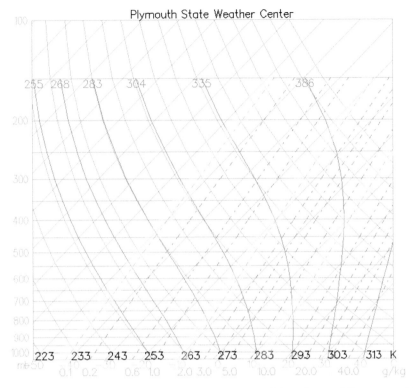

Figure 1.12. Mixing ratio isopleths are dashed, straight, diagonal lines stretching from lower left to upper right; some are highlighted in yellow. They are not *exactly* parallel, but would ultimately converge, if extended to a point well beyond the upper right corner of the diagram. The reason for this is discussed in Chapter 7. Mixing ratio [$g_{vapor}/kg_{dry\ air}$] scale is across the bottom of the diagram (highlighted in yellow).

Use Wikipedia to answer the following questions about thermodynamic diagrams. The web address is http://en.wikipedia.org.

5. What is the defining characteristic ("main feature") of a thermodynamic diagram?
6. What five different isopleths are usually shown on thermodynamic diagrams?
7. What are the names of three thermodynamic diagrams used by different weather services?
8. Search Wikipedia for "Emagram."
   • Who invented the Emagram?
   • When was it invented?
   • Where is it primarily used?
9. Search Wikipedia for "Tephigram."
   • Who invented the Tephigram?
   • When was it invented?
   • Where is it used?
   • What does "Tephi" mean?

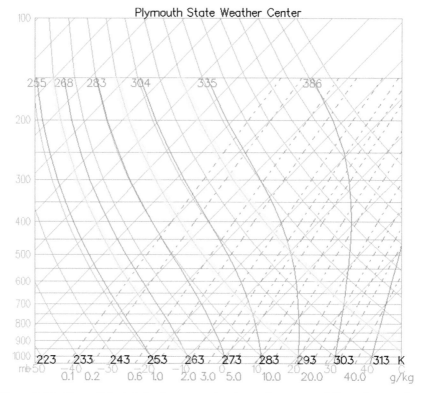

Figure 1.13. Dry adiabats (also known as isentropes) are curved lines that run from lower right to upper left; some are highlighted in yellow. They are approximately parallel to each other. Dry adiabats are discussed at length in Chapter 5.

10. Search Wikipedia for "Stuve Diagram."
    - Who invented the Stuve Diagram?
    - When was it invented?
    - Where is it used?
    - What defining characteristic of thermodynamic diagrams is sacrificed by using straight lines for p, T, and θ?
11. Search Wikipedia for "Skew-T log P Diagram."
    - Who invented the Skew-T?
    - When was it invented?
12. Show that:

$$dV = \alpha V_0 dT$$

evaluates to:

$$V' = V_0 \left(1 + \alpha T'\right)$$

when V is integrated from (constants) V_0 to V', and T is integrated from (constants) 0 to T'.

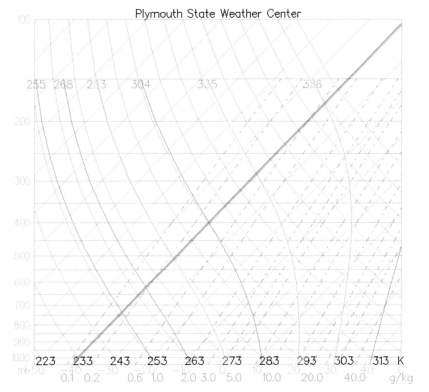

Figure 1.14. Moist adiabats are highly variable, curved lines that generally run from lower right to upper left; some are highlighted in yellow. Below temperatures of −40 °C (highlighted by heavy blue line), moist adiabats are essentially parallel to dry adiabats. The reason for this is discussed in Chapter 8.

13. In the following expression, p, α, and T are variables, and R_i is a constant:

$$p\alpha = R_i T$$

Take the derivative of both sides (using the chain rule), and show that:

$$p\,d\alpha + \alpha\,dp = R_i\,dT$$

14. Integrate both sides of the expression (from *initial* to *final* states):

$$\frac{dT}{T} = \frac{R_i}{c_p}\frac{dp}{p}$$

where T and p are variables, and R_i and c_p are constants. Show that the solution is:

$$ln\left(\frac{T_f}{T_i}\right) = ln\left[\left(\frac{p_f}{p_i}\right)^{\frac{R_i}{c_p}}\right]$$

15. Exponentiation (e^x) is the inverse function of the natural log ($ln\ x$). Use exponentiation to show that the solution to No. 14 can be written as:

$$T_f = T_i \left(\frac{p_f}{p_i} \right)^{\frac{R_i}{c_p}}$$

16. In the following expression, p, α, and T are variables, and R_i is a constant:

$$p\alpha = R_i T$$

Take the natural log of both sides, followed by the derivative of both sides, and show that:

$$\frac{dp}{p} + \frac{d\alpha}{\alpha} = \frac{dT}{T}$$

17. In the following expression, θ, T, and p are variables, and κ is a constant:

$$\theta = T_i \left(\frac{1000}{p_i} \right)^{\kappa}$$

Take the natural log of both sides, followed by the derivative of both sides (using the chain rule), and show that:

$$\frac{d\theta}{\theta} = \frac{dT}{T_i} - \kappa \frac{dp}{p}$$

18. In the following expression, e_s and T are variables, and l_v and R_d are constants:

$$\frac{de_s}{e_s} = \frac{l_v\ dT}{R_v T^2}$$

Integrate the LHS from e_{si} to e_{sf}, and the RHS from T_i to T_f, and show that:

$$ln\left[\frac{(e_s)_f}{(e_s)_i} \right] = \frac{l_v}{R_v} \left[\frac{1}{T_i} - \frac{1}{T_f} \right]$$

19. Integrate both sides of the following expression,

$$f'(z) = -\frac{g}{R_d (T_0 - \gamma_e z)} dz$$

using *udu* substitution on the RHS. Let $u = T_0 - \gamma_e z$, so that $du = -\gamma_e dz$. Show that the expression integrates to:

$$f(z) = ln\left[\left(\frac{T_0 - \gamma_e z}{T_0} \right)^{\frac{g}{R_d \gamma_e}} \right]$$

20. Solve the following system of equations for *dw/dt*:

$$\frac{dp}{dz} + \acute{\rho}g = -\acute{\rho}\frac{dw}{dt}$$

and:

$$\frac{dp}{dz} + \rho g = 0$$

2

Equations of State

2.1. Definition of an Ideal Gas

The real atmosphere is composed of a complex, ever-changing mixture[1] of gases that interact with each other, the surface(s) of the Earth, and the Earth's biosphere. There is even some interaction with the space environment. Attempting to compute something as simple as stability, while taking all of these variables into consideration, would be impractical. So we invent a fictional atmosphere (in fact, several of them, some of which are discussed in later chapters) that is much simpler than the real one. Our calculations are then performed on the fictional atmosphere rather than the real one, and we hope that the fiction is close enough to reality to make the effort useful.

The first simplifying assumption we make is that air is an *ideal gas* – that is, one with molecules that behave like simple billiard balls without an electrical charge. Most gases can be treated as "ideal"[2] when the density is low enough that the combined volume of the *molecules* of the gas is much smaller than the *total* volume of space the gas occupies. Further:[3]

- The molecules of the gas are in random motion (they move in all directions, and the same number of molecules move in any given direction).
- The only interactions between molecules occur when they collide with each other. They do not exert any other forces on each other, such as electrical attraction. Chemists refer to several variations of this electrical attraction as *Van der Waals*[4] forces.
- Collisions between molecules are *elastic* (i.e., they don't stick together), and these collisions occur at uniform speeds. The angle at which two molecules approach each other prior to a collision is the same as the angle they follow after collision.

Most of the laws discussed in this chapter refer specifically to ideal gases. Later in this chapter, there is a brief effort made to describe gases where Van der Waals forces *are* in effect, but it is not pursued further than that in this book.

2.2. Boyle's Law

In 1662, Anglo-Irish scientist Robert Boyle and his assistant, Robert Hooke, published the following relationship between pressure and volume for a gas:

$$pV = p'V' \tag{2.1}$$

where p and V describe the initial state, and p' and V' describe the final state. This relationship is true provided mass and temperature are held constant. Boyle and Hooke also noted that as a gas became hotter, its volume increased, but they were unable to quantify the relationship because there were no temperature scales at the time. The earliest temperature scales were not developed until more than seventy years later.[5]

2.3. Gas Laws of Gay-Lussac

In the early part of the nineteenth century, French chemist and physicist J. L. Gay-Lussac used laboratory experiments to empirically derive two laws describing the behavior of ideal gases. The first describes the changes that occur in *volume* when *temperature* changes, while pressure and mass are held constant. The second law describes the changes that occur in *pressure* when *temperature* changes, while holding volume and mass constant.[6]

Gay-Lussac's First Law. In 1802, Gay-Lussac published the following relation,[7] called the *Law of Volumes*, which he credited to an earlier, unpublished work by Jacques Charles:

$$dV = \alpha V_0 \, dT \tag{2.2}$$

where dV is the change in volume of an ideal gas, α is the *volume expansion coefficient* (constant pressure; 1/273 deg^{-1}), V_0 is the volume of the gas at $0\,°C$, and dT is the change in temperature. All quantities are in SI units.

Integrating the LHS from V_0 to V, and the RHS from 0 to T, results in:

$$V - V_0 = \alpha V_0 T \tag{2.3}$$

or:

$$V = \alpha V_0 T + V_0 \tag{2.4}$$

and, therefore:

$$V = V_0 \left(1 + \alpha T\right) \tag{2.5}$$

which shows that V is a linear function of T. Given than α is equal to 1/273 [deg^{-1}], the second term inside the parenthesis on the RHS goes to -1 if T is equal to -273 [°C]; therefore, the volume of the gas goes to zero. Mathematically,

$$V\left(-273°C\right) = V_0\left(1-1\right) = 0 \tag{2.6}$$

Equation (2.6) implies that the volume of an ideal gas goes to zero at a temperature of absolute zero. Clearly, this simplifying assumption can't be right (because the molecules of the gas occupy *some* volume of space), but at ordinary temperatures in the atmosphere, the error it introduces is relatively unimportant.

The absolute temperature scale can be defined using the Law of Volumes. Since the law implies that volume goes to zero if the temperature goes to -273 [°C], we can make this the absolute zero reference point and keep the incremental *differences* in temperature defined by the distance between the melting and boiling points of water at sea level (which is 100 Celsius degrees). This resets the melting and boiling points of water to 273 [K] and 373 [K], respectively.

Gay-Lussac's Law of Volumes can also be written:

$$\frac{V}{V'} = \frac{T}{T'} \tag{2.7}$$

which states that the ratio of the volumes before (unprimed) and after (primed) some change is equal to the ratio of the temperatures before and after the change. *The temperatures in (2.7) are on the absolute scale.*

Gay-Lussac's Second Law. Gay-Lussac is also credited[8] with a second gas law that we can call the *Law of Pressures*, which states:

$$dp = \beta p_0\, dT \tag{2.8}$$

where dp is the change in pressure of an ideal gas, β is the *pressure coefficient* (constant volume; 1/273 deg^{-1}), p_0 is the pressure of the gas at 0 °C, and dT is the change in temperature. All quantities are in SI units.

Integrating the LHS from p_0 to p, and the RHS from 0 to T, results in:

$$p - p_0 = \beta p_0 T \tag{2.9}$$

or:

$$p = \beta p_0 T + p_0 \tag{2.10}$$

and therefore:

$$p = p_0\left(1 + \beta T\right) \tag{2.11}$$

which shows that p is a linear function of T. Given than β is equal to 1/273 [deg^{-1}], the second term inside the parenthesis on the RHS goes to -1 if T is equal to -273 [$^\circ$C]; therefore, the pressure of the gas goes to zero. Mathematically:

$$p\left(-273\,^\circ\mathrm{C}\right) = p_0\left(1-1\right) = 0 \qquad (2.12)$$

Equation (2.12) implies that, like volume, the pressure of an ideal gas also goes to zero at a temperature of absolute zero. This simplifying assumption must also be wrong (because the molecules of the gas have mass and therefore weight), but at ordinary temperatures in the atmosphere, the error it introduces is also relatively unimportant.

Gay-Lussac's Law of Pressures can also be written:

$$\frac{p}{p'} = \frac{T}{T'} \qquad (2.13)$$

which states that the ratio of the pressures before (unprimed) and after (primed) some change is equal to the ratio of the temperatures before and after the change. *The temperatures in (2.13) are on the absolute scale.*

2.4. An Equation of State

The gas laws described by Boyle and Gay-Lussac can be combined into a more comprehensive gas law, describing the relationship between several state variables of an ideal gas.[9] In this case, the state variables related are pressure, temperature, and volume. To avoid transient conditions that might upset the relationships between the state variables defined by Boyle and Gay-Lussac, we assume that the gas is *never far from equilibrium*, and that any changes that *do* occur are small and gradual.[10]

The Boyle-Gay-Lussac Law. A first step toward developing a comprehensive *Ideal Gas Law* is obtained by changing all three state variables of an ideal gas in two stages. In the first stage, we hold volume constant and change the temperature from T to T'. According to Gay-Lussac's Law of Pressures (2.13),

$$\frac{p}{p'} = \frac{T}{T'}$$

Letting $p_1 = p'$ (the pressure reached after the first stage) and inverting both sides of (2.13) results in:

$$\frac{p_1}{p} = \frac{T'}{T} \qquad (2.14)$$

or,

$$p_1 = p\frac{T'}{T} \tag{2.15}$$

where p_1 on the LHS and T' on the RHS describe the final state after this stage, and p and T on the RHS describe the initial state. In the next stage, we hold the temperature constant and increase the volume from V to V', taking the gas to a final state described by p', V', and T'. According to Boyle's Law (2.1),

$$pV = p'V'$$

In this case, the initial pressure (before volume changes) is p_1, so

$$p_1V = p'V' \tag{2.16}$$

but (2.15) provides another way of writing p_1, and substituting the result of (2.15) into (2.16) results in:

$$\left(p\frac{T'}{T}\right)V = p'V' \tag{2.17}$$

Diving through by T' puts the all initial state variables on the LHS and all final state variables on the RHS:

$$\boxed{\frac{pV}{T} = \frac{p'V'}{T'}} \tag{2.18}$$

which is called the Boyle-Gay-Lussac Law. This relationship shows that the value of pV/T remains a constant for an ideal gas, provided mass and other properties of the gas are held constant.

The Ideal Gas Law. Equation (2.18) can also be written:

$$\frac{pV}{T} = R' \tag{2.19}$$

where R' is a constant dependent on mass and other individual properties of the gas. A unit analysis of the LHS of (2.19) shows that:

$$\left[\frac{pV}{T}\right] = \frac{(Pa)(m^3)}{K} \tag{2.20}$$

but *Pascals* (*Pa*) are not a fundamental unit. According to Table 1.4:

$$Pa \equiv \frac{kg}{ms^2} \tag{2.21}$$

Substituting this definition into (2.20) results in:

$$\left[\frac{pV}{T}\right] = \frac{(Pa)(m^3)}{K} = \left(\frac{kg}{ms^2}\right)\left(\frac{m^3}{K}\right) \tag{2.22}$$

After canceling the extraneous units, we are left with:

$$\left[\frac{pV}{T}\right] = \left(\frac{kgm^2}{s^2}\right)\left(\frac{1}{K}\right) \tag{2.23}$$

According to Table 1.4, the first parenthetical term on the RHS of (2.23) is the definition of a *Joule* (*J*) – a unit of energy, that is:

$$J \equiv \frac{kgm^2}{s^2} \tag{2.24}$$

so that:

$$\left[\frac{pV}{T}\right] = [R'] = \left(\frac{J}{K}\right) \tag{2.25}$$

According to (2.25), the quantity *pV/T*, which has units of J/K, is conserved *when an ideal gas undergoes changes that are never far from equilibrium.* Another way of saying "never far from equilibrium" is "reversible." As you'll see when we get to Chapter 6, (2.25) is a statement about the conservation of entropy, which is the province of the Second Law of Thermodynamics.

Multiplying (2.19) through by temperature brings us very close to a useful version of the Ideal Gas Law:

$$pV = R'T \tag{2.26}$$

Recall that one underlying condition for Boyle's and Gay-Lussac's gas laws was that mass was held constant. We can express this explicitly by rewriting the constant *R'* in (2.26) with:

$$R' = mR_i \tag{2.27}$$

where *m* is the total mass of the gas, and R_i is a constant dependent on the *other* individual properties of the gas. The units of R_i can be determined by another unit analysis, that is:

$$[R'] = [mR_i] \tag{2.28}$$

or:

$$\left(\frac{kgm^2}{s^2}\right)\left(\frac{1}{K}\right) = (kg)\left(\frac{m^2}{s^2}\right)\left(\frac{1}{K}\right) \tag{2.29}$$

where the LHS of (2.29) came from the RHS of (2.23). The relationship in (2.29) shows that, because mass has units of kg, the units of R_i must be:

$$[R_i] = \left(\frac{m^2}{s^2}\right)\left(\frac{1}{K}\right) \tag{2.30}$$

According to (2.24):

$$J \equiv \frac{kg\,m^2}{s^2}$$

which means that:

$$\frac{J}{kg} = \frac{m^2}{s^2} \tag{2.31}$$

or, energy per unit mass (specific energy) can also be expressed as m²/s². Using this, Equation (2.30) can be rewritten as:

$$[R_i] = \frac{J}{kg\,K} \tag{2.32}$$

Substituting (2.27) into (2.26), we arrive at our first practical version of the Ideal Gas Law:

$$\boxed{pV = mR_iT} \tag{2.33}$$

where p is in Pascals, V is in m³, m is in kg, R_i has units of J/(kg K), and T is in Kelvins. The Ideal Gas Law is one equation of state, relating the important state variables in an ideal gas. There are others (a few will be discussed at the end of this chapter), but (2.33) and additional forms that we'll derive next are sufficient for most meteorological applications. To reiterate, (2.33) applies to ideal gases or gases that can be approximated as ideal. Further, to apply, the gas in question must be homogeneous (having the same properties in all directions) and never far from internal equilibrium.

Let's look at a unit analysis of (2.33). The relationship in (2.25) shows that pV/T has units of J/K. Multiplying through by T leaves only Joules, or units of energy. This implies that the RHS of (2.33) must also have units of energy:

$$[mR_iT] = \left(\frac{kg}{1}\right)\left(\frac{J}{kg\,K}\right)\left(\frac{K}{1}\right) \tag{2.34}$$

After cancellation, Joules are all that remain. In other words, while the Boyle-Gay-Lussac Law (2.18) was a statement about the conservation of *entropy*, the

Ideal Gas Law is a statement about the conservation of *energy*. That's the province of the First Law of Thermodynamics, which we'll return to in Chapter 4.

2.5. Additional Forms of the Ideal Gas Law

There are several additional ways to write the Ideal Gas Law that are immediately useful in meteorology. In this section we'll derive four of them.

Forms involving density and specific volume. The first is obtained easily by dividing (2.33) through by volume:

$$p = \frac{mR_iT}{V} \qquad (2.35)$$

which can be rewritten:

$$p = \left(\frac{m}{V}\right)R_iT \qquad (2.36)$$

or:

$$\boxed{p = \rho R_i T} \qquad (2.37)$$

where ρ is density, and has the SI units of kg/m³. Next, we can divide through by density, resulting in:

$$p\frac{1}{\rho} = R_iT \qquad (2.38)$$

but the quantity $1/\rho$ is the same as specific volume (α; see Table 1.4 in Chapter 1) – that is:

$$\boxed{p\alpha = R_i T} \qquad (2.39)$$

Form involving number of moles. Let's define a new constant, the *Universal Gas Constant* (R^*), such that all individual gas constants can be computed by:

$$\boxed{R_i \equiv \frac{R^*}{M}} \qquad (2.40)$$

where M is the molar mass of the gas. Molar mass is defined in Chapter 1, Section 1.3, and has units of kg/kmol. Multiplying through by M yields:

$$R_i M = R^* \qquad (2.41)$$

We can perform a unit analysis on (2.41) to obtain the units of R^*, that is:

$$[R_i M] = \left(\frac{J}{kg\,K}\right)\left(\frac{kg}{kmol}\right) = \frac{J}{kmol\,K} \tag{2.42}$$

R^* is also the product of two constants, Avogadro's Number (N_A) and Boltzmann's Constant[11] (k), and is computed by:

$$\boxed{R^* = N_A k = \left(6.02 \times 10^{26}\,\frac{particles}{kmol}\right)\left(1.38 \times 10^{-23}\,\frac{J}{particle\,K}\right) = 8.31 \times 10^3\,\frac{J}{kmol\,K}}$$

$$\tag{2.43}$$

which we'll return to a bit later.

Substituting the identity in (2.40) into the definition of the Ideal Gas Law shown in (2.33), we obtain:

$$pV = m\left(\frac{R^*}{M}\right)T \tag{2.44}$$

which can be written:

$$pV = \left(\frac{m}{M}\right)R^* T \tag{2.45}$$

The quantity in parenthesis on the RHS of (2.45) has units of kmol, which you can prove by performing a straightforward unit analysis. In other words,

$$\boxed{pV = nR^* T} \tag{2.46}$$

where n is the number of kmols of substance in the sample of gas. Pressure (p) is in Pascals, V is in m³, R^* has units of J/(kmol K), and T is in Kelvins. This form of the Ideal Gas Law is probably more familiar to meteorology undergraduates who have taken a two- or three-semester sequence in basic university physics.

Form involving number density. Beginning with the form shown in (2.46), we can divide through by volume and multiply both sides by the number 1:

$$p = \frac{nR^* T}{V} \tag{2.47}$$

and:

$$(1)\,p = \left(\frac{N_A}{N_A}\right)\frac{nR^* T}{V} \tag{2.48}$$

Then we note that, if n is the number of kmols, then nN_A is the number of particles, and

$$\frac{nN_A}{V} = \frac{\text{number of particles}}{\text{volume}} \tag{2.49}$$

which is the *number density* (n_0). We can now rewrite (2.48) using this identity, obtaining:

$$p = \frac{n_0 R^* T}{N_A} \tag{2.50}$$

Now let's take a look at the quantity R^*/N_A in the RHS of (2.50). If N_A has units of particles/kmol, then $1/N_A$ has units of kmol/particle. Applying these in a unit analysis:

$$\left[\frac{R^*}{N_A}\right] = \left(\frac{J}{kmol\,K}\right)\left(\frac{kmol}{particle}\right) = \frac{J}{particle\,K} \tag{2.51}$$

This quantity is, in effect, the Universal Gas Constant for a single particle, which is the definition of Boltzmann's Constant (k), and completes the circle begun in Equation (2.43). Substituting this identity into (2.50) yields the last version of the Ideal Gas Law we are seeking in this section, that is:

$$\boxed{p = n_0 kT} \tag{2.52}$$

2.6. Additional Equations of State

No gas is really "ideal." Chemical reactions within the gas may change its composition, which, in turn, may alter the value of the individual gas constant (R_i). Gases may condense into liquids, or nearby liquids may evaporate and add to the mass of the gaseous system. The molecules in the gas take up space (they are not point particles), and Van der Waals forces (electrical attraction between molecules with separated positive and negative charge centers) cause the particles to behave differently than simple billiard balls. In the absence of a more complete theoretical expression to relate the state variables of a gas, empirically derived expressions (based on lab work) are the best solution. While this is less satisfying than an expression derived from first principles, it can be very accurate for the range of pressures, volumes, and temperatures studied in the lab.

Van der Waals' Equation of State. In the 1870s, Dutch scientist J. D. van der Waals first proposed the following expression that is more accurate over wider ranges of the state variables than the Ideal Gas Law:

$$\left(p + \frac{an^2}{V^2}\right)(V - nb) = nR^* T \tag{2.53}$$

Table 2.1. *Composition of air with Van der Waals constants a and b.*[i]

List is by decreasing concentrations for Earth's atmosphere. Water vapor is listed last because its concentration varies considerably. Percentages shown for N_2 through Xe are for *dry* air. Percentage shown for water vapor refers to *moist* air.

Substance	Chemical Symbol	Percentage of Air by Mass	Molar Mass (M_i) [kg/kmol]	a [× 10^{-3} Pa m^6/mol^2]	b [× 10^{-6} m^3/mol]
Nitrogen	N_2	75.47	28.02	140.4	39.13
Oxygen	O_2	23.20	32.00	137.4	31.83
Argon	Ar	1.28	39.94	136.3	32.19
Carbon Dioxide	CO_2	0.046	44.01	362.8	42.67
Neon	Ne	0.0012	20.18	21.35	17.09
Krypton	Kr	0.0003	83.80	234.9	39.78
Helium	He	0.00007	4.00	3.457	23.70
Xenon	Xe	0.00004	131.29	425	51.05
Water Vapor	H_2O	$\cong 0$ to $\cong 4$	18.02	551.9	30.49

[i] The Engineering Toolbox: Air Composition (2012); Espinola (1994); The Engineering Toolbox: Air Composition (2012).

which you should compare to (2.46).[12] Van der Waals' improved Equation of State attempts to correct for two of the largest effects that cause gases to behave "nonideally": electrical attractions between the molecules of the gas (term involving *a* on the LHS), and the volume of space taken up by the molecules (term involving *b*). The molecules are no longer treated as point particles behaving like simple billiard balls, as they were in the Ideal Gas Law. Values of the new constants *a* and *b* for the important gases in Earth's atmosphere are shown in Table 2.1.

Power series Equations of State. It's also possible to express the relationship between state variables using entirely empirical equations – that is, curve fits to laboratory data. For example, if we begin with the form of the Ideal Gas Law shown in (2.46):

$$pV = nR^*T$$

and then divide through by pV, we find that:

$$1 = \frac{nR^*T}{pV} \tag{2.54}$$

which would be true if the gas in question was really *ideal*. Of course, there are no real ideal gases, so we define a new parameter called the *compressibility factor* (Z), which for some gases at normal pressures, is *approximately* equal to 1. Given this definition, we rewrite (2.54) as

$$Z = \frac{nR^*T}{pV} \tag{2.55}$$

which is unitless. Laboratory data are then used to determine the values of the coefficients in the following power series (called a *virial expansion*):

$$Z = 1 + Bp + Cp^2 + Dp^3 + \cdots \tag{2.56}$$

Where p is pressure, and B, C, D... are the virial coefficients. There are other formulations using, for example, molar density as the independent variable on the RHS. Determining the best possible values of these coefficients is a subject of current research, but it is also beyond the scope of this book.

2.7. Ideal Gas Law for Dry Air and Water Vapor

The next problem is one of applying the Ideal Gas Law to Earth's atmosphere. One way of making the problem manageable is to assume that moist air is composed of two components: dry air and water vapor. We can then write separate equations of state for the two components and add the results as a simple linear sum.

Dalton's Law. That last point is an application of Dalton's Law,[13] which states:

$$p = \sum_{i=1}^{N} p_i = p_1 + p_2 + p_3 + \cdots \tag{2.57}$$

where p is the *total* pressure exerted by the mixture of gases, p_1, p_2, p_3, ... are the pressures of the individual gases, and N is the total number of different gases in the ensemble.

As a concrete example, let p be the standard sea-level pressure of 1013.25 [hPa]. For simplicity, round it off to 1000 [hPa]. Then let the p_is be the pressures resulting from the individual gases making up air, such as nitrogen, oxygen, argon, and so forth. In this case, p_1 (nitrogen) would be approximately 755 [hPa], p_2 (oxygen) 232 [hPa], and so forth, which comes from their percentages by mass (Table 2.1).

The total atmospheric pressure (p) from dry air and water vapor can be written:

$$p = p_d + p_v \tag{2.58a}$$

where p_d is the pressure resulting from the dry air component, and p_v is the pressure resulting from water vapor. Using more common symbols, it can be written as:

$$p = p_d + e \tag{2.58b}$$

where e symbolizes the vapor pressure. The task now is to write the Ideal Gas Law for the two separate components.

The Ideal Gas Law for dry air is given by:

$$\boxed{p_d = \rho_d R_d T} \tag{2.59}$$

or

$$\boxed{p_d \alpha_d = R_d T} \tag{2.60}$$

which are variations on (2.37) and (2.39). This, in turn, means we need R_d, which is the "individual" gas constant of the dry air ensemble.

Effective mean molar mass and individual gas constant for dry air. In Equation (2.40), we defined the individual gas constant by:

$$\boxed{R_i = \frac{R^*}{M}}$$

In the case of the *ensemble* of gases making up dry air, there is no single molar mass, but an *effective mean molar mass* of the entire ensemble, generally given the symbol $\bar{M}$, or specifically M_d (for the dry air ensemble), which must be computed. Once we have M_d, we can compute R_d by:

$$R_d = \frac{R^*}{M_d} \tag{2.61}$$

The value of $\bar{M}$ for a given ensemble of gases is computed by:

$$\boxed{\bar{M} = \frac{\sum_{i=1}^{N} m_i}{\sum_{i=1}^{N}\left(\frac{m_i}{M_i}\right)}} \tag{2.62}[14]$$

which is one variation on the idea of a *weighted average*. In this case, $\bar{M}$ is weighted by both the masses (m_i) and the molar masses (M_i) of the individual gases in the dry-air gas ensemble. N is the number of individual gases considered in the ensemble. If we assume a fixed total sample mass of 100 [kg], then the masses of the individual gases (m_i) shown in (2.62) can be replaced with the percentages by mass shown in Table 2.1. (The derivation of this relation is one of the practice problems at the end of this chapter.)

Using (2.62) and the data shown in Table 2.1, we can now compute M_d. The first four gases are all that are needed to estimate the correct value of M_d to accuracy of two decimal places. Rewriting (2.62) for this purpose results in:

$$M_d = \frac{\sum_{i=4}^{4} m_i}{\sum_{i=4}^{4} \frac{m_i}{M_i}} \cong \frac{m_{N_2} + m_{O_2} + m_{Ar} + m_{CO_2}}{\left(\frac{m_{N_2}}{M_{N_2}}\right) + \left(\frac{m_{O_2}}{M_{O_2}}\right) + \left(\frac{m_{Ar}}{M_{Ar}}\right) + \left(\frac{m_{CO_2}}{M_{CO_2}}\right)} \tag{2.63}$$

which has units of kg/(kg/(kg/kmol)), or kg/kmol. Substituting the values shown in Table 2.1:

$$M_d \cong \frac{75.47 + 23.20 + 1.28 + 0.046}{\left(\frac{75.47}{28.02}\right) + \left(\frac{23.20}{32.00}\right) + \left(\frac{1.28}{39.94}\right) + \left(\frac{0.046}{44.01}\right)} = 28.97 \left[\frac{kg}{kmol}\right] \tag{2.64}[15]$$

Substituting this result and the value of R^* from (2.43) into (2.61) yields:

$$\boxed{R_d = \frac{R^*}{M_d} = \frac{8310 \left[\dfrac{J}{kmol\,K}\right]}{28.97 \left[\dfrac{kg}{kmol}\right]} = 286.8 \left[\frac{J}{kg\,K}\right]} \tag{2.65}$$

For most purposes, $R_d = 287$ [J/(kg K)] is precise enough.

Ideal Gas Law for water vapor. Beginning with the forms of the Ideal Gas Law shown in (2.37) and (2.39):

$$p = \rho R_i T$$

and:

$$p\alpha = R_i T$$

we can write equations of state for water vapor, substituting the symbol e for water vapor pressure:

$$\boxed{e = \rho_v R_v T} \tag{2.66}$$

and:

$$\boxed{e\alpha_v = R_v T} \tag{2.67}$$

where the subscript v indicates that the values shown are for water vapor. Both forms involve the corresponding individual gas constant for water vapor, which can be easily computed using R^* and the value of M_i for vapor (M_v) shown in Table 2.1:

$$R_v = \frac{R^*}{M_v} = \frac{8310\left[\dfrac{J}{kmol\,K}\right]}{18.02\left[\dfrac{kg}{kmol}\right]} = 461.2\left[\frac{J}{kg\,K}\right] \tag{2.68}$$

Ratio of dry air and vapor constants. Using R_d and R_v, we can calculate a constant often used in meteorology. Epsilon (ε) is defined as the ratio of R_d to R_v, but it is also the ratio of M_v to M_d. Beginning with the definition:

$$\boxed{\varepsilon \equiv \frac{R_d}{R_v}} \tag{2.69}$$

we rewrite the two individual gas constants in terms of the R^* and the two molar masses and simplify the expression:

$$\varepsilon = \frac{\left(\dfrac{R^*}{M_d}\right)}{\left(\dfrac{R^*}{M_v}\right)} = \frac{\left(\dfrac{1}{M_d}\right)}{\left(\dfrac{1}{M_v}\right)} = \frac{M_v}{M_d} \tag{2.70}$$

Substituting the values we previously computed for M_v and M_d:

$$\varepsilon = \frac{M_v}{M_d} = \frac{18.02\left[\dfrac{kg}{kmol}\right]}{28.97\left[\dfrac{kg}{kmol}\right]} = 0.622 \tag{2.71}$$

which is unitless. The values of R_d and R_v can also be directly used in (2.71) to obtain the same result. We will use this ratio extensively in later chapters.

2.8. Summary of the Ideal Gas Law

The Ideal Gas Law is one equation of state. It is the simplest equation of state. More complete versions include Van der Waals's Equation of State. There are also empirically derived equations of state. The Ideal Gas Law can be applied to a gas, or an ensemble of gases, provided the parcel is of sufficiently low density, homogeneous, and in equilibrium. In this chapter, we derived several forms of the Ideal Gas Law. These were:

$$pV = nR^*T$$

where,

p = pressure [Pa]
V = volume [m^3]
n = number of moles [kmol]
R^* = Universal Gas Constant = 8310 [J/(kmol K)]
T = temperature [K]

$$pV = mR_iT$$

where,

m = mass of the sample [kg]
R_i = individual gas constant [J/(kg K)]

$$p = \rho R_iT$$

where,

ρ = density [kg/m^3]

$$p\alpha = R_iT$$

where,

α = specific volume [m^3/kg]

$$p = n_0kT$$

where,

n_0 = number density [particles/m^3]
k = Boltzmann's Constant = 1.38×10^{-23} [J/(particle K)]

For dry air, R_d (computed by $R_d = R^*/M_d$) is substituted for R_i, and M_d (computed using a weighted average) is substituted for M_i. We found that:

$$M_d = 28.97 \left[\frac{kg}{kmol} \right]$$

For water vapor, we have the forms:

$$e = \rho_v R_v T$$

where,

e = vapor pressure [Pa]
ρ_v = density of water vapor [kg/m^3]
R_v = individual gas constant for water vapor

$$e\alpha_v = R_v T$$

where,

α_v = specific volume of vapor [m³/kg]

and M_v was determined by adding up the molar masses of two hydrogen atoms and one oxygen atom:

$$M_v = 18.02 \left[\frac{kg}{kmol} \right]$$

We also computed the values of the individual gas constants for dry air and water vapor, which were:

$$R_d = \frac{R^*}{M_d} = 286.8 \left[\frac{J}{kg\ K} \right]$$

$$R_v = \frac{R^*}{M_v} = 461.2 \left[\frac{J}{kg\ K} \right]$$

as well as the ratio of the two individual gas constants:

$$\varepsilon = \frac{R_d}{R_v} = \frac{286.8 \left[\dfrac{J}{kg\ K} \right]}{461.2 \left[\dfrac{J}{kg\ K} \right]} = 0.622$$

Practice Problems

1. Write the Ideal Gas Law for pressure, density, temperature, and an individual gas constant, and indicate the correct units for each term in the equation, when referring to a specific gas or ensemble of gases. Show that the units balance in the equation.
2. Determine the specific gas constant for the following atmosphere:

Gas	Fraction by Mass [%]	Mole weight [kg kmol⁻¹]
CO_2	95.32	44.01
N_2	2.70	28.02
Ar	1.60	39.94
Kr	0.15	83.80
O_2	0.13	32.00
CO	0.07	28.01
H_2O	0.03	18.02
Ne	0.000 25	20.18

- Show your computations.
- If the surface pressure is 7 hPa, and the temperature is 1 °C, what is the surface density of the "air"?
- This atmosphere surrounds a planet in the Solar System. Which planet do you think it is?

3. Determine the specific gas constant for the Venusian atmosphere:

Gas	Fraction by Mass [%]	Mole weight [kg kmol^{-1}]
CO_2	96.500	44.01
N_2	3.472 4	28.02
SO_2	0.015 0	64.06
Ar	0.007 0	39.94
H_2O	0.002 0	18.02
CO	0.001 7	28.01
He	0.001 2	4.000
Ne	0.000 7	20.18

- Show your computations.
- If the "sea-level" pressure of the atmosphere in this problem is 92.1 bars, and the density is 65.5 kg m^{-3}, what is the equilibrium temperature? Express your answer in °C.

4. Use the Ideal Gas Law, and the U.S. standard atmosphere shown on a Skew-T, to calculate the mean density profile of Earth's atmosphere from 0 to 15 km, in 500 meter intervals. Assume dry air. Put the results into a table.

5. Plot the result of No. 4 on a graph with density (ρ) on the horizontal axis and altitude (z) on the vertical axis.

6. Calculate the temperature, density, and specific volume of 150 kg CO_2 confined in a volume of 85 m^3 at 800 hPa.

7. Calculate the temperature, density, and specific volume of 50 kg of water vapor confined in a volume of 1000 m^3 at 900 hPa.

8. Assume a 1-kg balloon filled with H_2 at temperature 232 Kelvins, pressure 7000 hPa, and density 0.7251 kg/m^3. What is its volume? Demonstrate two *different* methods to compute the answer.

9. Use the Ideal Gas Law and Van der Waals' Equation of State to complete the table in the following text. Compute the equilibrium temperature for a 1 mol sample of water vapor occupying volume 1 m^3. Express answers to four decimal places of precision. Comment on the relative benefit of using Van der Waals' equation over the Ideal Gas Law.

Pressure [Pa]	Pressure [hPa]	Temperature from with Ideal Gas Law [K]	Temperature from Van der Waals' Equation of State [K]
1	0.01		
10	0.1		
100	1		
1×10^3	10		
1×10^4	100	1 203.369 4	
2×10^4	200		
3×10^4	300		
4×10^4	400		
5×10^4	500		6 016.730 1
6×10^4	600		
7×10^4	700		
8×10^4	800		
9×10^4	900		
1×10^5	1 000		
1×10^6	10 000		
1×10^7	100 000		

10. Beginning with the Ideal Gas Law:

$$pV = mR_iT$$

we can write:

$$p_iV = m_iR_iT$$

for each gas in an ensemble. This assumes that the gases are well mixed, and occupy the same volume of space. Combine this with Dalton's Law, and write a complete derivation for the expression for the mean molar mass of the dry air ensemble:

$$M_d = \frac{\sum_{i=1}^{N} m_i}{\sum_{i=1}^{N} \left(\dfrac{m_i}{M_i} \right)}$$

3

Work, Heat, and Temperature

3.1. Forms of Energy

The early scientists working on thermodynamics problems were interested in understanding *energy* in its many forms, as well as how it changes from one to another. This concerned them for many different reasons, most of them practical. For example, Count Rumford[1] (Benjamin Thompson) was making munitions in Germany, and noticed that drilling a cannon barrel out of a solid column of brass caused the metal to get very hot. It occurred to him that the organized work done by the grinding machine was being converted into random heat.

Rumford realized that a rigorous understanding of the relationship between the *total* amount of energy in a system, the *work* the system performs, and *heat* was needed. Eventually, a simple, elegant statement was worked out that described the relationship in terms of *internal energy*. We'll return to that in Chapter 4. In this chapter, we'll develop quantitative ways of thinking about work, heat, and temperature, as well as mention what has sometimes been called "The Zeroth Law" of thermodynamics.

3.2. Work (W)

One way of describing the abstract concept of energy in practical terms is to call it "the ability to do work." Only *organized* energy can perform work, that is, there must be a *difference* between the amount of energy in one place and the amount in another that can be exploited. For example, all objects in the Earth's gravitational field possess potential energy, also known as the *energy of position*. To use an object's potential energy to perform work, there must be a way to exploit the potential energy difference between one place and another, such as using the difference in elevation between the top of a hill and the bottom of a valley. If the amount of potential energy is the same everywhere, such as in a flat area (Kansas), there's no way to do this. See Figure 3.1.

In the case of the ball on the top of the hill, gravity is the force responsible for performing the work of rolling the ball down the hill. In more general terms, work can

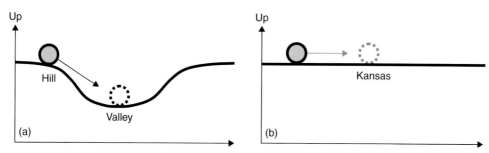

Figure 3.1. Two objects with the same gravitational potential energy. The objects in *a* and *b* are at the same elevation above the Earth's center of mass. The object on the left can exploit the potential energy *difference* between the top of the hill and the valley floor, and gravity can do work, causing the object to roll down the hill. There is no potential energy difference available for the object on the right, so gravity can't do any work: *The ball won't move.*

be thought of as the result of applying any force ($\vec{F}$) over a distance ($\overrightarrow{ds}$). A straightforward unit analysis shows the dot (scalar) product[2] $\vec{F} \bullet \overrightarrow{ds}$ has the units of work:

$$\left[\vec{F} \bullet \overrightarrow{ds}\right] = (N)(m) \tag{3.1}$$

Applying the definition of a Newton (Table 1.4) results in:

$$\left[\vec{F} \bullet \overrightarrow{ds}\right] = \left(kg\,\frac{m}{s^2}\right)(m) \tag{3.2}$$

or,

$$\left[\vec{F} \bullet \overrightarrow{ds}\right] = kg\,\frac{m^2}{s^2} \tag{3.3}$$

which is the definition of a Joule; that is, a unit of energy.

   Using this, we can define the *work* (*dW*), which is a scalar, done by a force ($\vec{F}$) acting on a mass over a distance ($\overrightarrow{ds}$) as:

$$\boxed{dW \equiv \vec{F} \bullet \overrightarrow{ds}} \tag{3.4}$$

The direction of the displacement $\overrightarrow{ds}$ may not be the same as the direction of the force causing it (see Figure 3.2). For example, let's return to the ball on the top of the hill. It can only roll diagonally down the side. The gravitational force acts on the ball, but this force is not pointing in the same direction as the ball can move. Gravity points straight "down." It can still do work on the mass, but only the component of the gravitational force acting in the same direction as the resulting motion of the ball does any work.

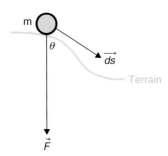

Figure 3.2 Gravitational force acting on a ball (with mass *m*) at a nonzero angle *θ*. Gravity ($\vec{F}$) is not pointing in the same direction as the resulting movement of the ball ($\vec{ds}$). Work is still done, but it is proportional to the component of the force acting in the direction of the movement.

Examining Figure 3.2, it is obvious that the force can do the *maximum* work if it is applied in the same direction as the test mass can move, but can't do *any* work if it's applied at a right (90°) angle. Put another way, it is proportional to the cosine of the angle *θ*, expressible with the more mathematically explicit form of (3.4):

$$dW = Fds\cos\theta \qquad (3.5)$$

where *θ* is the angle between the force and the direction of the displacement.

Next, let's apply this idea to a parcel of nonviscous[3] fluid, such as air, expanding into the surrounding environment. Beginning with the definition of pressure from Table 1.4:

$$p \equiv \frac{F}{A} \qquad (3.6)$$

we can write:

$$F = pA \qquad (3.7)$$

or,

$$dF = p\,dA \qquad (3.8)$$

If *dA* is a small area on the surface of the expanding parcel, and *p* is the pressure inside the parcel causing it to expand, then *dF* is the differential force exerted by the small area of the parcel's surface on the surrounding environment. Note that Newton's Third Law states that the environment also exerts force on the parcel's surface, of an equal magnitude, but in the opposite direction.

Equations (3.6) through (3.8) should lead to a mini-discussion about whether pressure is a vector or a scalar quantity (it is written as a scalar in these equations, as well as Table 1.4.). Because force *is* a vector, and we have defined pressure as force per unit area, it follows that pressure must be a vector as well. When considering the force acting on the surface area of a rigid mass, such as the force of your weight (i.e., gravity acting on your mass) on the floor of the room, pressure has a distinct direction (down). That makes it a *vector*. However, when pressure acts on the molecules of a fluid, such as air, it does so in all directions at the same time. That makes it a *scalar*. We'll adopt the latter convention now and stick with it.

Pressure in an ideal gas is a consequence of the random motion of the molecules in the parcel. As the temperature increases, the molecules move faster and exert a greater force per unit area (pressure) on each other and on the boundary of the fluid parcel. The direct relationship between temperature and the speed of the molecules in the parcel is discussed in the following text, and the direct relationship between temperature and pressure is clearly illustrated by the Ideal Gas Law, discussed in depth in Chapter 2.

Combining (3.8) with (3.4), we find that the work done by the expanding surface of the parcel is given by:

$$dW = p\overrightarrow{dA} \bullet \overrightarrow{ds} \tag{3.9}$$

which we can write as a dot-product, if we represent the small area dA as the vector $\overrightarrow{dA} = dA\,\widehat{dA}$, where the first part is the area [m²] of the surface element, and the second part is a unit vector perpendicular to the surface area. The explicit representation of this dot-product is:

$$dW = p\,dA\,ds\cos\theta \tag{3.10}$$

which is illustrated in Figure 3.3.

From Figure 3.3 and Equation (3.10), you can see that the quantity $dA\,ds\,\cos\theta$ is a small change in volume, shown as dV in the figure. With this we can rewrite (3.10) as:

$$\boxed{dW = p\,dV} \tag{3.11}$$

which states that the work (dW) done by the pressure (p) in the parcel is proportional to the parcel's change in volume (dV). Dividing through by mass, we obtain the intensive (or specific) work:

$$\boxed{dw = p\,d\alpha} \tag{3.12}$$

where dw is the specific work, and $d\alpha$ is the change in specific volume.

Sign conventions for work. We adopt the following conventions for the sign of the work, which you can prove to yourself are correct by examining (3.12). Because pressure can never be negative:

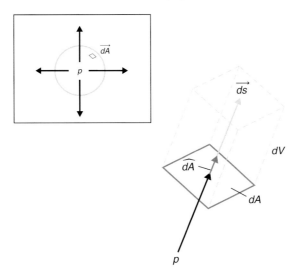

Figure 3.3. Pressure acting on the surface of an expanding parcel of air. A small patch on the surface of the parcel, represented by the vector $\vec{dA}$, moves through a distance $\vec{ds}$, as pressure acting on the inner surface causes the parcel to expand. The area dA traces out a small volume dV as it moves through distance ds. In this case, the angle between $\vec{dA}$ and $\vec{ds}$ is zero (i.e., the expansion occurs perpendicular to the surface of the parcel), but it doesn't have to be.

- if $d\alpha$ is positive (parcel expands), then dw is also positive, and we say that the *parcel* has done work on the *environment*.
- if $d\alpha$ is negative (parcel contracts), then dw is also negative, and we say that the *environment* has done work on the *parcel*.

Total work. The analysis shown in the preceding text illustrates the differential increment of work (dW or dw) that results when *constant* pressure causes a small change in volume (or specific volume). To obtain the *total* work done as the parcel expands (which you would probably be interested in, if this were an expanding piston inside a steam engine), differential work must be integrated from the beginning to final states of p and V (or α) in the system. In other words:

$$W = \int_i^f p\,dV \tag{3.13}$$

or,

$$w = \int_i^f p\,d\alpha \tag{3.14}$$

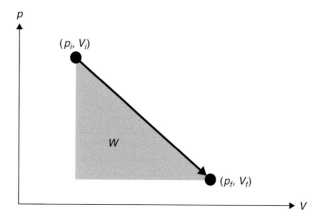

Figure 3.4. p-V diagram illustrating work. A parcel of air begins with an initial pressure p_i and volume V_i. As it expands to its final volume V_f, pressure drops to p_f. The *work* done by the pressure causing the expansion is equal to the area under the curve (shaded).

where (3.13) will compute the total *extensive* work taking the parcel from the initial to the final state and (3.14) will compute the total *specific* work. This can also be illustrated by examining a p-V diagram (see Figure 3.4).

Two things should immediately become evident. The first is that it would be rather difficult to perform the integrals shown in (3.13) and (3.14) as written, because, as Figure 3.4 clearly shows, p is not constant, so it can't be pulled out of the integral. We need another way to write the expressions inside the integrals that will allow us to perform these calculations analytically. To solve this, we will appeal to the Ideal Gas Law, but this imposes an additional condition: to use the Ideal Gas Law, we have to assume that the parcel is never far from equilibrium as it goes through these changes. This implies that it moves from its initial state to its final state in very small steps. Fortunately, the integral implies adding up a series of *infinitesimal* steps, so this requirement is satisfied. This will be discussed in greater detail later.

The second thing that should be obvious is that the total work done (equivalent to the shaded area in Figure 3.4) depends on the path the parcel takes from its initial to its final state. If it takes a different path than the simple straight line shown in the figure, the expansion will perform a different amount of work than is shown in this example. If you think this sounds familiar, you're right. Go back and reread Chapter 1, Section 1.9, which discusses line integrals. Section 1.10 illustrates line integrals on a p-V diagram.

3.3. Heat (Q)

When two bodies with different temperatures are brought into direct contact, the temperature difference between them eventually disappears. This was once described as the "flow" of a fluid, known as *caloric*, from the hotter to the cooler body. If you take a course in real analysis, you may have the opportunity to model this situation, and chances are you'll do this with a differential equation that still treats heat as a fluid that flows from one object to another.

We now understand that there is no such thing as caloric. Heat is nothing more than the random motion of the molecules of a substance – motion that, in theory, comes to a complete standstill when the temperature drops to zero on the absolute scale. Because absolute zero apparently doesn't occur anywhere in nature, there must be at least *some* random molecular motion everywhere in the universe, even in the coldest parts of deep space. As the temperature of an object increases, the random motion (velocity) of its molecules must also increase. In classical physics, there is no upper limit to either temperature or the velocity obtained by the molecules of a hot substance.

Conservation of heat. Heat, as a form of energy, must be conserved, so if two objects with different temperatures are put next to each other, they will exchange heat in such a way that, after a long period of time, the temperature difference disappears and the objects reach *thermal equilibrium*. Before equilibrium is reached, heat will "flow" from the hot to the cold object, or, one object will gain heat while the other loses heat. Since the total amount of heat is conserved:

$$\Delta Q_1 + \Delta Q_2 = 0 \qquad (3.15)$$

where the ΔQ_1 is the heat *lost* by the hot object, and ΔQ_2 is the heat *gained* by the cold object. If heat is conserved, these two must add up to zero.

The amount of heat stored in an object depends on three things:

• Temperature,
• Mass, and
• The details of its molecular structure.

Hotter, more massive objects can deliver a great deal more heat than cooler, less massive objects. This also illustrates an important difference between *heat* and *temperature*. Because heat content depends on the mass as well as the temperature, a small object can have an extremely high temperature, but contain very little heat, while a large, massive object can contain a huge amount of heat, without necessarily having a high temperature.

Specific heat (*c*). We define the relationship between change in heat content (ΔQ) and change in temperature (ΔT) by:

$$\boxed{\Delta Q \equiv cm\Delta T} \qquad (3.16)$$

or:

$$\boxed{dQ \equiv cm\,dT} \qquad (3.17)$$

where m is the mass of the object, and c is a *specific heat coefficient* that describes the heat transfer characteristics of the particular substance, but is independent of mass. This coefficient parameterizes the details of the object's molecular structure mentioned in the preceding text. It is approximately constant over narrow temperature ranges. Specific heat has units of J/(kg K), which can be shown by manipulating (3.17) to obtain:

$$c = \frac{dQ}{m\,dT} \tag{3.18}$$

and then plugging in the units for heat, mass, and temperature into the RHS.

Definition of a calorie. A *calorie* is defined as the incremental amount of heat (dQ) necessary to raise the temperature (dT) of one gram (m) of pure water from 14.5 to 15.5 °C at sea level. Put another way, the specific heat of pure, liquid water at a temperature of 15 °C and a pressure of 1013.25 hPa is 1 [cal/(g °C)]. Calories are not SI units, but they can be converted to Joules (the SI unit for energy) by:

$$1.000 \text{ cal} = 4.186 \text{ J} \tag{3.19}$$

To obtain the value of c for a given substance in SI units, we must also convert the mass units from g to kg and use the fact that a temperature change of 1 °C is equivalent to a temperature change of 1 K. Combined, the conversion is given by:

$$1.000 \left[\frac{\text{cal}}{\text{g °C}}\right] \times \left[\frac{4.186\,\text{J}}{1.000\,\text{cal}}\right] \times \left[\frac{1000\,\text{g}}{1.000\,\text{kg}}\right] \times \left[\frac{1\,°\text{C}}{1\,\text{K}}\right] = 4186 \left[\frac{\text{J}}{\text{kg K}}\right] \tag{3.20}$$

Sample values of c in both types of units are shown in Tables 3.1 and 3.2.

Heat capacity (C). A related quantity is called *heat capacity*, which *is* dependent on the mass of the object, and can found by multiplying (3.18) through by mass:

$$C \equiv cm = \frac{dQ}{dT} \tag{3.21}$$

Heat capacity has units of J/K, so would necessarily have to apply to objects with a known mass. For the majority of the work in this text, we'll stick with specific heat.

Final temperature of an ensemble of objects. Using the identity shown in (3.16), we can now rewrite the heat exchange described in (3.15) by:

$$c_1 m_1 \Delta T_1 + c_2 m_2 \Delta T_2 = 0 \tag{3.22}$$

where c_1 and c_2 are the specific heat coefficients of the substances making up the two objects [J/(kg K)], m_1 and m_2 are the masses of the two objects [kg], and ΔT_1 and ΔT_2

Table 3.1. *Specific heat coefficients for pure liquid water and ice.*[i]

Note the discontinuity between liquid water and ice that occurs at $0\,^\circ C$, resulting from the difference in their molecular structures.

Substance	Temperature [°C]	c [cal/(g °C)]	c [J/(kg K)]
Pure liquid water	50.0	0.999	4182
	45.0	0.999	4181
	40.0	0.998	4179
	35.0	0.998	4178
	30.0	0.998	4178
	25.0	0.999	4181
	20.0	0.999	4182
	15.0	1.000	4186
	10.0	1.002	4193
	5.0	1.004	4203
	0.0	1.006	4215
Pure water ice	0.0	0.490	2050
	−5.0	0.484	2027
	−10.0	0.478	2000
	−15.0	0.471	1972
	−20.0	0.464	1943
	−25.0	0.457	1913
	−30.0	0.450	1882
	−35.0	0.442	1851
	−40.0	0.434	1818
	−45.0	0.426	1785[ii]
	−50.0	0.418	1751

[i] Hess (1959); The Engineering Toolbox: Water – Thermal Properties (2012); The Engineering Toolbox: Ice – Thermal Properties (2012).
[ii] Estimated from values for c at −40.0 and −50.0 °C.

are the differences between the final (equilibrium) and initial temperatures of the two objects [Kelvins]. Because we know that the equilibrium temperatures for both objects are the same, we can rewrite (3.22) as:

$$c_1 m_1 \left(T_f - T_1 \right) + c_2 m_2 \left(T_f - T_2 \right) = 0 \qquad (3.23)$$

where T_1 and T_2 are the *initial* temperatures of the two objects. It must also be true that:

$$c_1 m_1 \left(T_f - T_1 \right) = -c_2 m_2 \left(T_f - T_2 \right) \qquad (3.24)$$

Table 3.2. *Specific heat coefficients for sea water.*[i]

All values in the table have units of J/(Kg K). Some values of c for liquid sea water are listed for temperatures below 0 °C, because very salty water (salinity of 35‰ or greater) freezes at about −2 °C at normal sea-level pressure.

Temperature [°C]	Salinity [‰][ii]								
	0	5	10	15	20	25	30	35	40
30.0	4178	4152	4126	4100	4074	4048	4023	3999	3976
25.0	4181	4153	4126	4099	4072	4046	4020	3995	3972
20.0	4182	4154	4126	4098	4071	4045	4018	3993	3968
15.0	4186	4157	4128	4099	4071	4043	4016	3990	3965
10.0	4193	4161	4130	4100	4071	4042	4015	3988	3963
5.0	4203	4168	4135	4105	4072	4042	4014	3986	3959
0.0	4215	4179	4142	4107	4074	4043	4013	3985	3957
−2.0				No Data				3984	3957

[i] Cox and Smith (1959).

[ii] Parts per thousand, vs. "%," which indicates parts per hundred. ‰ is also sometimes indicated by "ppt," and can be considered the portion of the mass of the sea water made up of salt, by mass, thus has units of $g_{salt}/kg_{seawater}$. In a more rigorous study of sea water, "physical salinity units" (psu) would replace both. Some of the zero-salinity values of c in this table were adjusted slightly to match values shown in Table 3.1.

which is another way of stating that the heat lost by m_1 is equal in magnitude, but opposite in sign, to the heat gain by m_2. Applying the distributive law results in:

$$c_1 m_1 T_f - c_1 m_1 T_1 = -c_2 m_2 T_f + c_2 m_2 T_2 \qquad (3.25)$$

which can then be regrouped as:

$$c_1 m_1 T_f + c_2 m_2 T_f = c_1 m_1 T_1 + c_2 m_2 T_2 \qquad (3.26)$$

or

$$T_f \left(c_1 m_1 + c_2 m_2 \right) = c_1 m_1 T_1 + c_2 m_2 T_2 \qquad (3.27)$$

and finally:

$$T_f = \frac{\left(c_1 m_1 \right) T_1 + \left(c_2 m_2 \right) T_2}{\left(c_1 m_1 + c_2 m_2 \right)} \qquad (3.28)$$

which, like (2.62), is a variation on the theme of weighted averages. In this case, the values being averaged are the two initial temperatures, T_1 and T_2. The *weights* are the products of the respective masses and specific heat coefficients of the two substances, that is, their heat capacities. For the final temperature of an *ensemble* of objects, we can use the more general form:

$$T_f = \sum_{i=1}^{N} \frac{(c_i m_i) T_i}{c_i m_i} = \sum_{i=1}^{N} \frac{W_i T_i}{W_i}$$

(3.29)

where T_f is the equilibrium temperature of the ensemble of objects, c_i and m_i are the specific heat coefficients and masses of the individuals objects, W_i is the weight applied to each initial temperature, T_i is the initial temperature of the objects, and N is the total number of objects in the ensemble. Note that, for this equation to work, all temperatures must be in Kelvins, and c_i and m_i must be in the SI units discussed previously (or equivalent CGS units). In addition to solid objects, (3.28) and (3.29) will also work for computing the final temperature of a mixture of fluids.

3.4. Temperature

The relationship described by (3.29) is another way of stating a condition, originally recognized by Fowler[4] in 1939, and now called the Zeroth Law of Thermodynamics: *Two objects in thermal equilibrium with a third object must also be in thermal equilibrium with each other.*[5] From here a somewhat circular definition of temperature can be stated, which is that *temperature is the property of an object that determines whether it is in thermal equilibrium with other objects.*

Temperature can also be defined by appealing to an equilibrium relation that states:

$$\frac{1}{2} m \langle v \rangle^2 = \frac{3}{2} N k T$$

(3.30)

where m is the total mass of the object [kg], $\langle v \rangle$ is the root-mean-square (RMS) speed[6] of the random molecular motion of the molecules in the substance of the object [m/s], N is the total number of particles in the object, k is Boltzmann's Constant (1.38×10^{-23} [J/(particle K)]), and T is the object's *thermodynamic* (or *equilibrium*) *temperature* [K]. A parcel of air where (3.30) applies is in thermodynamic equilibrium, which is a state where no mechanical, chemical, or thermal changes occur.[7] This is a reasonable approximation for air parcels below the turbopause – that is, within about 100 km of the Earth's surface.

The LHS of (3.30) is the mean kinetic energy of the molecules in a sample of ideal gas and is responsible for the pressure in the parcel. (We will redefine the RHS of (3.30) as the *internal energy* of a parcel in the next chapter.) At a temperature of absolute zero, (3.30) states that all random molecular motion in the parcel must stop. The Ideal Gas Law and Gay-Lussac's Law of Pressures (see Chapter 2) further state that the pressure must *also* go to zero when the temperature is absolute zero.

For (3.30) to be true, the RHS must also have units of kinetic energy, which can be proven using a straightforward unit analysis. The factor of 3/2 has no units, so it can be ignored, leaving:

$$[NkT] = (particles)\left(\frac{J}{particle\,K}\right)(K) \tag{3.31}$$

which simplifies to Joules – a unit of energy.

The thermodynamic temperature can be explicitly stated by rearranging (3.30). First, we see that both sides have a common factor of 1/2, so that:

$$m\langle v\rangle^2 = 3NkT \tag{3.32}$$

Next, we simply divide through by the factor of *3Nk*, resulting in:

$$T = \frac{m\langle v\rangle^2}{3Nk} \tag{3.33}$$

You should prove to yourself that the RHS of (3.33) has units of Kelvins by performing a unit analysis.

Degrees of freedom.[8] The factor of 3/2 on the RHS of (3.30) applies only in the case of a *monatomic gas* (i.e., a simple molecule consisting of one atom). The 3 in the numerator of this term represents the *degrees of freedom* (*f*), which, interpreted spatially, is a description of the different dimensions the molecule can move through – two horizontal and one vertical dimension. It can also be thought of as the ways in which a gas can store energy or the ways in which energy is distributed into the motion of the gas's molecules. The *Equipartition Principle* states that, if a gas is in thermodynamic equilibrium, the kinetic energy of its molecules is distributed equally among their degrees of freedom.

For a monatomic gas, such as Ar or Ne, the three degrees of freedom refer to the three spatial dimensions in which it can *translate*. For *diatomic* molecules at ordinary temperatures, *f* increases to 5, since, in addition to translating in the *x*, *y* and *z* dimensions, the atoms in the molecule can rotate around in each other in two different 2-D planes. For nonlinear *triatomic* molecules, such as CO_2 and H_2O, the atoms can rotate about each other in *three* different planes, increasing *f* to 6. At high temperatures, complex molecules can also *vibrate* (expand and contract) along their associated molecular bonds, increasing *f* to even higher values. (Vibrational degrees of freedom make only a small contribution to *f* at ordinary temperatures in the Earth system.) Table 3.3 summarizes *f* for some of the important gases in the Earth's atmosphere.

Using *f*, (3.30) can be rewritten in the more general form, which states:

$$\boxed{\frac{1}{2}m\langle v\rangle^2 = \frac{f}{2}NkT} \tag{3.34}$$

Table 3.3. *Degrees of freedom (f) for gases in Earth's atmosphere.*

These values include degrees of freedom for translation and rotation but not degrees of freedom resulting from vibration.

Gas	f
N_2	5
O_2	5
Ar	3
CO_2	6
Ne	3
Kr	3
He	3
Xe	3
H_2O	6

and the thermodynamic temperature (3.33) becomes:

$$T = \frac{m\langle v \rangle^2}{fNk} \tag{3.35}$$

Measuring temperature (thermometry). In order for *temperature* to be a useful concept, it must be measurable by either *in situ* or *remote* means. The former implies direct contact between the object being measured and another object (the sensor), which are allowed to remain in contact until they reach thermal equilibrium. Remote sensing involves standing off and detecting some emissions (or back-scattered energy) from the object whose temperature is being assessed, from which temperature can be inferred (such as through Planck's blackbody relation). Remote sensing is the subject of a separate course and is beyond the scope of this book.

The principle behind *in situ* temperature sensors is fairly straightforward: We build a device with some *thermometric property* that changes in a predictable, systematic way when the temperature of the device changes. But the Zeroth Law of Thermodynamics, expressed in (3.29), that is:

$$T_f = \sum_{i=1}^{N} \frac{\left(c_i m_i\right) T_i}{c_i m_i}$$

provides a warning: The final temperature (T_f) of the two bodies (the object being measured, and the sensor placed into contact with it) will be a weighted average of the initial temperatures (T_i) of the two bodies. The final temperature is also dependent on the masses and specific heat coefficients of both bodies in the ensemble. That is, *the temperature reported by the sensor at equilibrium is also dependent on the mass, specific heat, and initial temperature of the sensor itself.* For this reason, an ideal temperature sensor has very low mass and specific heat (meaning a very low heat capacity), so it exerts a minimum influence on the final temperature of the two-body ensemble, or:

$$T_f = \frac{\left(c_o m_o\right)T_o + \left(c_s m_s\right)T_s}{\left(c_o m_o + c_s m_s\right)} \cong \frac{\left(c_o m_o\right)T_o}{\left(c_o m_o\right)} = T_o \tag{3.36}$$

where the subscript *o* refers to the object being measured, and the subscript *s* refers to the properties of the temperature sensor.

There are several thermometric properties that can be used to build temperature sensors. One commonly exploited is *volume* expansion, which is used in mercury and alcohol-based thermometers. As the fluid in the spherical reservoir of the thermometer changes temperature, its volume also changes in a predictable way, and a narrow column of the fluid is pushed upward into a vertical scale. *Linear* expansion is used in bimetallic thermometers, which are coils made up of two different metals bonded together. As the temperature changes, the two metals expand or contract linearly in different ways, causing the coil to become tighter or looser. A needle can be attached to the bimetallic coil so that it sweeps across a circular scale as the shape of the coil changes. A third example is electrical resistance. *Thermistors* are made of a semiconducting material whose resistance decreases exponentially as the temperature increases linearly. They are inexpensive and accurate enough at ordinary temperatures for most meteorological applications. There are several more designs in use in different settings, and the choice between them is usually made depending on (1) the amount of money the builder wishes to spend and (2) the precision and accuracy needed. For more about temperature sensors, see Brock and Richardson (2001).

In all cases, the sensor response must be calibrated to a specific *temperature scale* for it to be useful. This usually means fixing the position of the indicator needle (or other sensor response) at two or more known temperatures and then using some kind of interpolation and extrapolation to estimate the sensor response at other temperatures. The function relating the sensor response to the temperature scale is most often a polynomial, such as:

$$T = ax \tag{3.37a}$$

$$T = a_1 + a_2 x \tag{3.37b}$$

$$T = a_1 + a_2 x + a_3 x^2 \tag{3.37c}$$

$$T = a_1 + a_2 x + a_3 x^2 + H.O.T. \tag{3.37d}$$

where T is the temperature, x is the amount of displacement of the fluid or needle (or, in general, any kind of sensor variation) from a reference point, and a or a_i are coefficients determined when the sensor is calibrated. *H.O.T.* refers to *higher-order terms*, some or all of which may be discarded. Some polynomials also use *cross terms* – that is, multiples and powers of two kinds of response.

The first relation, (3.37a), is a simple linear relationship, and, because it lacks an explicit intercept, it assumes that there is zero sensor response when the temperature is zero. One example of this is the absolute temperature scale. If x is the volume of an ideal gas, and this volume goes to zero, then the temperature is defined as zero. Lower temperatures (implying negative volumes) are not possible. (See *Gay-Lussac's First Law*, Chapter 2, Section 2.3.) The absolute temperature scale's slope is actually determined using two other reference points: the melting and boiling temperature of pure water at sea level. The constant a is fixed so that the difference between melting and boiling is divided into one hundred subdivisions called Kelvins.

The second relation, (3.37b), is also a simple linear relationship, but has an intercept (a_1) that does not need to be set to zero. The slope (a_2) is selected based on the number of subdivisions the inventor of the scale wishes to have between two reference points. One example is the Fahrenheit temperature scale. Fahrenheit's lower reference point was originally set to the temperature of an equal mixture of salt, water, and ammonium chloride, which stabilizes at a fixed point, that he defined as "zero." Following a suggestion from Isaac Newton, his upper reference point was set to the internal body temperature of an adult human (96 degrees).[9] There are ninety-six divisions between his upper and lower reference points, which is an integer multiple of eight, the latter being the preferred method of dividing larger portions (rather than into tenths) during the eighteenth century. Fahrenheit used this difference to determine the slope of his temperature scale and then to quantify the melting and boiling temperatures of water at 32 and 212 degrees, respectively.[10] The intercept (a_1) on Fahrenheit's temperature scale corresponds to absolute zero and is equal to −459.67 degrees. (Additional temperature scales, and the conversions between them, are discussed in Chapter 1.) Relations with higher-order polynomials, such as (3.37c) and (3.37d), are usually reserved for laboratory work, or in the factory when calibrating research-grade instrumentation.

3.5. Summary of Work, Heat, and Temperature

Organized energy can be defined as the ability to do work, and in this chapter, we defined work as either force applied over a distance, or pressure resulting in a change in volume. Heat was defined as energy stored as random motion in the molecules of a substance. Mathematically, we have:

$$dW \equiv \vec{F} \bullet \vec{ds}$$

where,
  dW = element of extensive work [J]
  $\vec{F}$ = vector force [N]
  $\vec{ds}$ = vector element of displacement [m]

$$dW = F\,ds\cos\theta$$

where,
  F = scalar magnitude of force [N]
  ds = scalar element of displacement [m]
  θ = angle between force and displacement [rad]

$$dF = p\,dA$$

where,
  dF = element of scalar force [N]
  p = pressure acting on fluid parcel [Pa]
  dA = element of parcel surface area [m²]

$$dW = p\,\vec{dA} \bullet \vec{ds}$$

where,
  $\vec{dA}$ = vector element of surface area [m²]
  $\vec{ds}$ = vector element of displacement [m]

$$dW = p\,dA\,ds\cos\theta$$

where,
  θ = angle between unit vector $\hat{dA}$ and displacement $\vec{ds}$ [rad]

$$dW = p\,dV$$

where,
  dV = element of extensive volume [m³]

$$dw = p\,d\alpha$$

where,
  dw = element of specific work [J/kg]
  $d\alpha$ = element of specific volume [m³/kg]

$$W = \int_i^f p\,dV$$

where,

　　W = total extensive work performed [J]

$$w = \int_i^f p\,d\alpha$$

where,

　　w = total specific work performed [J/kg]

$$\Delta Q = cm\,\Delta T$$

where,

　　ΔQ = finite element of extensive heat [J]
　　c = specific heat coefficient [J/(kg K)]
　　m = mass [kg]
　　ΔT = finite temperature change [K]

$$dQ = cm\,dT$$

where,

　　dQ = infinitesimal change of extensive heat [J]
　　dT = infinitesimal temperature change [K]

The "specific heat" and "heat capacity" of an object were defined by:

$$c \equiv \frac{dQ}{m\,dT}$$

where,

　　c = the specific heat (coefficient) [J/(kg K)]

and,

$$C \equiv \frac{dQ}{dT}$$

where,

　　C = the heat capacity of the object [J/K]

We defined heat as a conserved quantity, and when two objects of different masses, specific heat coefficients, and initial temperatures are put into direct contact, we found that:

$$T_f = \frac{(c_1 m_1)T_1 + (c_2 m_2)T_2}{(c_1 m_1 + c_2 m_2)}$$

where,

   T_f = final (equilibrium) temperature [K]
   c_1, c_2 = specific heat coefficients [J/(kg K)]
   m_1, m_2 = masses [kg]
   T_1, T_2 = initial temperatures [K]

which can be written in a more general form, describing the equilibrium temperature of an *ensemble* of objects (the Zeroth Law of Thermodynamics):

$$T_f = \sum_{i=1}^{N} \frac{(c_i m_i)T_i}{c_i m_i}$$

where,

   T_f = final (equilibrium) temperature [K]
   c_i = specific heat coefficients [J/(kg K)]
   m_i = masses [kg]
   T_i = initial temperatures [K]

Temperature was also defined in terms of the mean kinetic energy of the particles in the parcel of a gas by:

$$\frac{1}{2} m\langle v \rangle^2 = \frac{f}{2} NkT$$

where,

   m = total mass of molecules in gas [kg]
   $\langle v \rangle$ = RMS velocity of molecules in gas [m/s]
   f = degrees of freedom [integer]
   N = Number of molecules in parcel
   k = Boltzmann's Constant = 1.38×10^{-23} J/(particle K)
   T = thermodynamic temperature [K]

or explicitly by:

$$T = \frac{m\langle v \rangle^2}{fNk}$$

"Measured" temperature was shown to be a special case of the Zeroth Law, and the temperature reported by the sensor was shown to approach the true temperature of the object being measured, as the specific heat and mass of the sensor approached zero, namely:

$$T_f = \frac{(c_o m_o)T_o + (c_s m_s)T_s}{(c_o m_o + c_s m_s)} \cong \frac{(c_o m_o)T_o}{(c_o m_o)} = T_o$$

where,

T_f = "measured" temperature [K]

c_o, m_o = specific heat and mass of object [MKS]

c_s, m_s = specific heat and mass of sensor [MKS]

T_o = initial temperature of object [K]

T_s = initial temperature of sensor [K]

Practice Problems

1. How many degrees of freedom are there for a monatomic gas?
2. How many degrees of freedom are there for a diatomic gas?
3. The molecules in two different gas samples have the same RMS velocity. One of the samples consists of pure argon. The other consists of pure diatomic oxygen. Assuming m/N is the same in both cases, which gas has a higher temperature? Explain.
4. Two different gas samples have the same temperature. One of the samples consists of dry air. The other consists of water vapor. Assuming m/N is the same in both cases, which gas has the higher RMS velocity? Explain.
5. Calculate the work done by gravity acting on a mass of 10 kg, causing the mass to move through a distance of 100 m at an angle of 30° to the direction of the gravitational force.
6. Calculate the work done by gravity acting on a mass of 1000 kg, causing the mass to move through a distance of 1 km at an angle of 45° to the direction of the gravitational force. Then perform a unit analysis to show that the work has units of Joules.
7. Earth's gravity (at sea level) moves a 1000 kg object through a distance of 1 km, performing work equal to 2.539×10^6 Joules. Determine the angular distance between the direction of the gravitational force and the direction of the object's movement. Express your answer in degrees.
8. A downward pressure of 1000 hPa causes a disk of area 10 m^2 to move directly downward at a rate of 1 m/s. Calculate the work done by the pressure on the disk after 10 s. Then perform a unit analysis to show that the work has units of Joules.
9. Calculate the specific work done by the environment on a spherical parcel of air at 10^5 Pa over a 10-sec period, if the initial radius of the parcel is 1 m, and it decreases at a rate of 10 percent every second.
10. A constant pressure of 10^5 Pa moves a thin disk through a distance of 10 m, performing work equal to 2.5×10^6 Joules. Determine the radius of the disk.
11. A constant pressure of 10^5 Pa is applied to a 1 kg parcel of dry air with initial temperature of 10 °C. Determine the work done *by the environmental pressure* if the volume of the parcel is reduced to 0.75 m^3.
12. A mercury thermometer with initial temperature 10 °C and mass 10 grams is placed into contact with a block of pure water ice with temperature −5 °C and mass 10 kg. Compute the temperature that the thermometer will indicate after it has come into thermal equilibrium with the ice. (Note: Assume the thermometer is made of liquid mercury, which has a specific heat of about 138 J/(kg K).)

13. One kg of pure liquid water at 0 °C is placed into an insulated tank containing 5 kg of seawater of salinity 3 percent and initial temperature 17.5 °C. The system is allowed to reach equilibrium. Calculate the final temperature of the system. Express your answer in °C.

14. 5 kg of liquid seawater with initial temperature 0 °C and salinity 3 percent (by mass) is mixed with pure water with initial temperature 15 °C. The system reaches equilibrium at +5 °C. Determine the initial mass of the pure water and the final salinity of the mixture.

15. 1000 kg of seawater at 17.5 °C and initial salinity of 3 percent is mixed with 500 kg of pure water with initial temperature 5 °C. Determine the final temperature [°C] and salinity [%] of the mixture.

4

The First Law of Thermodynamics

4.1. The First Law

Energy can neither be created nor destroyed, but can only be changed from one form to another. That simple, profound statement is the basis of the First Law of Thermodynamics. This law is often thought of as so fundamental that it's simply called *the First Law*. Stated that way it's also imprecise. If we're going to make use of the First Law, we have to restate it mathematically. To do so, we'll think about energy in three ways: internal energy, heat, and work.

4.2. Internal Energy and the First Law

If heat is added to an air parcel by conduction, radiation or any other means, then the parcel's *internal energy* (U) will increase, or the parcel will perform work (by expanding its volume), or both (see Figure 4.1). This can happen with either an *open* or *closed* parcel – both allow the passage of energy from the surrounding environment into the parcel – but not an *isolated* parcel.

By *internal energy*, we mean the kinetic energy of the molecules in the sample of air. In Chapter 3, we defined this with Equation (3.30), which states that, for an ideal gas:

$$\frac{1}{2}m\langle v \rangle^2 = \frac{3}{2}NkT$$

where m is the total mass of the molecules in an object [kg], $\langle v \rangle$ is the root-mean-square (RMS) speed of the random molecular motion of the molecules in the substance of the object [m/s], N is the total number of molecules in the object, k is Boltzmann's Constant (1.38×10^{-23} J/(particle K)), and T is the object's thermodynamic temperature [K]. The factor of 3 on the RHS is the number of degrees of freedom (f), which was discussed in Chapter 3. At ordinary temperatures, f varies according to the number of atoms in the molecules of the gas. For molecules with a single atom, f is 3. For

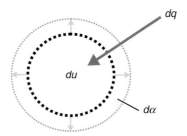

Figure 4.1. Addition of heat (dq) to an open or closed parcel. Parcel may respond by experiencing a change in internal energy (du), perform work by expanding ($d\alpha$), or both.

molecules with two atoms, such as O_2, f is 5 (see Table 3.3.) *Joule's Law*[1] is the RHS of this equation and states that the internal energy of an ideal gas depends only on its temperature. It is expressed directly by:

$$U = \frac{f}{2}NkT$$

(4.1)[2]

where U is the internal energy of the object and has units of Joules.

The First Law can be written as the relationship between heat added, *changes* in internal energy, and work performed by the parcel:

$$dQ = dU + dW$$

(4.2)

where dQ is the extensive (mass dependent) heat added to the parcel, dU is the change in extensive internal energy, and dW is the extensive work performed. All three terms have units of Joules. Dividing through by the mass of the parcel yields the intensive (specific) form of the relationship:

$$dq = du + dw$$

(4.3)

where all three terms have units of J/kg. Adopting the expression for differential specific work (dw) shown in (3.12), we can write this somewhat more explicitly as:

$$dq = du + pd\alpha$$

(4.4)

This is usually rearranged to solve for the change in specific internal energy, resulting in:

$$du = dq - pd\alpha$$

(4.5)

4.3. Enthalpy (*h*)

Enthalpy[3] is a thermodynamic function of specific internal energy (*u*), pressure (*p*), and specific volume (*α*), defined by:

$$h = u + p\alpha \tag{4.6}$$

where all three terms have units of J/kg. (A similar equation could be written using the extensive, mass-dependent variations on these terms.) A unit analysis confirms that the second term on the RHS has units of energy per unit mass, that is:

$$[p\alpha] = (Pa)\left(\frac{m^3}{kg}\right) \tag{4.7}$$

Substituting the definition of a Pascal from Table 1.4 into the first set of parenthesis on the RHS results in:

$$[p\alpha] = \left(\frac{kg}{ms^2}\right)\left(\frac{m^3}{kg}\right) = \left(kg\frac{m^2}{s^2}\right)\left(\frac{1}{kg}\right) = \frac{J}{kg} \tag{4.8}$$

The differential change in specific enthalpy is obtained by:

$$dh = d(u + p\alpha) \tag{4.9}$$

or:

$$dh = du + d(p\alpha) \tag{4.10}$$

which requires that you apply the chain rule[4] to the second term on the RHS, resulting in:

$$dh = du + p\,d\alpha + \alpha\,dp \tag{4.11}$$

Equation (4.4) shows that the first two terms on the RHS of (4.11) are equal to the change in heat (*dq*), so that we can rewrite *dh* as:

$$dh = dq + \alpha\,dp \tag{4.12}$$

which can be solved for *dq*, resulting in:

$$dq = dh - \alpha\,dp \tag{4.13}$$

If a process is *isobaric* (occurring at a constant pressure), then $dp = 0$, and the change in specific enthalpy is equal to the change in specific heat in the parcel, that is:

$$dq = dh \tag{4.14}$$

4.4. Specific Heats of an Ideal Gas[5]

If heat is added to a parcel, the parcel's internal energy will increase, and it may also perform some work on the surrounding environment by expanding. The Ideal Gas Law shows that there are three state variables that must be considered as the gas changes. The form of the Ideal Gas Law shown in (2.33) illustrates this nicely:

$$pV = mR_i T$$

where p is pressure [Pa], V is volume [m³], m is the mass of the parcel [kg], T is its temperature [K], and R_i is the individual gas constant of the parcel [J/(kg K)]. If we assume that mass is held constant (i.e., it is *not* an open parcel), then the only three variables that can vary are p, V, and T.

The current problem is to see how T varies with p and V *separately*. For example, if p is held constant and V is allowed to change, how will T change? Conversely, how will T change if V is held constant and p is allowed to change? *A more useful question is:* how much heat must be added to a parcel of air (or any other gas) to raise its temperature by 1 °C? We'll have one answer for heat added at constant *volume*, and another answer for heat added at a constant *pressure*.

We can begin with the two methods we've already derived for writing dq, shown in (4.4) and (4.13):

$$dq = du + p\,d\alpha \tag{1*}$$

$$dq = dh - \alpha\,dp \tag{2*}$$

which we'll call (1*) and (2*) for now. The term for the change in internal energy in the first equation, du, is rather vague, and we need a more explicit way of writing it if we're going to make practical use of it. The Ideal Gas Law tells us that, with mass held constant, if we know any two of the state variables, we can compute the third. In other words, we really only need *two* of them. With this in mind, let's begin with the following two assumptions:

$$u = f(T,\alpha) \tag{4.15}$$

$$h = f(T,p) \tag{4.16}$$

or, the internal energy (u) of a parcel is a function of its temperature (T) and specific volume (α), and the enthalpy (h) of a parcel is a function of its temperature and pressure (p). Taking the total differentials of (4.15) and (4.16) results in:

$$du = \left(\frac{\partial u}{\partial T}\right)_\alpha dT + \left(\frac{\partial u}{\partial \alpha}\right)_T d\alpha \tag{4.17}$$

$$dh = \left(\frac{\partial h}{\partial T}\right)_p dT + \left(\frac{\partial h}{\partial p}\right)_T dp \qquad (4.18)$$

which you should compare to (1.15).

If we substitute the identity for du from (4.17) into (1^*), we have:

$$dq = \underbrace{\left(\frac{\partial u}{\partial T}\right)_\alpha dT + \left(\frac{\partial u}{\partial \alpha}\right)_T d\alpha}_{du} + pd\alpha \qquad (4.19)$$

Combining like terms results in:

$$\boxed{dq = \left(\frac{\partial u}{\partial T}\right)_\alpha dT + \left[\left(\frac{\partial u}{\partial \alpha}\right)_T + p\right]d\alpha} \qquad (4.20)$$

Similarly, we can substitute the identity for dh from (4.18) into (2^*), resulting in:

$$dq = \underbrace{\left(\frac{\partial h}{\partial T}\right)_p dT + \left(\frac{\partial h}{\partial p}\right)_T dp}_{dh} - \alpha dp \qquad (4.21)$$

Once again, combining like terms results in:

$$\boxed{dq = \left(\frac{\partial h}{\partial T}\right)_p dT + \left[\left(\frac{\partial h}{\partial p}\right)_T - \alpha\right]dp} \qquad (4.22)$$

Changes in temperature at constant volume. One of the two questions we are trying to answer is *how much heat must be added to a parcel of an ideal gas <u>at constant volume</u> to raise its temperature by 1 °C?* The relationship shown in (4.22) doesn't have a term for volume, but (4.20) does. In fact, if volume is held constant, then $d\alpha = 0$ and (4.20) becomes:

$$dq = \left(\frac{\partial u}{\partial T}\right)_\alpha dT \qquad (4.23)$$

Dividing through by dT results in:

$$\frac{dq}{dT}_\alpha = \left(\frac{\partial u}{\partial T}\right)_\alpha \qquad (4.24)$$

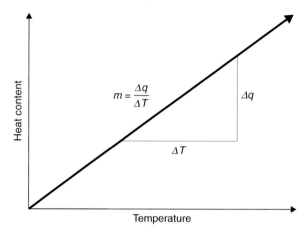

Figure 4.2. Illustration of Equation 4.24. The specific heat coefficient (*c*) of an object is the slope (*m*) of a straight line (shown here with finite differences), indicating how the object's heat content increases (decreases) as its temperature increases (decreases). The origin of the graph corresponds to a temperature of absolute zero and a heat content of zero.

which is a statement about the rate of change of heat with respect to the change in temperature – that is, the slope of a linear equation (Figure 4.2). We'll define the relationship in (4.24) as c_v, the *specific heat at constant volume*, that is:

$$c_v \equiv \frac{dq}{dT}_\alpha = \left(\frac{\partial u}{\partial T} \right)_\alpha \tag{4.25}$$

which has units of J/(kg K).

We can now answer the question posed in the preceding text, by rearranging (4.25) to obtain:

$$\boxed{dq = c_v dT} \tag{4.26}$$

where both sides of the equation have units of J/kg. A bit later we'll see how to compute values of c_v for individual gases, or the dry air gas ensemble. For a parcel with nonunit mass, we can use the mass-dependent (extensive) form to compute the total amount of heat required to raise its temperature by a small amount:

$$dQ = c_v m \, dT \tag{4.27}$$

which you should compare to Equation (3.17) in Chapter 3. Both sides of (4.27) have units of Joules. Joule's Law, shown in (4.1), states that internal energy (*u*) of an ideal gas is a function of its temperature only (plus three constants), so that the partial derivative on the RHS of (4.25) is, in fact, the total derivative of *u*. This means we can rewrite this as:

$$c_v = \frac{du}{dT} \tag{4.28}$$

so that, for changes in heat occurring at constant volume, we also have:

$$\boxed{du = c_v dT} \tag{4.29}$$

which is a fact we will return to later. Using (4.29), we can write a more explicit form of the First Law. Beginning with (4.4), which stated:

$$dq = du + p\,d\alpha$$

and substituting the identity for *du* shown in (4.29), we now obtain:

$$\boxed{dq = c_v dT + p\,d\alpha} \tag{4.30}$$

The form of the First Law shown in (4.30) is an explicit statement about how heat added to a parcel (per unit mass) is partitioned into (a) changes in its internal (kinetic) energy (the first term on the RHS), and (b) the work the parcel performs on its surrounding environment by expanding (the second term on the RHS). As long as we know pressure, and the changes in temperature and specific volume, all that's needed to compute a numerical answer now is the value of the constant c_v.

Changes in temperature at constant pressure. The second question asked was *how much heat must be added to a parcel of an ideal gas <u>at constant pressure</u> to raise its temperature by 1 °C?* The relationship shown in (4.22) has a term for pressure, and states that if pressure is held constant ($dp = 0$), we have:

$$dq = \left(\frac{\partial h}{\partial T}\right)_p dT \tag{4.31}$$

or:

$$\frac{dq}{dT}_p = \left(\frac{\partial h}{\partial T}\right)_p \tag{4.32}$$

which, like (4.24), is the slope of a linear equation. We'll define this as c_p, the *specific heat at constant pressure*, that is:

$$c_p \equiv \frac{dq}{dT}_p = \left(\frac{\partial h}{\partial T}\right)_p \tag{4.33}$$

which we can rearrange to obtain:

$$\boxed{dq = c_p dT} \tag{4.34}$$

Equation (4.34) describes the resulting change in temperature in heat is added at constant pressure.

Relationship between c_p and c_v. The Ideal Gas Law – by providing a relationship between pressure, volume, and temperature – implies that there ought to be a relationship between c_p and c_v, and we can derive this relationship by returning to the form of the First Law expressed in (4.30):

$$dq = c_v dT + p\,d\alpha \tag{3*}$$

which we'll call (3*) for now. The objective is to manipulate this form of the First Law so that we can find another way of writing the specific heat at constant pressure (c_p) that involves c_v. The first step is to focus on the second term on the RHS. We start by taking the total differential of $p\alpha$:

$$d(p\alpha) = p\,d\alpha + \alpha\,dp \tag{4.35}$$

which is obtained by the chain rule. This can be rearranged to yield:

$$p\,d\alpha = d(p\alpha) - \alpha\,dp \tag{4.36}$$

The next step begins with the form of the Ideal Gas Law shown in (2.39), namely:

$$p\alpha = R_i T$$

and taking the total differential of both sides:

$$d(p\alpha) = d(R_i T) \tag{4.37}$$

or, since is R_i a constant,

$$d(p\alpha) = R_i dT \tag{4.38}$$

Substituting the identity for $p\,d\alpha$ in (4.36) into (3*) results in:

$$dq = c_v dT + \underbrace{d(p\alpha) - \alpha\,dp}_{p\,d\alpha} \tag{4.39}$$

and substituting the identity for $d(p\alpha)$ shown in (4.38) into (4.39):

$$dq = c_v dT + \underbrace{R_i dT}_{d(p\alpha)} - \alpha\,dp \tag{4.40}$$

Next, we can combine like terms in (4.40), resulting in:

$$dq = (c_v + R_i)dT - \alpha\,dp \tag{4.41}$$

If pressure is held constant as we add the heat dq, then $dp = 0$, and we have:

$$dq = (c_v + R_i)dT \qquad (4.42)$$

or:

$$\frac{dq}{dT_p} = c_v + R_i \qquad (4.43)$$

But in (4.33), we have already established that:

$$\frac{dq}{dT_p} = c_p$$

so we now also know that:

$$\boxed{c_p = c_v + R_i} \qquad (4.44)$$

Combining (4.41) and (4.44) results in:

$$\boxed{dq = c_p dT - \alpha\, dp} \qquad (4.45)$$

which is another way of describing the result of adding heat. Note that the first term on the RHS is *not* internal energy, and the second term is *not* work, so that (4.45) is *not* a form of the First Law.

Computing numerical values of c_p and c_v. We begin by assuming heat is added to a parcel of air at constant volume. According to (4.29),

$$du = c_v dT$$

which we can rewrite for the extensive change in internal energy (dU), by multiplying through by the mass (m) of the parcel:

$$dU = c_v m\, dT \qquad (4.46)$$

We also have Joule's Law (4.1), which states:

$$U = \frac{f}{2} NkT$$

Taking the differential of both sides, we obtain:

$$dU = \frac{f}{2} Nk\, dT \qquad (4.47)$$

since f, N, and k are all constants. Comparing (4.46) and (4.47), we find that:

$$\frac{f}{2} N k\, dT = c_v m\, dT \tag{4.48}$$

or:

$$\frac{f}{2} N k = c_v m \tag{4.49}$$

Next, we can compare (2.33) and (2.46), which state:

$$pV = m R_i T$$

and

$$pV = n R^* T$$

From this comparison, we can see that:

$$\boxed{m R_i = n R^*} \tag{4.50}$$

where R_i is the individual gas constant for the substance in the parcel [J/(kg K)], n is the number of kmols of particles in the parcel, and R^* is the Universal Gas Constant (8.31×10^3 J/(kmol K)). From this, we see that:

$$m = \frac{n R^*}{R_i} \tag{4.51}$$

But, from (2.40), we also know that:

$$R_i = \frac{R^*}{M}$$

which means that:

$$M = \frac{R^*}{R_i} \tag{4.52}$$

Combining (4.51) and (4.52) results in:

$$m - nM \tag{4.53}$$

where M is the molar mass of the gas [kg/kmol]. We can substitute this identity into (4.49):

$$\frac{f}{2}Nk = c_v nM \tag{4.54}$$

Solving for c_v results in:

$$c_v = \frac{fNk}{2nM} \tag{4.55}$$

We replace N with nN_A, where n is the number of kmols, and N_A is Avagadro's Number $(6.02 \times 10^{26}$ particles/kmol), obtaining:

$$c_v = \frac{fnN_A k}{2nM} \tag{4.56}$$

or:

$$c_v = \frac{fN_A k}{2M} \tag{4.57}$$

Next, we appeal to (2.43), which states that:

$$R^* = N_A k$$

and which allows us to write the numerator on the RHS of (4.57) as:

$$c_v = \frac{fR^*}{2M} = \frac{f}{2}\left(\frac{R^*}{M}\right) \tag{4.58}$$

But (2.40) shows that this is:

$$\boxed{c_v = \frac{f}{2}R_i} \tag{4.59}$$

Last, we can obtain the basis of c_p, the specific heat at constant pressure, by combining information from (4.44) and (4.59), resulting in:

$$c_p = \left(c_v + R_i\right) = \left(\frac{f}{2}R_i + R_i\right) = \left(\frac{f}{2}R_i + \frac{2}{2}R_i\right)$$

or simply:

$$\boxed{c_p = \left(\frac{f+2}{2}\right) R_i}$$
(4.60)

According to (2.65), R_i for dry air (R_d) = 286.8 [J/(kg K)]. We can assume that dry air is primarily a diatomic gas (N_2 and O_2 make up 98.67 percent of dry air by mass; see Table 2.1), with 5 degrees of freedom (Table 3.3). Plugging these into (4.59) and (4.60), we at last obtain:

$$c_v^{air} = \left(\frac{5}{2}\right) 286.8 \left[\frac{J}{kg\,K}\right] = 717.0 \left[\frac{J}{kg\,K}\right]$$
(4.61)

$$c_p^{air} = \left(\frac{7}{2}\right) 286.8 \left[\frac{J}{kg\,K}\right] = 1003.8 \left[\frac{J}{kg\,K}\right]$$
(4.62)

A similar calculation can be performed for water vapor, which has an individual gas constant of 461.2 [J/(kg K)] (2.68) and is a triatomic gas with six degrees of freedom (Table 3.3):

$$c_v^{vapor} = \left(\frac{6}{2}\right) 461.2 \left[\frac{J}{kg\,K}\right] = 1383.6 \left[\frac{J}{kg\,K}\right]$$
(4.63)

$$c_p^{vapor} = \left(\frac{8}{2}\right) 461.2 \left[\frac{J}{kg\,K}\right] = 1844.8 \left[\frac{J}{kg\,K}\right]$$
(4.64)

Variations in specific heat with increasing temperature. Both c_p and c_v are functions of the degrees of freedom (f) in a gas, and as the temperature of the gas increases, the vibration of the molecules in the gas begins making an important contribution to f. This implies that the values of the specific heat coefficients should increase with increasing temperature, implying that, as the temperature of a gas increases, it takes progressively more and more heat to raise its temperature by another degree.

This effect is small at ordinary temperatures in Earth's atmosphere, but it is definitely real, and some effort has been spent trying to determine the analytical means of computing c_p and c_v at any given temperature. These computations are beyond the scope of this book, but Table 4.1 lists some sample values of both specific heats for dry air between 250 and 1000 K. As you can see from the table, the values of the specific heats computed in (4.61) through (4.64) should be taken as correct at 0 °C, but as you can also see, both of them only vary by about 1 percent between 0 °C and 100 °C. For this reason, they can be treated as constants without much loss of accuracy.

Table 4.1. *Specific heats for dry air as a function of temperature.*[i]

Increasing temperature adds additional degrees of freedom, resulting in increased specific heat coefficients at constant pressure and constant volume. The values of both specific heats at $0\,°C$ are included for comparison. (Values in parenthesis have been estimated from available data.)

Temperature [K]	Temperature [°C]	c_p [J/(kg K)]	c_v [J/(kg K)]
250	−23.2	1003	716
273.2	0	1003.8	717.0
300	26.8	1005	718
350	76.8	1008	721
400	126.8	1013	726
450	176.8	1020	733
500	226.8	1029	742
550	276.8	1040	753
600	326.8	1051	764
650	376.8	1063	776
700	426.8	1075	788
750	476.8	1087	800
800	526.8	1099	812
850	576.8	(1110)	(823)
900	626.8	1121	834
950	676.8	(1132)	(845)
1000	726.8	1142	855

[i] Hilsenrath et al. (1955).

4.5. Latent Heat: A Special Case of Specific Heat

In Chapter 3, we derived an expression to describe the conservation of energy when two different *solid* objects were placed into direct contact, or when two different *liquids* were mixed together. In this chapter, we've described the conservation of energy when heat is added to a parcel of *gas*, either at constant volume or at constant pressure. What we haven't considered is the possibility of *phase changes* when discussing changes in the internal energy of water, such as when enough heat is added to a block of ice to cause it to melt, or when heat is added to a sample of liquid until it boils (evaporates).

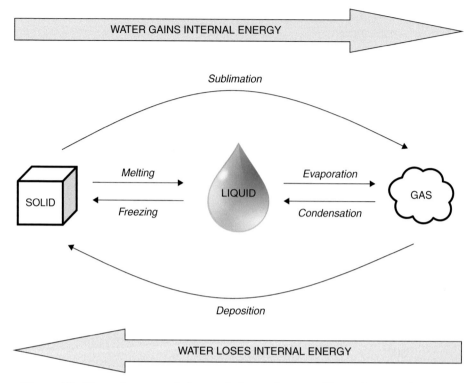

Figure 4.3. Phase changes and changes in internal energy. Phase changes take water from one of its *states* to another. Upper three phase changes (sublimation, melting, and evaporation) result in water substance *gaining* internal energy in the form of latent (hidden) heat. Lower three phase changes (deposition, freezing, and condensation) result in water substance *losing* internal energy.

You learned about phase changes in your introductory meteorology course, but here is a brief reminder of the different phase changes taking a substance from one state to another (also see Figure 4.3). We focus on water because of its prime importance in meteorology, although the following applies to all substances. The three ordinary *states*[6] of matter are solid, liquid, and gas. There is a distinct molecular configuration with each of these states. In solids, molecules are locked into a lattice structure that prevents them from moving with respect to each other. (They can still vibrate if they become hot.) In liquids, the lattice structure disappears, but the molecules are still bound to each other by electrical attraction. Without the lattice structure, the molecules can move around with respect to each other, but they can't easily escape each other's electrical attraction. In gases, each molecule has enough kinetic energy to overcome the electrical attraction of its neighbors.

Solids, therefore, maintain a fixed shape because of the lattice structure, and represent the lowest internal energy state of a substance. Liquids are not bound into a specific shape, but, at a constant pressure and temperature, they will maintain a fixed volume.[7] Liquids represent an intermediate level of internal energy of the substance.

(Water is unusual, in that its liquid state is denser than its solid state. This is why ice floats. The solid state is denser than the liquid state for most other substances.) In gases, the molecules break free from each other and spread out until they fill the available volume. Gases are the highest ordinary state of internal energy.

Phase changes are processes that take a substance from one state to another. Each substance undergoes phase changes at particular temperatures and pressures. Phase change processes that take a substance toward the right in Figure 4.3 represent a gain in internal energy for the system and a loss of energy to the system's surrounding environment. Phase change processes that move a substance toward the left represent a loss in internal energy for the system, and an energy gain for the environment.

If heat is added to a system, and the system undergoes a phase change, the increase in internal energy becomes entirely associated with the change in molecular configuration, rather than an increase in temperature. In other words, the relationship described by (4.26):

$$dq = c_v dT$$

does *not* apply. Because the added heat does not change the temperature, we call it hidden, or *latent* heat. If the heat added is enough to complete the phase change, we define it as the *latent heat coefficient* (l) of a substance, which, in its *generic* form, is defined by:

$$l \equiv dq_T \qquad (4.65)$$

where the subscript T indicates heat added at a constant temperature. Because phase changes occur at a constant temperature, the units of l are J/kg. The following sign conventions are adopted:

- If dq_T is positive (the system *gains* heat), and the system undergoes a phase change that takes it toward the right in Figure 4.3, we say that the latent heat of the substance in the system has *increased*.
- If dq_T is negative (the system *loses* heat), and the system undergoes a phase change that takes it toward the left in Figure 4.3, we say that the latent heat of the substance in the system has *decreased*.
- In more general terms, if dq_T is positive, then l is positive as well.

Latent heats of specific phase changes. We have three different possibilities:

- Latent heat of fusion (l_f) $\equiv$ specific heat of melting or freezing (dq_f).
- Latent heat of vaporization (l_v) $\equiv$ specific heat of evaporation or condensation (dq_v).
- Latent heat of sublimation (l_s) $= l_f + l_v \equiv$ specific heat of sublimation and deposition (dq_s).

In general, l_v decreases as temperature increases, dropping from 2.50×10^6 J/kg at $0\,°C$, to 2.25×10^6 J/kg at $100\,°C$. As liquid water gets hotter, it takes less latent heat to make it evaporate. This makes intuitive sense. (Boiling is a special case of evaporation that occurs when the vapor pressure is equal to the total atmospheric pressure.

This is discussed further in Chapter 7 and Appendix 5.) But l_f does the opposite, *increasing* as the temperature increases toward 0 °C. As liquid water gets colder and colder (it *can* occur with temperatures below freezing),[8] it must give up less and less latent heat to drop into a crystal-lattice configuration. Because the energy necessary to completely liberate a water molecule from the electrical attraction of its neighbors is much greater than the energy required to merely break down the crystal-lattice structure of ice, l_v is approximately an order of magnitude larger than l_f. And, by definition, sublimation can only occur at temperatures below freezing, representing a single jump from a solid to a gaseous state, so the value of l_s is the linear sum of l_v and l_f.

Some sample values for the latent heats of water are shown in Table 4.2. The latent heat of vaporization can also be estimated with the following empirical relationship:

$$l_v = l_{v0} - \left(2369 T_c\right) \tag{4.66}$$

where l_{v0} is the value of l_v at 0 °C [J] (see Table 4.2), and T_c is the temperature [°C]. Equation (4.66) gives a reasonable approximation (within a few percent) of l_v for temperatures between −30 °C and +30 °C.

4.6. Summary of the First Law, Specific Heat, and Latent Heat

In this chapter, we derived several forms of the First Law of Thermodynamics, which relates the heat added to a parcel to its subsequent change in internal energy and performance of work. Some of these forms were:

$$dq = du + dw$$

where,

    dq = change in specific heat content [J/kg]
    du = change in specific internal energy [J/kg]
    dw = change in specific work [J/kg]

$$dq = du + \underbrace{pd\alpha}_{dw}$$

where,

    p = pressure [Pa]
    $d\alpha$ = change in specific volume [m³/kg]

$$dq = \underbrace{c_v dT}_{du} + \underbrace{pd\alpha}_{dw}$$

where,

    c_v = specific heat at constant volume [J/(kg K)]
    dT = change in temperature [K]

Table 4.2. *Latent heats for water substance.*[i]

Numbers shown indicate amount of latent heat water must gain to go from a lower to a higher energy state, *or* the amount of heat water must lose to go from a higher to a lower internal energy state. (Some values have been interpolated from available data.)

Temperature [°C]	l_f [× 10^6 J/kg]	l_v [× 10^6 J/kg][ii]	l_s [× 10^6 J/kg]
50		2.3893	
45		2.3945	
40		2.4062	
35		2.4183	
30		2.4300	
25		2.4418	
20		2.4535	
15		2.4656	
10		2.4774	
5		2.4891	
0	0.3337	2.5008	2.8345
−5	0.3228	2.5128	2.8356
−10	0.3119	2.5247	2.8366
−15	0.3004	2.5366	2.8370
−20	0.2889	2.5494	2.8383
−25	0.2764	2.5622	2.8386
−30	0.2638	2.5749	2.8387
−35	0.2498	2.5890	2.8388
−40	0.2357	2.6030	2.8387
−45	0.2196	2.6189	2.8385
−50	0.2035	2.6348	2.8383

[i] Tsonis (2007).
[ii] Laboratory experiments have shown that very small droplets of water can remain liquid at temperatures as cold as about −40 °C.

We also two additional relations for specific heat:

$$dq = dh - \alpha\, dp$$

where,

dh = change in specific enthalpy [J/kg]
α = specific volume [m³/kg]
dp = change in pressure [Pa]

$$dq = c_p dT - \alpha dp$$

where,[9]

    c_p = specific heat at constant pressure [J/(kg K)]
    dT = change in temperature [K]
    α = specific volume [m³/kg]
    dp = change in pressure [Pa]

When heat is added to a system, the magnitude of the resulting increase in its temperature depends on whether the heat is added at constant volume or constant pressure. If heat is added at constant *volume*, we have a relation that says:

$$dq = c_v dT$$

where,

    c_v = specific heat at constant volume [J/(kg K)]

and if heat is added at constant *pressure*, we have a relation that says:

$$dq = c_p dT$$

where,

    c_p = specific heat at constant pressure [J/(kg K)]

Both specific heat coefficients are functions of the individual gas constant of the gas (or ensemble) in the parcel, given by:

$$c_v = \frac{f}{2} R_i$$

$$c_p = \left(\frac{f+2}{2} \right) R_i$$

where,

    f = degrees of freedom for the molecules (discussed in Chapter 3)
    R_i = individual gas constant [J/(kg K)]

Associated values of these constants for dry air and water vapor (at $0\,°C$) are:

$$c_v^{air} = 717.0 \left[\frac{J}{kg\ K} \right]$$

$$c_p^{air} = 1003.8 \left[\frac{J}{kg\ K} \right]$$

$$c_v^{vapor} = 1383.6 \left[\frac{J}{kg\ K} \right]$$

$$c_p^{vapor} = 1844.8 \left[\frac{J}{kg\ K} \right]$$

These values increase as temperature increases, although the effect is small at ordinary temperatures. Table 4.1 lists sample values for c_p and c_v of air between 250 and 1000 Kelvins, and indicates that, for ordinary temperatures, both specific heats can be treated as constants without a significant loss of accuracy.

If heat is added to (or subtracted from) a substance, such as water, resulting in a phase change, the resulting change in internal energy is focused on altering the molecular structure of the substance, not on changing the temperature. We define this as latent (hidden) heat, and we call the amount of heat required to cause a *complete* phase change in a substance its latent heat coefficient. We define three latent heat coefficients: one for melting and freezing (l_f), one for evaporation and condensation (l_v), and one for sublimation and deposition (l_s). The last one is a linear sum of the first two. Table 4.2 lists values for all three for the range $-50\,°C$ to $+50\,°C$. For temperatures in the range from $-30\,°C$ to $+30\,°C$, the following empirical relation provides a satisfactory approximation of l_v:

$$l_v = l_{v0} - (2369 T_c)$$

where,

l_{v0} = latent heat of vaporization at $0\,°C$ [J/kg]
T_c = temperature [$°C$]

Practice Problems

1. Write c_p as a function of c_v and R_i for *any* gas.
2. Write c_p and c_v as a function of R_i for a diatomic gas.
3. Compute c_v and c_p for a diatomic gas with molar mass 28.02 kg/kmol, at temperature $0\,°C$.
4. Compute R_i for a triatomic gas with molar mass 44.01 kg/kmol, at temperature $0\,°C$.
5. How many molecules of pure air are in a 1000-kg sample at $0\,°C$?
6. Write the First Law of Thermodynamics in terms of internal energy, work, and heat.
7. Write the First Law of Thermodynamics in terms of changes in specific volume and changes in temperature.
8. Compute the internal energy of a 10-mol sample of triatomic gas at 300 K.
9. Compute the *change* in internal energy of a 1-kmol sample of dry air if its temperature is raised by $10\,°C$.
10. Compute the amount of latent heat taken up when 5 kg of liquid water evaporates.
11. Compute the amount of latent heat released when 50 kg of liquid water freezes.

12. Compute the change in the heat content of a 1-kg sample of *air* if it heats up by 10 Kelvins at a constant pressure 850 hPa, and expands by 10 m³.

13. A sample of pure, dry air has initial temperature of 15 °C at 1000 hPa. Compute the change in enthalpy if 10 Joules of heat are added to the sample at a constant specific volume, changing the pressure by 10 hPa.

14. How much would the temperature of a 10-kg sample of dry air change if 3000 Joules of heat were added at constant *pressure*?

15. Compute the change in temperature if 3000 Joules of heat are added to a 10-kg sample of dry air at constant *volume*. How much heat would be required to raise the temperature of a 10-kg sample of water vapor by the same amount?

5

Adiabatic Processes

5.1. Introduction

In the last chapter, we derived several forms of the First Law, which describes how energy is conserved. In general, we saw that if heat is added to a parcel (which implies it is either *open* or *closed*, but not *isolated*), the parcel's internal energy may increase, or the parcel may expand (performing work), or both. In this chapter, we consider energy conservation in an *isolated* parcel – that is, one that does not exchange mass or energy with its surrounding environment. This doesn't mean that an isolated parcel has no interaction with the environmental air. The surrounding air can still perform work on an isolated parcel by squeezing it, and, the parcel can perform work on the environment by expanding.

The application we're interested in here is vertical motion, and the temperature changes that occur in an isolated parcel as a result (Figure 5.1). This occurs because a rising parcel moves from a high-pressure environment (near the Earth's surface) to a low-pressure environment (aloft). As the pressure falls, the parcel expands, performing work. The energy to perform the work is taken from the parcel's internal energy, which we saw is a function of its temperature. This means that, as the parcel's internal energy is converted to work, the temperature inside the parcel falls. Or, in summary, expansion results in cooling. The reverse is also true: a descending parcel is compressed (work is performed on it by the environment), causing its temperature to rise.

The reason we care about this is that when an isolated parcel is lifted aloft by an external mechanism (such as a front, or an obstacle to air flow, such as a mountain), it may wind up with a different temperature than its surrounding environment at the same altitude. If a lifted parcel is warmer than the surrounding air, this implies a lower density, which makes the parcel *buoyant*. A parcel in this condition will continue to rise on its own, even after the external lifting force is removed, and it may very well continue to climb upward until it reaches the lower stratosphere. Provided there is some moisture available, this situation is likely to produce thunderstorms and other forms of severe convective weather.

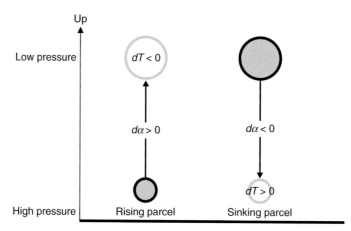

Figure 5.1. Isolated air parcel rising and descending. Rising parcels move from high pressure to low pressure, causing them to expand ($d\alpha > 0$), perform work ($dw > 0$), and expend internal energy ($du < 0$). The decrease in internal energy implies a decrease in temperature ($dT < 0$). The opposite is true for descending parcels.

Before we begin formal derivations of the governing equations, we need two definitions, both taken from the AMS Glossary of Meteorology:[1]

- *Diabatic:* A thermodynamic process in which a parcel exchanges energy with its environment by virtue of the temperature difference between them.
- *Adiabatic:* A process in which a parcel does *not* exchange energy with its environment.

The former implies an *open* or *closed* parcel, and the latter implies an *isolated* parcel. When we model vertical motion, we simplify the equations by assuming isolated parcels. This means that our equations for vertical motion revolve around the idea of adiabatic changes.

Initially, we'll assume that the parcels in question are either totally dry, or at least, that any water in them remains in a vapor state, and that its presence doesn't significantly change the characteristics of the air sample in the parcels. Later, we'll allow for latent heat released when the water vapor in the parcel condenses, *adding* heat, and the latent heat absorbed when liquid water in the parcel evaporates, *subtracting* heat.

5.2. Adiabatic Processes, the First Law, and Enthalpy

Beginning with the form of the First Law described by (4.30), we have:

$$dq = c_v dT + p\, d\alpha$$

where dq is the specific heat added to (or subtracted from) a parcel [J/kg], c_v is the constant-volume specific heat coefficient of the gas (or gas ensemble) in the parcel [J/(kg K)], dT is the change in temperature [K], p is pressure [Pa], and $d\alpha$ is the change in specific volume [m³/kg]. The first term on the RHS corresponds to the change in specific internal energy [du], and the second term on the RHS is the specific work performed by (or on) the parcel [dw]. If the parcel is isolated, and there are no phase changes, then there can be no change in the parcel's specific heat, and, by definition, $dq = 0$. In other words, for adiabatic processes,

$$\boxed{0 = c_v dT + p d\alpha}$$ (5.1)

or

$$\boxed{-c_v dT = p d\alpha}$$ (5.2)

which implies:

- If the parcel expands ($d\alpha > 0$), work is performed on the environment ($dw > 0$),[2] and the parcel temperature will decrease ($dT < 0$).
- If the parcel is compressed ($d\alpha < 0$), the environment performs work on the parcel ($dw < 0$), and the parcel temperature will increase ($dT > 0$).

We also have the specific heat relation expressed by (4.45), which states:

$$dq = c_p dT - \alpha dp$$

where c_p is the constant-pressure specific heat coefficient of the gas (or gas ensemble) of the parcel [J/(kg K)], α is the specific volume [m³/kg], and dp is change in pressure [Pa]. Again, if the process is adiabatic, then by definition $dq = 0$, and we have:

$$0 = c_p dT - \alpha dp$$ (5.3)

or:

$$\boxed{c_p dT = \alpha dp}$$ (5.4)

which implies that:

- If the pressure in the parcel increases ($dp > 0$) in a fixed volume, the parcel temperature will also increase ($dT > 0$).
- If the pressure in the parcel decreases ($dp < 0$) in a fixed volume, the parcel temperature will also decrease ($dT < 0$).

Adiabatic changes in specific enthalpy. From the definition of specific enthalpy, provided by (4.6), we have:

$$h = u + p\alpha$$

where h is the specific enthalpy [J/kg], u is the specific internal energy [J/kg], p is pressure [Pa], and α is the specific volume [m³/kg]. By taking the differential of both sides, and comparing the result to the form of the First Law given by (4.4), we found the identity specified by (4.12):

$$dh = dq + \alpha\,dp$$

where dh is the differential change in specific enthalpy [J/kg]. If a process is adiabatic, then $dq = 0$, and the relationship becomes:

$$dh = \alpha\,dp \qquad\qquad (5.5)$$

which, by comparing to (5.4), also implies that:

$$dh = c_p\,dT \qquad\qquad (5.6)$$

in adiabatic processes.

5.3. Mathematical Models of Adiabatic Variation

The scenario we want to model is an isolated parcel with an initial temperature, starting in an environment with some known pressure. The parcel is then moved to a different part of the environment, with a *different* known pressure. *How will the parcel's temperature change?* The Ideal Gas Law, in the form given by (2.39), states:

$$p\alpha = R_i T$$

where p is the pressure [Pa], α is the specific volume [m³/kg], R_i is the individual gas constant of the substance in the parcel [J/(kg K)], and T is the temperature [K]. The relationship indicates a direct relationship between pressure and temperature, *provided the specific volume remains constant*. If the parcel is moved from a high-pressure environment (such as near sea level), to a low-pressure environment, the temperature must also fall. Unfortunately, the volume will certainly change as well. In this example, the parcel will expand – the volume will increase – making the final temperature a more difficult matter to determine.

Temperature changes as a function of pressure. In mathematical terms, the present objective is to derive an expression that will allow you to compute a final temperature, if you know the initial temperature, the initial pressure, and the final pressure. Somehow the specific volume must be either controlled or eliminated from the problem altogether. Equation (5.4) has terms that relate temperature to pressure, and states that:

$$c_p\,dT = \alpha\,dp$$

This expression also contains a term for the specific volume (α), but not for the *change* in specific volume that must occur if a parcel moves from an environment with one pressure, to an environment with a different pressure. To solve the problem, we have to rewrite the term on the RHS of the equation without specific volume.

Fortunately, we can appeal to the Ideal Gas Law again. This forces us to adopt the implicit assumption that the gas in the parcel is never far from equilibrium, which we can satisfy by assuming that movement from the initial to the final condition is done in infinitesimally small steps. Solving (2.39) for specific volume yields:

$$\alpha = \frac{R_i T}{p} \tag{5.7}$$

which we can then substitute into (5.4), resulting in:

$$c_p dT = \frac{R_i T}{p} dp \tag{5.8}$$

Next, we rearrange (5.8) so that all terms involving temperature are on the LHS, and all terms involving pressure are on the RHS. We also move all constants[3] to the RHS:

$$\frac{dT}{T} = \frac{R_i}{c_p} \frac{dp}{p} \tag{5.9}$$

which is the relationship for an *infinitesimal* change. Surely, moving a parcel from 1000 hPa to 500 hPa wouldn't be considered "infinitesimal," so we have to integrate the expressions on both sides from the initial to the final condition:

$$\int_{T_i}^{T_f} \left(\frac{dT}{T} \right) = \int_{p_i}^{p_f} \left(\frac{R_i}{c_p} \frac{dp}{p} \right) \tag{5.10}$$

where the subscripts *i* and *f* indicate the *initial* and *final* conditions, respectively, on both sides of the equation. The only unknown is the final temperature (T_f), so that's what we want to isolate. We can pull the two constants out of the integral on the RHS, giving us:

$$\int_{T_i}^{T_f} \left(\frac{dT}{T} \right) = \frac{R_i}{c_p} \int_{p_i}^{p_f} \left(\frac{dp}{p} \right) \tag{5.11}$$

which is a straightforward integral to perform. From your first-semester calculus course, you know that:

$$A \int_{x_i}^{x_f} \left(\frac{dx}{x} \right) = A \left[ln\left(x_f \right) - ln\left(x_i \right) \right] = A \, ln \left(\frac{x_f}{x_i} \right) = ln \left[\left(\frac{x_f}{x_i} \right)^A \right] \tag{5.12}$$

where A is a constant, and $ln(x)$ is the natural log of x.[4] Applying this identity to (5.11) yields:

$$ln(T_f) - ln(T_i) = \frac{R_i}{c_p} \left[ln(p_f) - ln(p_i) \right] \tag{5.13}$$

or:

$$ln\left(\frac{T_f}{T_i} \right) = ln\left[\left(\frac{p_f}{p_i} \right)^{\frac{R_i}{c_p}} \right] \tag{5.14}$$

Next, we exponentiate[5] both sides of (5.14), which results in:

$$\exp\left[ln\left(\frac{T_f}{T_i} \right) \right] = \exp\left(ln\left[\left(\frac{p_f}{p_i} \right)^{\frac{R_i}{c_p}} \right] \right) \tag{5.15}$$

or:

$$\left(\frac{T_f}{T_i} \right) = \left(\frac{p_f}{p_i} \right)^{\frac{R_i}{c_p}} \tag{5.16}$$

All that remains is to solve for the final temperature, that is:

$$\boxed{T_f = T_i \left(\frac{p_f}{p_i} \right)^{\frac{R_i}{c_p}}} \tag{5.17}$$

We began with the goal of defining temperature changes as a function of changing pressure, and found a relationship that involves the constants R_i and c_p, where, interestingly, the latter describes changes occurring at constant pressure. We can further simplify the relation in (5.17) by defining κ as the ratio of the individual gas constant to the specific heat at constant pressure, that is,

$$\boxed{\kappa \equiv \frac{R_i}{c_p}} \tag{5.18}$$

but from (4.60), we know that:

$$c_p = \left(\frac{f + 2}{2} \right) R_i$$

where f is the number of degrees of freedom for the molecule in question, equal to 5 for a diatomic molecule (such as N_2), and 6 for a triatomic molecule (such as H_2O).[6] We can substitute this identity into (5.18), obtaining:

$$\kappa = \frac{R_i}{\frac{(f+2)}{2}R_i} = \frac{\frac{2}{2}R_i}{\frac{(f+2)}{2}R_i} \tag{5.19}$$

or:

$$\kappa = \frac{2}{(f+2)} \tag{5.20}$$

which has a value of about 0.286 for air (dominated by diatomic N_2 and O_2), and 0.250 for triatomic water vapor. Substituting κ into (5.17) yields:

$$\boxed{T_f = T_i \left(\frac{p_f}{p_i}\right)^{\kappa}} \tag{5.21}$$

which describes how the temperature of a parcel of air changes with pressure, as it undergoes an adiabatic change. To obtain correct results, the initial temperature must be in Kelvins, which means the final temperature comes out in Kelvins. The pressures can be expressed in any one of the units discussed in Equations (1.4a) through (1.4f), provided they are both expressed in the same units.

Visualizing the T-p relationship on a thermodynamic diagram. The relationship shown in (5.17) and (5.21) can be more easily visualized by illustrating it on a thermodynamic diagram. In Chapter 1, we looked at three variations on these diagrams: A simple 2-D diagram (with volume on the horizontal axis and pressure on the vertical axis; Figure 1.7), a 3-D diagram (adding temperature – the third state variable included in the Ideal Gas Law; Figure 1.8), and an operational variation called a Skew-T Log P (Figure 1.9).

We will look at Skew-Ts again toward the end of this chapter, but for now, we can use the simplest form (the 2-D p-V diagram) to see how temperature changes along an adiabat (see Figure 5.2). A parcel begins in a high-pressure area with initial p_i, initial volume V_i, and initial temperature T_i. It follows an adiabat (Γ) toward a low-pressure area aloft ($dp < 0$), expanding as it moves upward ($dV > 0$) and cooling as its internal energy is converted to work ($dT < 0$). The parcel achieves a final condition of p_f, V_f, and T_f.

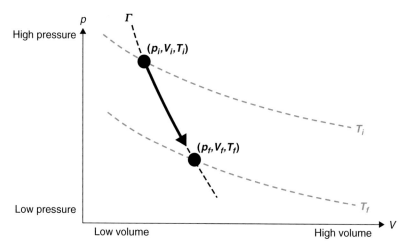

Figure 5.2. Adiabatic temperature changes on a p-V diagram. An isolated parcel moves from a high-pressure to a low-pressure environment, causing it to expand and cool. Adiabatic cooling occurs along adiabats labeled Γ (dashed black line). Isotherms are dashed gray lines, where $T_f < T_i$. (Compare to Figure 1.7.)

Additional ways of expressing the T-p relationship. There are additional ways of expressing adiabatic temperature changes as a function of pressure. Beginning with (5.17):

$$T_f = T_i \left(\frac{p_f}{p_i} \right)^{\frac{R_i}{c_p}}$$

we raise both sides of the equation to the power of c_p:

$$\left(T_f \right)^{c_p} = \left[T_i \left(\frac{p_f}{p_i} \right)^{\frac{R_i}{c_p}} \right]^{c_p} \qquad (5.22)$$

or:

$$\left(T_f \right)^{c_p} = \left(T_i \right)^{c_p} \left(\frac{p_f}{p_i} \right)^{R_i} \qquad (5.23)$$

or:

$$\left(T_f \right)^{c_p} = \left(T_i \right)^{c_p} \frac{p_f^{R_i}}{p_i^{R_i}} \qquad (5.24)$$

and finally:

$$\frac{T_f{}^{c_p}}{p_f{}^{R_i}} = \frac{T_i{}^{c_p}}{p_i{}^{R_i}} \tag{5.25}$$

which states that the ratio of the quantities T^{c_p} and p^{R_i} remains constant in adiabatic processes. This can be further generalized by:

$$\frac{T^{c_p}}{p^{R_i}} = \text{constant} \tag{5.26a}$$

or:

$$\left(T^{c_p}\right)\left(p^{-R_i}\right) = \text{constant} \tag{5.26b}$$

Poisson's Equation. We define *potential temperature* (θ) as the final temperature a parcel of air would have if brought adiabatically from any pressure to 1000 hPa. It may imply either *upward* motion (with the parcel originating at pressures higher than 1000 hPa) or *downward* motion (with the parcel originating at pressures below 1000 hPa). Potential temperature is computed by a special case of (5.21) known as *Poisson's Equation,*[7] given by:

$$\boxed{\theta = T_i \left(\frac{1000}{p_i}\right)^{\kappa}} \tag{5.27}$$

where θ replaces T_f on the LHS, and T_i and p_i are the temperature and pressure of the parcel at its point of origin. As written in (5.27), p_i must be expressed in hPa. κ is defined in (5.18) through (5.20).

Temperature changes as a function of specific volume. In Equation (5.1), we defined the following relationship (which originates with the First Law, and assumes $dq = 0$):

$$0 = c_v dT + p \, d\alpha$$

where c_v is the specific heat at constant volume [J/(kg K)], dT is the change in temperature [K], p is pressure [Pa], and $d\alpha$ is the change in specific volume [m³/kg]. Because pressure is not constant ($dp \neq 0$), if a parcel is moved adiabatically from one elevation to another, and we want a general expression (not just one that will work for *horizontal* adiabatic changes), we need another way of writing pressure. Once again we appeal to the form of the Ideal Gas Law expressed in (2.39), and solve for p:

$$p = \frac{R_i T}{\alpha} \tag{5.28}$$

which we then substitute into (5.1), yielding:

$$0 = c_v dT + \frac{R_i T}{\alpha} d\alpha \qquad (5.29)$$

This result is encouraging, because – knowing that we'll probably have to integrate again – we have terms for T, dT, α, and $d\alpha$ in (5.29). Next, we subtract the second term on the RHS of (5.29) from both sides of the equation, and divide by T and c_v, obtaining:

$$\frac{dT}{T} = -\frac{R_i}{c_v} \frac{d\alpha}{\alpha} \qquad (5.30)$$

Next, we integrate both sides of (5.30) from the initial (i) to the final (f) state:

$$\int_{T_i}^{T_f} \frac{dT}{T} = -\frac{R_i}{c_v} \int_{\alpha_i}^{\alpha_f} \frac{d\alpha}{\alpha} \qquad (5.31)$$

and use the calculus identity:

$$-\int_{x_i}^{x_f} x\,dx = \int_{x_f}^{x_i} x\,dx \qquad (5.32)$$

to reverse the order of the limits of integration on the RHS of (5.31), resulting in:

$$ln(T_f) - ln(T_i) = \frac{R_i}{c_v}\left[ln(\alpha_i) - ln(\alpha_f)\right] \qquad (5.33)$$

or:

$$ln\left(\frac{T_f}{T_i}\right) = ln\left[\left(\frac{\alpha_i}{\alpha_f}\right)^{\frac{R_i}{c_v}}\right] \qquad (5.34)$$

Finally, by exponentiating both sides, as we did in (5.15), we obtain:

$$\frac{T_f}{T_i} = \left(\frac{\alpha_i}{\alpha_f}\right)^{\frac{R_i}{c_v}} \qquad (5.35)$$

Multiplying through by the initial temperatures yields:

$$\boxed{T_f = T_i \left(\frac{\alpha_i}{\alpha_f}\right)^{\frac{R_i}{c_v}}} \qquad (5.36)$$

which resembles (5.21), with three important differences. The first is that the relationship is between temperature and specific volume, not temperature and pressure. The second regards the fractional term inside the parenthesis on the RHS. Note that the *initial* specific volume is in the *numerator*, and the *final* specific volume is in the *denominator*. In (5.21), the *final* pressure was in the numerator, and the *initial* was in the denominator. The last is the exponent of the parenthetical argument on the RHS. In (5.21), this was R_i/c_p, which we defined as κ (5.18). (Again, interestingly, we began with the goal of deriving adiabatic temperature changes with changes in volume, and the result involves c_v, a constant used to describe changes occurring at *constant* volume.) In the exponent of (5.36), we have:

$$\frac{R_i}{c_v} = \frac{R_i}{\left(\dfrac{f}{2}R_i\right)} = \frac{\left(\dfrac{2}{2}R_i\right)}{\left(\dfrac{f}{2}R_i\right)} = \frac{2}{f} \tag{5.37}$$

where, as in (5.19) and (5.20), f is the number of degrees of freedom for the molecules in the parcel, and the identity used for c_v comes from (4.59). For a diatomic gas, $f = 5$, and the value of the ratio in (5.37) is 0.400. For a triatomic gas, $f = 6$, and the ratio is 0.333. Using this identity for the exponent, we can rewrite (5.36) somewhat more simply as:

$$\boxed{T_f = T_i \left(\frac{\alpha_i}{\alpha_f}\right)^{\frac{2}{f}}} \tag{5.38}$$

but you must be careful not to confuse the f in the exponent (where it means the number of degrees of freedom) with the f subscript in T and α (where it means the "final" state).

Let's define another constant γ, by the relation:

$$\boxed{\gamma \equiv \frac{c_p}{c_v}} \tag{5.39}$$

which, like κ, can be rewritten in terms of the number of degrees of freedom for the gas in question, using the identities shown in (4.59) and (4.60):

$$\gamma = \frac{c_p}{c_v} = \frac{\left(\dfrac{f+2}{2}\right)R_i}{\left(\dfrac{f}{2}\right)R_i} = \frac{f+2}{f} \tag{5.40}$$

The exponent in (5.36) is R_i/c_v, but, (4.44) states that:

$$c_p = c_v + R_i$$

Combining this with (5.39) yields:

$$\gamma = \frac{c_v + R_i}{c_v} \tag{5.41}$$

or:

$$\gamma = \frac{c_v}{c_v} + \frac{R_i}{c_v} \tag{5.42}$$

or finally:

$$\gamma = 1 + \frac{R_i}{c_v} \tag{5.43}$$

so that:

$$\frac{R_i}{c_v} = \gamma - 1 \tag{5.44}$$

which can then be used to rewrite (5.36):

$$\boxed{T_f = T_i \left(\frac{\alpha_i}{\alpha_f} \right)^{\gamma-1}} \tag{5.45}$$

For diatomic gases, where $f = 5$ (such as N_2 and O_2), and $\gamma - 1$ is exactly 0.400. For triatomic gases, such as water vapor and ozone, $f = 6$, and $\gamma - 1$ is about 0.333. The method shown in Figure 5.2 can also be used to visualize the relationships shown in (5.36), (5.38), and (5.45), and by following a procedure similar to that used for (5.22) through (5.26), additional forms of the relation can be derived.

Specific volume changes as a function of pressure. We now close the loop of the three state variables in the Ideal Gas Law, and derive a relationship that shows how the specific volume of a parcel changes as it is taken adiabatically from one known pressure, to another known pressure. *The objective now is to eliminate temperature from the problem.*

Beginning with the form of the Ideal Gas Law given in (2.39):

$$p\alpha = R_i T$$

we take the natural log of both sides:

$$ln(p\alpha) = ln(R_iT) \tag{5.46}$$

Next we remember two identities for logarithms. The first states that:

$$ln(xy) = ln(x) + ln(y) \tag{5.47}$$

and the second one states:

$$d\left[ln(x)\right] = \frac{dx}{x} \tag{5.48}$$

Applying (5.47) to (5.46) yields:

$$ln(p) + ln(\alpha) = ln(R_i) + ln(T) \tag{5.49}$$

Next, we take the derivative of (5.49), and applying the identity in (5.48), which results in:

$$\frac{dp}{p} + \frac{d\alpha}{\alpha} = \frac{dR_i}{R_i} + \frac{dT}{T} \tag{5.50}$$

The numerator of first term on the RHS of (5.50) is the differential of a constant (R_i), which, like the derivative of a constant, is equal to zero, so that term can be eliminated. This leaves:

$$\frac{dp}{p} + \frac{d\alpha}{\alpha} = \frac{dT}{T} \tag{5.51}$$

Next, recall (5.30), which stated that:

$$\frac{dT}{T} = -\frac{R_i}{c_v} \frac{d\alpha}{\alpha}$$

and notice that (5.51) now gives us another way to write dT/T. Substituting the LHS of (5.51) into the LHS of (5.30) results in:

$$\frac{dp}{p} + \frac{d\alpha}{\alpha} = -\frac{R_i}{c_v} \frac{d\alpha}{\alpha} \tag{5.52}$$

which succeeds in eliminating temperature from the relationship. If we multiply through by c_v, we obtain:

$$c_v\left(\frac{dp}{p} + \frac{d\alpha}{\alpha}\right) = -R_i \frac{d\alpha}{\alpha} \tag{5.53}$$

or:

$$c_v \frac{dp}{p} + c_v \frac{d\alpha}{\alpha} = -R_i \frac{d\alpha}{\alpha} \tag{5.54}$$

Next, we add the RHS to both sides of the equation:

$$c_v \frac{dp}{p} + c_v \frac{d\alpha}{\alpha} + R_i \frac{d\alpha}{\alpha} = 0 \tag{5.55}$$

and combine like terms, resulting in:

$$c_v \frac{dp}{p} + \left(c_v + R_i\right)\frac{d\alpha}{\alpha} = 0 \tag{5.56}$$

The parenthetical term on the LHS may look familiar to you, and if you go back and check (4.44) you'll find:

$$c_p = c_v + R_i$$

which means we can substitute this identity into (5.56) and obtain:

$$c_v \frac{dp}{p} + c_p \frac{d\alpha}{\alpha} = 0 \tag{5.57}$$

Our next task is to integrate these terms and obtain a closed solution for changes in specific volume as a function of pressure. We begin by moving the second term in (5.57) to the RHS:

$$c_v \frac{dp}{p} = -c_p \frac{d\alpha}{\alpha} \tag{5.58}$$

Next, integrate as before, taking advantage of the fact that c_v and c_p are approximately constant:

$$c_v \int_{p_i}^{p_f} \frac{dp}{p} = -c_p \int_{\alpha_i}^{\alpha_f} \frac{d\alpha}{\alpha} \tag{5.59}$$

resulting in:

$$c_v ln\left(\frac{p_f}{p_i}\right) = -c_p ln\left(\frac{\alpha_f}{\alpha_i}\right) \tag{5.60}$$

Dividing through by c_p, and exponentiating both sides yields:

$$-\frac{c_v}{c_p} ln\left(\frac{p_f}{p_i}\right) = ln\left(\frac{\alpha_f}{\alpha_i}\right) \tag{5.61}$$

or:

$$exp\left[ln\left[\left(\frac{p_f}{p_i}\right)^{-\frac{c_v}{c_p}}\right]\right] = exp\left[ln\left(\frac{\alpha_f}{\alpha_i}\right)\right] \tag{5.62}$$

or:

$$\left(\frac{p_f}{p_i}\right)^{-\frac{c_v}{c_p}} = \left(\frac{\alpha_f}{\alpha_i}\right) \tag{5.63}$$

Last, we multiply through by the initial specific volume (α_i) and obtain:

$$\boxed{\alpha_f = \alpha_i \left(\frac{p_f}{p_i}\right)^{-\frac{c_v}{c_p}}} \tag{5.64}$$

in which we notice an interesting outcome (with respect to the constants involved) similar to the results found in (5.17) and (5.36). In this case we have a relationship describing adiabatic *changes* to both volume and pressure, that involves constants c_v and c_p, both of which assume these variables are *constant*.

Recalling the definition of the constant γ, shown in (5.39):

$$\gamma \equiv \frac{c_p}{c_v}$$

We see that the exponent in (5.64) is the negative inverse of γ, which by comparing to (5.40) can be written as:

$$-\frac{c_v}{c_p} = -\frac{1}{\gamma} = -\left(\frac{f}{f+2}\right) \tag{5.65}$$

For diatomic gases, where $f = 5$ (such as N_2 and O_2), $-1/\gamma$ is about -0.714. For tri-atomic gases, such as water vapor and ozone, $f = 6$, $-1/\gamma$ is exactly -0.750. The identity in (5.65) can also be used to rewrite (5.64) by:

$$\boxed{\alpha_f = \alpha_i \left(\frac{p_f}{p_i}\right)^{-\frac{1}{\gamma}}} \tag{5.66}$$

As with the temperature-specific volume adiabatic relationship, the method shown in Figure 5.2 can also be used to visualize the relationships shown in (5.64) and (5.66), and you can follow a procedure similar to that used for (5.22) through (5.26) to derive additional forms of the specific volume-pressure adiabatic relation.

5.4. Adiabatic Temperature Change as a Function of Height

All three adiabatic relationships derived in the preceding text involved variations in the three state variables included in the Ideal Gas Law. These relationships have their uses, such as when computing *Convectively Available Potential Energy* (CAPE), which we will return to in Chapter 12. But there are many meteorological applications, such as computing a stability using index or the *Brunt–Väisälä Frequency* (N^2), that involve the variation in temperature with *elevation*. Adiabatic temperature changes with elevation are one kind of *lapse rate*. In this section, we'll derive the *Dry Adiabatic Lapse Rate* (Γ_D), which is the rate of adiabatic cooling for an unsaturated parcel. Later, we'll derive the *Moist* (or *Pseudo*) *Adiabatic Lapse Rate* (Γ_M), which is applicable to saturated parcels.

To begin this, we have to introduce the idea of *hydrostatics*, which, as its name implies, originates with the dynamic analysis of the behavior of bodies of water, and concludes that vertical motions in a large body are very small (thus, *static*). We can apply this concept to the atmosphere by recognizing that it is only an approximation, which definitely does not apply in regions of strong upward or downward vertical motion, such as thunderstorms. On large scales (meso-α or larger[8] – i.e., at least 200 km across), this is a pretty good assumption: vertical motions are two or three orders of magnitude smaller than horizontal motions.

In Table 1.4, we defined pressure as force over area, that is:

$$\vec{p} \equiv \frac{\vec{F}}{A} \tag{5.67}$$

where A is the area [m²] over which the force $\vec{F}$ is applied. For a little while, we'll stick with the vector force then switch to the scalar definition, as discussed in Chapter 3. Because $\vec{F} = m\vec{a}$ (also from Table 1.4),

$$\vec{p} = \frac{m\vec{a}}{A} \tag{5.68}$$

where $\vec{p}$ is the vector pressure acting on the area [Pa], m is the mass of the object sitting on area A [kg], and $\vec{a}$ is the acceleration acting on this mass [m/s²]. According to Table 1.4,

$$\rho = \frac{m}{V} \tag{5.69}$$

where ρ is the density of the object [kg/m³], and V is the volume of the object [m³]. This means:

$$m = \rho V \tag{5.70}$$

which we substitute into (5.62) to obtain:

$$\vec{p} = \frac{\rho V \vec{a}}{A} \tag{5.71}$$

Let's put this in the Earth context. The acceleration at work "near" the Earth's surface – *which describes most of the atmosphere* – is "little g," that we used for a unit analysis problem in Chapter 1, Equations (1.6) through (1.11). Little g has a magnitude of about 9.81 m/s², although this varies by a few percent on the Earth's surface, depending on latitude and altitude above sea level. Because gravity points "down," we can rewrite (5.71) in its scalar form as:

$$p = -\frac{\rho V g}{A} \tag{5.72}$$

If the object in question is a column of air with a cross-sectional area equal to A, then its total volume is given by:

$$V = zA \tag{5.73}$$

where z is the vertical coordinate, describing the distance between the top of the atmosphere and the point where the pressure is being evaluated. All points below the top of the atmosphere are at negative z values. We can ignore the negative sign associated with z values in the atmosphere, and assume that the relation shown in (5.73) represents the volume of the air column between the top of the atmosphere and some point within the atmosphere. The largest negative z value would correspond to the surface of the Earth (see Figure 5.3). This means that:

$$z = \frac{V}{A} \tag{5.74}$$

Making this substitution, (5.72) becomes:

$$\boxed{p = -\rho g z} \tag{5.75}$$

which states that as height increases, pressure decreases. According to (5.75), at the top of the atmosphere, where $z = 0$, the pressure is zero. As z becomes increasingly negative, shifting the point of evaluation further and further down into the atmosphere, pressure increases. When z reaches its maximum negative value, resting on the Earth's surface, pressure is at a maximum. This analysis reflects its "hydro" origins, where $z = 0$ is often used to indicate the surface of the ocean.

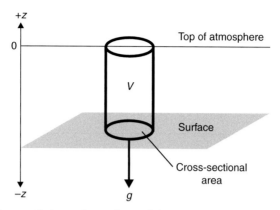

Figure 5.3 Column of air on the surface of the Earth. Total volume of the column
is equal to its height (z) multiplied by its cross-sectional area (A). Gravity (g) points
"down," causing the mass of the column to exert pressure on the surface of the Earth.

To complete this analysis, we take the total differential of both sides, and assume
that (over small vertical distances) ρ and g are constants:

$$dp = -\rho g dz \tag{5.76}$$

A little rearranging yields the hydrostatic relation we need to continue the analysis
of lapse rate:

$$\boxed{\frac{dp}{dz} = -\rho g} \tag{5.77}$$

Let's manipulate (5.77) a little further. First, we divide through by density and multiply both sides by dz, resulting in:

$$\frac{1}{\rho} dp = -g\, dz \tag{5.78}$$

Next, we use the identity from Table 1.4 that states:

$$\rho = \frac{1}{\alpha} \therefore \alpha = \frac{1}{\rho} \tag{5.79}$$

changing (5.78) to:

$$\alpha\, dp = -g\, dz \tag{5.80}$$

but (5.4) states that:

$$c_p dT = \alpha\, dp$$

Substituting the LHS of (5.4) for the LHS of (5.80) yields:

$$c_p dT = -g dz \qquad (5.81)$$

Dividing through by c_p and dz results in:

$$\boxed{\frac{dT}{dz} = -\frac{g}{c_p}} \qquad (5.82)$$

which describes the rate at which air temperature changes with height, near the surface of the Earth. A unit analysis will convince you that the RHS of (5.82) has units of K/m. Note that this result is unique to *Earth*, because it depends on both *this* planet's gravitational pull (g), and on the composition of *this* planet's atmosphere (which is parameterized in c_p).

Definition of the Dry Adiabatic Lapse Rate. We define the *Dry Adiabatic Lapse Rate* (Γ_D) as the rate at which an unsaturated air parcel cools (warms) when it's moved upward (downward) adiabatically. The term *lapse rate* implies a rate of *decrease*, so we define Γ_D as *positive* if the temperature *decreases* with increased height. To be precise:

$$\boxed{\Gamma_D \equiv -\frac{dT}{dz} = \frac{g}{c_p}} \qquad (5.83)$$

Numerical values for Γ_D can be computed by plugging in the known values of g and c_p the latter from Chapter 4 (4.62):

$$\boxed{\Gamma_D = \frac{g}{c_p} = \frac{9.81 \left[\dfrac{m}{s^2}\right]}{1003.8 \left[\dfrac{J}{kg\,k}\right]} = 0.00977 \left[\frac{K}{m}\right]} \qquad (5.84)$$

We usually convert this to K/km to make it a more manageable number, but don't forget that, when performing physical calculations, you have to use the form shown in (5.84), which has the correct *SI* units. Multiplying by a factor of (1000 m/km) results in:

$$\boxed{\Gamma_D = 0.00977 \left[\frac{K}{m}\right] \times \frac{1000}{1}\left[\frac{m}{km}\right] = 9.77\left[\frac{K}{km}\right] \cong 10\left[\frac{K}{km}\right]} \qquad (5.88)$$

where the approximate value shown on the far RHS is close enough for most applications.

There is an important point that ought to be made about the Dry Adiabatic Lapse Rate. We derived the relationship shown in (5.83) and (5.84) to show how a parcel of air cools as it is lifted vertically in the atmosphere, but we did it by beginning with the Hydrostatic Relation shown in (5.77), which assumes that there *are no* important vertical motions. This apparent contradiction can be resolved (or, to the cynic, *papered over*) by assuming that the vertical motions governed by the dry rate really *are* small, that is, *infinitesimal*. Buried in the relationship defining c_p are appeals to the Ideal Gas Law, which demand that, not only are changes *infinitesimal*, but further, these changes never take the gas very far from equilibrium. That tells us that, for (5.83) and (5.84) to apply, the vertical lifting process must be done in very small steps, which are *integrated* to compute the final temperature.

Visualizing the Dry Adiabatic Lapse Rate on a thermodynamic diagram. Γ_D can also be manipulated graphically on a thermodynamic diagram, such as a Skew-T, to compute the temperature of a rising or a sinking parcel. Because every isopleth on the Skew-T represents either a constant (such as a constant value of temperature or pressure) or an equation, the act of correctly drawing lines on the diagram is equivalent to driving one of the equations shown in the preceding text.

Let's return to the examples illustrated in Figure 5.1. One was a rising parcel, originating near the Earth's surface, and the other was a descending parcel, originating aloft. In Figure 5.4, we see the examples again, but now we can perform real computations. The rising parcel is shown on the left, originating at a temperature of 0 °C and a pressure of 1000 hPa, and rising adiabatically to 500 hPa. The path it follows upward (along one of the dry adiabats) is shown with a heavy black arrow. Because it expands as it rises, it is shown with a larger diameter at 500 hPa than it had at 1000 hPa. A thinner, gray arrow points from its position at 500 hPa to the Celsius temperature scale at the bottom of the diagram and indicates a final temperature of about −50 °C. The relations shown in (5.83) and (5.84) support this, since 500 hPa is about 5 km above 1000 hPa, and the value of Γ_D is about 10 K/km (5.88). It takes about one second to perform this calculation graphically, which makes it very convenient in an operational setting. If you want an extremely precise answer, you can use (5.21) and plug in the known parameters (with temperatures in Kelvins):

$$T_f = T_i\left(\frac{p_f}{p_i}\right)^\kappa = 273.16\left(\frac{500}{1000}\right)^{0.286} = 224.04\,K = -49.12\,°C$$

You can compute the change in the rising parcel's specific volume using the relation shown in (5.66), and assuming an initial specific volume of 1 kg/m³:

$$\alpha_f = \alpha_i\left(\frac{p_f}{p_i}\right)^{-\frac{1}{\gamma}} = (1.00)\left(\frac{500}{1000}\right)^{-0.714} = 1.64\left[\frac{m^3}{kg}\right]$$

From this result we can see that the parcel expanded by about 64 percent.

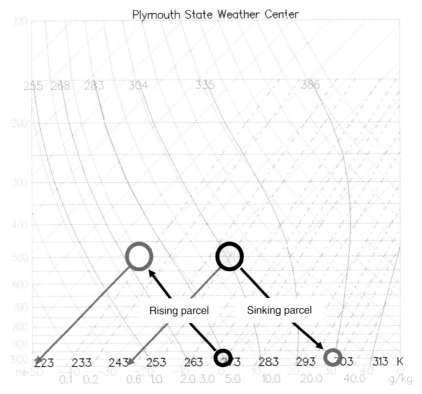

Figure 5.4. Track of parcel rising and falling adiabatically on a Skew-T Log P diagram. Dry adiabats (Γ_d) are gently curving diagonal lines originating in the lower right, and sloping toward the upper left. Isotherms (T) are straight diagonal lines originating at lower left, and sloping toward upper right – scale is in °C (brown) and K (black). Isobars (p) are horizontal lines – scale is in hPa (or mb). Vertical elevation scale (not shown) is *linear*. (Background image courtesy of Plymouth State University meteorology program, 2012.)

The descending parcel is shown on the right side of Figure 5.4, originating at a temperature of about −25 °C and a pressure of 500 hPa and descending adiabatically to 1000 hPa. The path it follows downward (along one of the dry adiabats) is shown with a heavy black arrow. Because it contracts as it sinks, it is shown with a smaller diameter at 1000 hPa than it had at 500 hPa. A thinner, gray arrow points from its position at 500 hPa to the Celsius temperature scale at the bottom of the diagram. Its final temperature is about 30 °C. Once again, you can use the equations derived in the preceding text to compute precise answers if needed. Note that, by bringing the parcel to 1000 hPa, we have also computed its potential temperature (5.27).

Two additional points. Looking at the Skew-T shown in Figure 5.4, you should notice a couple more things about the Dry Adiabatic Lapse Rate. The first is that the dry adiabats are not straight lines, as you would expect if the quantity g/c_p were really a constant. The curvature in the Γ_D isopleths is caused by the diagonally tilted isotherms, which, although scaled linearly, are not at a right angle to the height scale (which is also linear). Additional nonlinearity in the Dry Adiabatic Lapse Rate comes from vertical variations in g (see Equation 1.8) and temperature-driven variations in c_p (see Table 4.1).

The second point, which is probably more important, can be clearly illustrated by looking at the sinking parcel in the figure. In the example, the parcel begins with an *in situ* (actual) temperature of about −25 °C. It then descends dry adiabatically to 1000 hPa, achieving a final temperature of about 30 °C. But its 1000 hPa temperature is also, by definition, its potential temperature (θ), which is described by Equation 5.27. Notice that a parcel originating at any other elevation on this parcel's path would have the same potential temperature, whether it began at 340 hPa with an *in situ* temperature of −50 °C, or at 610 hPa with an *in situ* temperature of −10 °C. *For this reason, we state that dry adiabats are also lines of constant potential temperature.* We'll prove this, as well as the fact that dry adiabats are lines of constant *entropy* (*isentropes*), in Chapter 6.

5.5 Summary of Adiabatic Processes

In this chapter, we derived several adiabatic relationships, which related changes in the temperature of an isolated parcel to changes in its pressure, volume, or altitude. We also derived a relationship showing how its volume changes as a function of pressure. Some of these relationships were:

$$T_f = T_i \left(\frac{p_f}{p_i} \right)^{\frac{R_i}{c_p}}$$

where,

   T_f = final temperature [K]
   T_i = initial temperature [K]
   p_f = final pressure [Pa]
   p_i = initial pressure [Pa]
   R_i = individual gas constant [J/(kg K)]
   c_p = specific heat at constant pressure [J/(kg K)]

$$T_f = T_i \left(\frac{p_f}{p_i} \right)^{\kappa}$$

where,

   $\kappa = R_i/c_p = 2/(f+2)$
   f = degrees of freedom

$$\theta = T_i \left(\frac{1000}{p_i} \right)^{\kappa}$$

where,

θ = potential temperature [K]

1000 = reference pressure [hPa]

p_i = initial pressure [hPa]

$$T_f = T_i \left(\frac{\alpha_i}{\alpha_f} \right)^{\frac{R_i}{c_v}}$$

where,

α_i = initial specific volume [m³/kg]

α_f = final specific volume [m³/kg]

c_v = specific heat at constant volume [J/(kg K)]

$$T_f = T_i \left(\frac{\alpha_i}{\alpha_f} \right)^{\frac{2}{f}}$$

$$T_f = T_i \left(\frac{\alpha_i}{\alpha_f} \right)^{\gamma - 1}$$

where,

$\gamma = c_p/c_v = (f+2)/f$

$$\alpha_f = \alpha_i \left(\frac{p_f}{p_i} \right)^{-\frac{c_v}{c_p}}$$

$$\alpha_f = \alpha_i \left(\frac{p_f}{p_i} \right)^{-\frac{1}{\gamma}}$$

$$\frac{dT}{dz} = -\frac{g}{c_p}$$

where,

dT/dz = vertical temperature lapse rate [K/m]

g = gravitational acceleration [m/s²]

$$\Gamma_D \equiv -\frac{dT}{dz} = \frac{g}{c_p}$$

where,

Γ_D = Dry Adiabatic Lapse Rate [K/m]

Values of constants κ, $(\gamma - 1)$ and $-1/\gamma$ for dry air and water vapor are:

$$\kappa^{air} = \frac{2}{(5+2)} \cong 0.286$$

$$\kappa^{vapor} = \frac{2}{(6+2)} = 0.250$$

$$(\gamma-1)^{air} = \left[\frac{(5+2)}{5} - 1\right] = 0.400$$

$$(\gamma-1)^{vapor} = \left[\frac{(6+2)}{6} - 1\right] \cong 0.333$$

$$\left(-\frac{1}{\gamma}\right)^{air} = -\frac{5}{(5+2)} \cong -0.714$$

$$\left(-\frac{1}{\gamma}\right)^{vapor} = -\frac{6}{(6+2)} = -0.750$$

The Dry Adiabatic Lapse Rate had a value of:

$$\Gamma_D = 0.00977\left[\frac{K}{m}\right] \times \frac{1000}{1}\left[\frac{m}{km}\right] = 9.77\left[\frac{K}{km}\right] \cong 10\left[\frac{K}{km}\right]$$

Finally, we showed that dry adiabats were also lines of constant potential temperature.

Practice Problems

1. A 1 kg spherical weather balloon filled with diatomic hydrogen (H_2) rises adiabatically from 1000 hPa to 1 hPa, where it bursts. Its initial volume is 0.5 m³. Determine its final volume.
2. If the initial temperature of the balloon in the previous question is 5 °C, what is its temperature just before bursting? Express your answer in °C. Does this seem realistic?
3. What is the specific work performed by the balloon on the environment during ascent?

4. Perform a unit analysis on your answer to question 3 and show that the units are J/kg.
5. Determine the density of the balloon just before it bursts.
6. Plot the temperature profile below on a Skew-T. Do you see a low-level temperature inversion? What kind of inversion do you think it is? What is the pressure level of the tropopause?

Station: KGYX
Latitude: 43.90
Longitude: −70.25
Elevation: 125.00

ELEV	PRES mb	HGHT m	TEMP C
SFC	990	125	−6.3
2	983	181	−3.3
3	969	295	−2.5
4	939	545	−4.1
5	925	664	−4.3
6	914	758	−4.3
7	886	1003	−6.3
8	864	1199	−7.9
9	850	1326	−8.5
10	828	1529	−10.3
11	797	1823	−10.1
12	770	2088	−9.5
13	700	2817	−14.7
14	643	3454	−20.1
15	620	3723	−21.9
16	596	4012	−24.5
17	576	4261	−24.1
18	536	4782	−27.9
19	500	5280	−29.7
20	481	5555	−30.9
21	419	6513	−39.3
22	409	6678	−39.3
23	400	6830	−40.3
24	362	7507	−44.9
25	355	7638	−45.6
26	305	8642	−50.7
27	300	8750	−50.9
28	284	9104	−51.7
29	266	9527	−51.5
30	255	9801	−49.9

7. Calculate the potential temperature (θ) profile two ways: by using a dry adiabat (Γ_d), and by using Poisson's Equation. Enter your answers in the following table. How does the potential temperature profile differ from the *in situ* temperature profile?

Pressure [hPa]	θ from Γ_d [°C]	θ from Poisson's Equation [K]	θ from Poisson's Equation [°C]
990			
983			
969			
939			
925			
914			
886			
864			
850			
828			
797			
770			
700			
643			
620			
596			
576			
536			
500			
481			
419			
409			
400			
362			
355			
305			
300			
284			
266			
255			

8. Calculate the potential density (ρ_{1000}) profile of this atmosphere, assuming the potential temperature (θ) values determined in column 3 of the preceding table, a pressure of 1000 hPa, and no water vapor. Enter your answers in the following table. Where do you find the largest negative vertical density gradient ($\Delta\rho_{1000}/\Delta z$)? How does this relate to the temperature inversion in the plotted profile?

Pressure [hPa]	ρ_{1000} [kg m^{-3}]	z [km]	$\Delta\rho_{1000}/\Delta z$ [(kg m^{-3}) / km]
990		0.125	
983		0.181	
969		0.295	
939		0.545	
925		0.664	
914		0.758	
886		1.003	
864		1.199	
850		1.326	
828		1.529	
797		1.823	
770		2.088	
700		2.817	
643		3.454	
620		3.723	
596		4.012	
576		4.261	
536		4.782	
500		5.280	
481		5.555	
419		6.513	
409		6.678	
400		6.830	
362		7.507	
355		7.638	
305		8.642	
300		8.750	
284		9.104	
266		9.527	
255		9.801	

9. Use the results of No. 8 to say whether a surface-based parcel would be able to reach an altitude of 1 km, without the assistance of an external lifting force. Explain your conclusions.

10. Derive the "Dry Adiabatic Lapse Rate" (Γ_D) for Jupiter. Use $g_j = GM/R^2$, where G is the universal gravitational constant (6.673×10^{-11} Nm^2kg^{-2}), M is the mass of Jupiter (1.899×10^{27} kg), and R is the radius of the planet (71.49×10^6 m). You may ignore the slight offset to gravity caused by the rotation of the planet, and because Jupiter's atmosphere is dominated by H_2, you may assume it behaves as a diatomic gas. The composition of Jupiter's atmosphere is as follows:

Gas	Fraction by Mass [%]	Mole Weight [kg kmol^{-1}]
H_2	86.00	2.00
He	13.00	4.00
CH_4	0.500	16.04
H_2O	0.500	18.02

Express your answer in K/km. How does it compare to Earth's Dry Adiabatic Lapse Rate?

6

The Second Law of Thermodynamics

6.1. Introduction

The First Law was a statement about the conservation of energy, which in the twentieth century (thanks to the work of Einstein), was further generalized into a statement of conservation of energy and *mass* (which he showed were not independent of each other). Put simply, it now states that *the sum totality of mass and energy in the universe is a constant*. But there is another idea in thermodynamics, related but separate, that when translated into English, says *you can't get something for free – you always have to pay somehow*. And payment is in the form of waste material of some sort. The *Second Law of Thermodynamics*, like the First Law, is often simply called the Second Law, because of the deceptively simple yet fundamental ideas that it captures.

Every so often, someone claims to have built a machine (or claims to know how to build a machine) that runs on its own batteries so efficiently that it can recharge the batteries, and have enough energy left over to perform useful work, without consuming any fuel from an outside source. Once started, the machine perpetually recharges itself, and need never be refueled or even switched off. It will run forever. For this reason these mythical devices are known as *perpetual motion machines*. They do not exist and can never exist.

A much more prosaic problem, examined and finally understood in the nineteenth century, was the inability to build steam engines that converted all the heat generated in their boilers to mechanical work. In spite of the engineering genius applied, engines always dump some heat into the atmosphere. Their efficiency is always less than 100 percent, and in fact is *much less* than 100 percent.

The Second Law explains both of these phenomena, and it also applies to the atmosphere. Instead of steam engines or the pistons of an internal combustion engine, our "engine" consists of a parcel of air heated by the surface of the Earth, causing it to expand and perform work, which manifests itself as "weather." This means we need to understand the Second Law quantitatively to effectively model the atmosphere and forecast the weather. But before we start deriving Second Law relationships, let's define three classes of thermodynamic process:[1]

- *Natural* (or *irreversible*): A process in which there are significant departures from equilibrium – that is, one where the Ideal Gas Law *does not* apply. Some examples include the diffusion of one gas into another (resulting in a mixture) or the conduction of heat from a hot object to a colder object across a finite[2] temperature gradient (resulting in two objects with the same temperature). Neither of these can be undone without doing work powered by an external source.
- *Unnatural* (or *reversible*): A process that moves a system through a succession of equilibrium states, departing only infinitesimally from equilibrium in each step, so that the Ideal Gas Law *does* apply. A system moving through a reversible process from one state to another can be pushed backward along the same path and returned to its original state. These processes do not violate the First Law, but they don't occur in nature either.
- *Antinatural* (**impossible**):[3] Another kind of process that does not occur in nature, such as the spontaneous separation of a mixture of gases into its constituent gases, or the flow of heat from a cold object to a hotter object.

A key idea in Second Law physics is that *irreversible* processes can be modeled with a series of infinitesimal *reversible* steps, implying integration, and permitting us to use the Ideal Gas Law where needed. This is a bit of a dodge, but it makes it possible to solve the equations, and it provides a pretty good model of the atmosphere.

6.2. Entropy

Entropy is a quantity originally defined to describe the loss of energy as *waste heat* in steam engines. After a great deal of investigation, it was found that *some* energy waste is inescapable, and not merely the result of less-than-ideal design. *It applies to all machines*. In the late nineteenth century, Boltzmann[4] redefined entropy as the *amount of disorder* in a system, which must always increase in natural (irreversible) systems, unless work (powered by an outside source) is performed on the system.

It is Boltzmann's definition that has led to a great deal of confusion. One example of this confusion is the claim, made by some, that the emergence of life on Earth – a highly organized state, from less organized "nonlife" – is evidence of the deliberate hand of an outside agency, because it appears to imply a spontaneous reduction in entropy. From a thermodynamic point of view, the biological system includes Earth and *the Sun*. The Sun provides the work to drive biology from less organized to more organized states, and collects the resulting disorder in its thermonuclear core. (Eventually, this waste material will cause the Sun's thermonuclear reactions to cease, and the Sun will collapse into a white dwarf.) A considerable amount of waste heat is also deposited directly into the Earth system.[5]

Entropy is defined mathematically by the general expression:

$$\boxed{ds \geq \frac{dq}{T}} \tag{6.1}$$

where ds is the change in specific entropy [J/(kg K)], and dq is the specific heat [J/kg] "dispersed" at temperature T [K].

Reversible processes. For an infinitesimal step of a reversible process,

$$ds = \frac{dq}{T} \tag{6.2}$$

and, for the total change in entropy, this must be integrated from the initial (i) to the final (f) state, that is:

$$\Delta s = \int_i^f ds = \int_i^f \frac{dq}{T} \tag{6.3}$$

where T may or may not be a constant.

Variations in entropy with adiabatic processes. A special case of *reversible process* is an *adiabatic process*, which, as we are about to show, results in zero change in entropy.[6] We begin with the relation shown in (4.45), which states:

$$dq = c_p \, dT - \alpha \, dp$$

and rewrite the expression by dividing through by T, and substituting (6.2) into the LHS, resulting in:

$$ds = \frac{dq}{T} = \frac{c_p \, dT}{T} - \frac{\alpha \, dp}{T} \tag{6.4}$$

We need another way to write temperature in the denominator of the second term on the RHS of (6.4). Because adiabatic processes are reversible, and occur in infinitesimal steps never far from equilibrium, we can appeal to the Ideal Gas Law, beginning with the form described in (2.39):

$$p\alpha = R_i T$$

and solving for T, which results in:

$$T = \frac{p\alpha}{R_i} \tag{6.5}$$

where T is temperature [K], p is pressure [Pa], α is specific volume [m³/kg], and R_i is the individual gas constant for the gas in the parcel [J/(kg K)]. Inverting (6.5) results in:

$$\frac{1}{T} = \frac{R_i}{p\alpha}$$

(6.6)

which we then substitute into the second term on the RHS of (6.4):

$$ds = \frac{c_p\,dT}{T} - \frac{\alpha R_i\,dp}{p\alpha}$$

(6.7)

or, after canceling α:

$$\boxed{ds = c_p\frac{dT}{T} - R_i\frac{dp}{p}}$$

(6.8)

Now recall *Poisson's Equation* (5.27), which stated that (for adiabatic processes):

$$\theta = T_i\left(\frac{1000}{p_i}\right)^{\kappa}$$

where $\kappa \equiv R/c_p$, T_i is a parcel's initial temperature [K], p_i is its initial pressure [hPa], and θ is its potential temperature (the temperature it would have if brought dry adiabatically to 1000 hPa) [K]. Dropping the subscript i and taking the natural log of both sides results in:

$$ln(\theta) = ln\left[T\left(\frac{1000}{p}\right)^{\kappa}\right]$$

(6.9)

or:

$$ln(\theta) = ln(T) + ln\left[\left(\frac{1000}{p}\right)^{\kappa}\right]$$

(6.10)

or further:

$$ln(\theta) = ln(T) + \kappa\,ln\left(\frac{1000}{p}\right)$$

(6.11)

which can be rewritten as:

$$ln(\theta) = ln(T) - \kappa\,ln\left(\frac{p}{1000}\right)$$

(6.12)

and finally:

$$ln(\theta) = ln(T) - \kappa ln(p) + \kappa ln(1000) \tag{6.13}$$

Note that the third term on the RHS of (6.13) is a constant. If we take the derivative of both sides, we obtain:

$$\frac{d\theta}{\theta} = \frac{dT}{T_i} - \kappa \frac{dp}{p} + \kappa \frac{d(1000)}{1000} \tag{6.14}$$

or, because the derivative of a constant is zero,

$$\frac{d\theta}{\theta} = \frac{dT}{T_i} - \kappa \frac{dp}{p} \tag{6.15}$$

We now substitute the definition of κ, resulting in:

$$\frac{d\theta}{\theta} = \frac{dT}{T} - \frac{R_i}{c_p} \frac{dp}{p} \tag{6.16}$$

and then multiply through by c_p:

$$c_p \frac{d\theta}{\theta} = c_p \frac{dT}{T_i} - R_i \frac{dp}{p} \tag{6.17}$$

Adding the second term on the RHS to both sides of the equation, we obtain:

$$\boxed{c_p \frac{d\theta}{\theta} + R_i \frac{dp}{p} = c_p \frac{dT}{T}} \tag{6.18}$$

We then substitute the identity for $c_p \dfrac{dT}{T}$ given in (6.18) into (6.8), which gives us:

$$ds = \underbrace{\left[c_p \frac{d\theta}{\theta} + R_i \frac{dp}{p} \right]}_{c_p \frac{dT}{T}} - R_i \frac{dp}{p} \tag{6.19}$$

or:

$$ds = c_p \frac{d\theta}{\theta} + R_i \frac{dp}{p} - R_i \frac{dp}{p} \tag{6.20}$$

Noting that the second and third terms on the RHS of (6.20) cancel, we are left with:

$$\boxed{ds = c_p \frac{d\theta}{\theta}} \tag{6.21}$$

which states that, if potential temperature is constant ($d\theta = 0$), then there is also no change in entropy ($ds = 0$). For this reason, dry adiabats (Γ_D) are not only lines of constant potential temperature,[7] but are also lines of constant entropy, and are therefore also called *isentropes*.

Irreversible processes. The entropy of an *isolated* system undergoing a *natural* process must always increase. If one integrates the total change in entropy from the initial (*i*) to the final (*f*) state,

$$\Delta s > \int_i^f \frac{dq}{T} \tag{6.22}$$

Variations in entropy with diabatic temperature increases: With *closed* or *open* parcels,[8] the environment can diabatically add heat. If heat is added to a substance, it may result in a temperature increase, a phase change, or both. For the entropy change resulting from an increase in temperature, we appeal to the definition of heat provided in (3.17), namely:

$$dQ \equiv cm\,dT$$

where dQ is the heat input required [J] to increase the temperature by a small amount dT [K] of an object with mass m [kg], and c is the specific heat of the object [J/(kg K)]. Dividing through by mass yields:

$$dq = c\,dT \tag{6.23}$$

which we can then substitute into (6.22):

$$\Delta s > \int_i^f \frac{c\,dT}{T} \tag{6.24}$$

We can compute a *minimum* value of the specific entropy change for an object by integrating from the initial to the final temperature, and assuming c is a constant over a small temperature range:

$$\Delta s \geq c \int_{T_i}^{T_f} \frac{dT}{T} \tag{6.25}$$

The solution to the integral on the RHS of (6.25) is:

$$\Delta s \geq c\left[ln\left(T_f\right) - ln\left(T_i\right) \right] \tag{6.26}$$

or:

$$\boxed{\Delta s \geq c\,ln\left(\frac{T_f}{T_i}\right)}$$ (6.27)

If we want the total change in entropy for a nonunit mass, we can simply multiply by the mass *m* to obtain:

$$\boxed{\Delta S \geq cm\,ln\left(\frac{T_f}{T_i}\right)}$$ (6.28)

where both sides of (6.28) have units of [J/K].

Variations in entropy with phase changes: For entropy changes resulting from a phase change, $dq = l$ (4.65), and l is the latent heat of fusion, vaporization, or sublimation (see Table 4.2 for latent heat values of water substance). Substituting this definition into (6.22) results in:

$$\Delta s > \int_i^f \frac{l}{T}$$ (6.29)

Table 4.2 shows that l is constant at a fixed temperature, so – because T is also constant – there is nothing to integrate, and for the minimum change in specific entropy we have:

$$\boxed{\Delta s \geq \frac{l}{T}}$$ (6.30)

For the total entropy change in a nonunit mass, we multiply through by mass again, obtaining:

$$\boxed{\Delta S \geq \frac{lm}{T}}$$ (6.31)

Example: Let's say we have a block of pure water ice with a mass of 100 grams and an initial temperature of −20 °C. The question we want to answer is: *What is the total change in entropy for the water substance if we add heat until its temperature increases to zero Celsius, the ice melts, the temperature then increases to 100 °C, and the water boils?*

Step 1: Raise the temperature of the ice from −20 °C to 0 °C. For this step, apply (6.28) using:

$c \cong 2000$ J/(kg K)[9]
$m = 100$ g $= 0.100$ kg
$T_i = -20\,°C = 253.16$ K
$T_f = 0\,°C = 273.16$ K

Therefore:

$$\Delta S_1 \geq \left(2000\left[\frac{J}{kg\,K}\right]\right)(0.100[kg])\,ln\left(\frac{273.16[K]}{253.16[K]}\right) = 15.21\left[\frac{J}{K}\right]$$

Step 2: Melt the ice at a constant temperature of $0\,°C$. For the second step, apply (6.31) using:

   $m = 100\ g = 0.100\ kg$
   $l_f = 3.337 \times 10^5$ J/kg[10] (*positive* because substance goes from a lower to a higher-energy state[11])
   $T = 0\,°C = 273.16\ K$

The relationship becomes:

$$\Delta S_2 \geq \frac{\left(3.337\times10^5\left[\frac{J}{kg}\right]\right)(0.100[kg])}{(273.16[K])} = 122.16\left[\frac{J}{K}\right]$$

Step 3: Raise the temperature of the liquid water from $0\,°C$ to $100\,°C$. Because there are no phase changes in this step, we once again apply (6.28) using:

   $c \cong 4200$ J/(kg K)[12]
   $m = 100\ g = 0.100\ kg$
   $T_i = 0\,°C = 273.16\ K$
   $T_f = 100\,°C = 373.16\ K$

Therefore:

$$\Delta S_3 \geq \left(4200\left[\frac{J}{kg\,K}\right]\right)(0.100[kg])\,ln\left(\frac{373.16[K]}{273.16[K]}\right) = 131.02\left[\frac{J}{K}\right]$$

Step 4: Boil the water at a constant temperature of $100\,°C$. For the final step, apply (6.31) using:

   $m = 100\ g = 0.100\ kg$
   $l_v = 2.25 \times 10^6$ J/kg[13] (*positive*; water substance moves from lower to higher-energy state)
   $T = 100\,°C = 373.16\ K$

And the last step is written:

$$\Delta S_4 \geq \frac{\left(2.25\times10^6\left[\frac{J}{kg}\right]\right)(0.100[kg])}{(373.16[K])} = 602.96\left[\frac{J}{K}\right]$$

Step 5: Sum up the total change in entropy to the water system by:

$$\Delta S = \sum_{i=1}^{N} \Delta S_i = \Delta S_1 + \Delta S_2 + \Delta S_3 + \Delta S_4 = \underbrace{(15.21)}_{\text{Temp Increase}} + \overbrace{(122.16)}^{\text{Melting}} + \underbrace{(131.02)}_{\text{Temp Increase}} + \overbrace{(602.96)}^{\text{Boiling}}$$

$$= 871.35 \left[\frac{J}{K} \right]$$

A look at these numbers gives a little more insight into the nature of entropy. Raising the temperature of the block of ice only added about fifteen units of entropy, but melting it added nearly ten times that much. Similarly, raising the temperature of the liquid water by 100 degrees Celsius added about the same amount of entropy as melting it, but boiling it added more than 600 units – five times as much as raising its temperature from melting to the boiling point.

Boltzmann defined entropy as the *amount of disorder* in a system. The results of the preceding thought experiment tells us that warming up a block of ice by 20 Celsius degrees adds a relatively small amount of disorder. Because the molecules are locked into a crystal-lattice structure, the additional heat energy can only go into the *vibrational* degrees of freedom of the water molecules.[14] Once you begin breaking the crystal structure by melting the ice, the molecules become highly mobile with respect to each. Although they are still electrically bound to each other (surface tension and internal viscosity), and may be confined (by gravity) to a basin, they can now move a great deal more. The added *translational* degrees of freedom of the molecules equates to a greater amount of disorder, or entropy, in the system. Adding additional heat and raising the temperature of the liquid by another 100 degrees causes the molecules to vibrate, translate, and rotate even more, adding even more disorder. Finally, boiling the water liberates the molecules from their electrical attraction to each other, and they freely expand to fill the volume of the vessel containing them. This represents an enormous addition of disorder, because the molecules will no longer be predictably contained in the region of minimum gravitational potential energy (the lowest points available in the container) or bound to each other.

Net entropy changes: Systems and their environments. The entropy of an *isolated* system must either remain the same or increase, but – because no heat is exchanged – the entropy of the surrounding environment is not affected. For a nonisolated system, such as a *closed* or *open* parcel, the *combined* entropy of the parcel and its surrounding environment must always increase during natural processes:

$$\sum \left(\Delta s_{\text{system}} + \Delta s_{\text{environment}} \right) > 0 \qquad (6.32)$$

which implies that the individual entropies of the parcel and environment may increase or decrease, as long as the *sum* of the entropy change is a positive number.

Should there be a decrease in the parcel's entropy, like what would occur during a phase change from a higher to a lower energy state, or during the diabatic transfer of heat from the parcel to the environment, this will be offset by a somewhat larger increase in the environment's entropy.

6.3. The Carnot Cycle

In the early nineteenth century, Carnot was studying the physics of steam engines. He defined the *reversible* and *irreversible* processes described in the preceding text, and he also defined *cyclic processes*, one of which is now known as the *Carnot Cycle*.[15] A cyclic process is one in which the state of the working substance (water, in the case of steam engines) changes, but, after a series of steps, it is returned to its original state. The changes in the cyclic process are all reversible and occur in a way that the working substance is never far from equilibrium. At any point in the cycle, the direction of the changes can be reversed, and the working substance will return to its original state.

In a cyclic process, the final temperature of the working substance is equal to its initial temperature, so that its internal energy (u) is unchanged ($du = 0$) when integrated over all steps of the cycle; that is:

$$\oint du = 0 \qquad\qquad (6.33)$$

which you should compare to (1.21) in Chapter 1. If we keep this in mind and compare this to the form of the First Law shown in (4.3), namely:

$$dq = du + dw$$

then we also see that, for a cyclic process:

$$\boxed{\oint dq = \oint dw} \qquad\qquad (6.34)$$

when integrated through the full cycle, which seems to imply that *all* heat expended (dq) can be converted to work (dw) performed by the engine; that is, we have an engine with a 100 percent efficiency. As we will see a bit later, this conclusion is mistaken.

Heat engines and atmospheric parcels. Why does this matter? What does a steam engine have to do with the behavior of Earth's atmosphere? Carnot created an abstract version of steam engines called *heat engines*, which perform work by transferring heat from a high-temperature reservoir to a low-temperature reservoir. Leaving out the details of the machine's mechanical operations, Carnot's heat engines are idealized versions of the real thing, and thus are always more efficient that real steam engines. There is no internal friction from gears to slow anything down or waste energy.

The abstraction can also be applied to a parcel of air, which, when heated by the high-temperature surface of the Earth, expands (performing work), and may rise as high as the low-temperature tropopause, where it dumps the excess heat. Instead of the piston of a steam engine, our "piston" is a parcel of air. Instead of water (steam), the working substance is dry air. Notice that, right away, we're assuming that the parcel is *closed* (capable of diabatically exchanging heat with its surroundings) when in touch with the surface of the Earth, as well as when it's in contact with the tropopause. But we also assume that it becomes *isolated* when moving vertically. We're also going to assume away the possibility of condensation, evaporation, or any other phase changes involving water that may release or absorb latent heat, changing the temperature of the parcel. And we're going to ignore (1) the possibility that temperature inversions may slow or stop the ascent of the parcel and (2) all other complexities (such as horizontal winds) that might change the parcel's behavior. Using this model, which is unquestionably a highly simplified description of the real world, we can nonetheless gain some understanding of how energy moves around in the atmosphere. This will ultimately lead to a better understanding of stability and convection.

The Carnot Cycle for an atmospheric parcel. The Carnot Cycle followed by an air parcel consists of four steps (shown schematically in Figure 6.1):

1. In the first step, the (*closed*) air parcel is in contact with the surface of the Earth (high-temperature reservoir), where it absorbs heat diabatically and undergoes *isothermal expansion*, taking it from point *a* to point *b*.
2. In the second step, the (now *isolated*) parcel rises *adiabatically* to the tropopause, taking it from point *b* to point *c*.
3. In the third step, the parcel (*closed* again) undergoes *isothermal contraction* as it diabatically dumps its heat into the stratosphere (low-temperature reservoir), taking it from point *c* to point *d*.
4. In the final step, the (*isolated* again) parcel sinks *adiabatically* back to the surface of the Earth, taking it from point *d* back to point *a*.

Figure 6.2 illustrates these steps on a p-V diagram, which you should compare to Figures 1.7 and 5.2. If this parcel were the piston in a steam engine, we would call the combination of steps 1 and 2 the "forward stroke," and the combination of steps 3 and 4 the "return stroke."

Our first task, in obtaining a quantitative description of an air parcel undergoing a Carnot Cycle between the surface of the Earth and the tropopause, is to develop precise mathematical expressions for the work performed (either by or on the parcel) and the heat transferred (either into or out of the parcel). The following steps correspond to those shown in Figures 6.1 and 6.2.

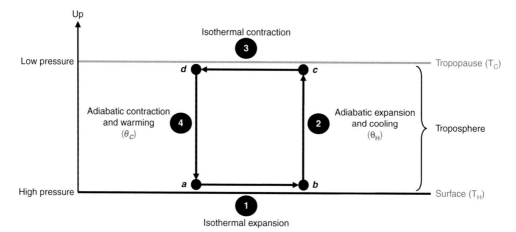

Figure 6.1. Schematic diagram for an atmospheric parcel following the Carnot Cycle. A parcel of air begins in state *a*, then proceeds around a closed loop to states *b*, *c*, and *d*, and back to *a*. Step 1 (indicated by "1" in a black circle) is isothermal expansion on the Earth's surface (at temperature T_H), increasing potential temperature from θ_C to θ_H. In Step 2, the parcel rises adiabatically to the tropopause, expanding and cooling, and decreasing temperature from T_H to T_C. Step 3 is an isothermal contraction at T_C, decreasing potential temperature from θ_H to θ_C. Step 4 is adiabatic contraction and warming, increasing temperature from T_C to T_H. Steps 1 and 2 constitute the *forward stroke*; steps 3 and 4 constitute the *return stroke*. (Compare to Figure 6.2.)

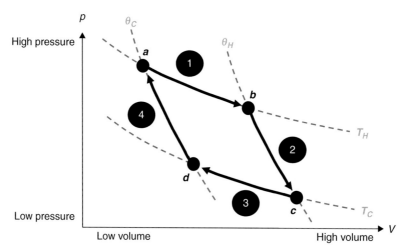

Figure 6.2. p-V diagram for an atmospheric parcel following the Carnot Cycle. A parcel of air begins in state *a* (pressure $= p_a$, Volume $= V_a$), then proceeds around a closed loop to states *b*, *c*, and *d*, and back to *a*. Steps shown (numbers in black circles) are the same as described in Figure 6.1. T_H and T_C indicate isotherms, where $T_H > T_C$. θ_H and θ_C indicate adiabats, which are lines of equal potential temperature (where $\theta_H > \theta_C$), as well as isentropes. (Compare to Figures 5.2. and 6.1.)

Step 1: Isothermal expansion on Earth's surface. Heat is transferred diabatically from the environment (high-temperature reservoir) to the parcel, but does *not* result in a temperature increase in the parcel ($\therefore dT = 0$). According to (4.29):

$$du = c_v \, dT$$

which implies that, because the temperature of the parcel remains constant in this step, there is no change in internal energy either ($\therefore du = 0$). We also have the form of the First Law shown in (4.4), which states:

$$dq = du + p \, d\alpha$$

Combining the former conclusion with the latter implies that all the heat added to the parcel on the Earth's surface goes into performing work ($dw = p \, d\alpha$); see (3.12). In other words, we can write:

$$dq = dw = p \, d\alpha \tag{6.35}$$

which must be integrated to compute the total work done during Step 1:

$$q_1 = w_{a-b} = \int_a^b p \, d\alpha \tag{6.36}$$

If you examine Figure 6.2, you'll see that pressure (p) does *not* remain constant during the first step, so we need another way to express p to perform the integral. Because this is one part of a Carnot Cycle, which assumes that the parcel is never far from equilibrium, we can appeal to the Ideal Gas Law for a way to eliminate p. First we'll generalize this analysis a little, factoring the mass back into each term, that is:

$$Q_1 = W_{a-b} = \int_a^b p \, dV \tag{6.37}$$

where Q_1 is the heat *absorbed* by the parcel during Step 1 (between points a and b in Figures 6.1 and 6.2), and W_{a-b} is the work the parcel performs as it expands (dV). Next, to rewrite p, we can use the form of the Ideal Gas Law shown in (2.46), which says:

$$pV = nR^*T$$

where n is the number of moles of material in the parcel, and R^* is the Universal Gas Constant, equal to 8.31×10^3 [J/(kmol K)]. Solving for pressure, we obtain:

$$p = \frac{nR^*T}{V} \tag{6.38a}$$

or, because the parcel is on the surface (where $T = T_H$):

$$p = \frac{nR^*T_H}{V} \tag{6.38b}$$

which can then be substituted into the integral on the RHS of (6.37), yielding:

$$Q_1 = W_{a-b} = \int_a^b \left(\frac{nR^*T_H}{V} \right) dV \tag{6.39}$$

or:

$$Q_1 = W_{a-b} = nR^*T_H \int_a^b \frac{dV}{V} \tag{6.40}$$

because n, R^*, and T_H are all constants. Noting that $nR^* = mR_d$ (4.50),[16] and performing the integral, we have:

$$\boxed{Q_1 = W_{a-b} = mR_d T_H \, ln\left(\frac{V_b}{V_a} \right)} \tag{6.41}$$

which describes how the heat absorbed by the parcel from Earth's surface (Q_1) causes it to expand and perform work (W_{a-b}).

Step 2: Adiabatic expansion and cooling from Earth's surface to the tropopause. According to the form of the First Law shown in (5.1):

$$0 = c_v \, dT + p \, d\alpha$$

for an adiabatic process (because $dq = 0$), or, as shown in (5.2):

$$p \, d\alpha = -c_v \, dT$$

Writing this in terms of the extensive properties (i.e., multiplying through by mass), we have:

$$p \, dV = -c_v m \, dT \tag{6.42}$$

or:

$$dW = -c_v m \, dT \tag{6.43}$$

We want the total work performed by the parcel during the second step of the cycle (between points b and c in Figures 6.1 and 6.2), so we must integrate both sides of (6.43). The RHS is integrated from T_H (the initial condition, on the surface) to T_C (the final condition, at the tropopause), resulting in:

$$W_{b-c} = -c_v m \left(T_C - T_H \right) \tag{6.44}$$

which (by distributing the negative sign into the parenthesis on the RHS) can also be written as:

$$W_{b-c} = c_v m \left(T_H - T_C \right)$$ (6.45)

where W_{a-b} is the work performed during this step of the cycle, T_H is the temperature of Earth's surface (high-temperature reservoir), and T_C is the temperature of the tropopause (low-temperature reservoir).

Step 3: Isothermal contraction at the tropopause. The heat absorbed by the parcel at Earth's surface (that caused it to expand) is dumped into the base of the stratosphere, causing the parcel to contract isothermally. Once again, $dT = 0$ and therefore $du = 0$, so that:

$$Q_3 = W_{c-d} = \int_c^d p \, dV$$ (6.46)

where Q_3 is the heat transferred from the parcel to the base of the stratosphere, and W_{c-d} is the work performed during Step 3 (between points c and d in Figures 6.1 and 6.2).

Making a substitution similar to that used in (6.39), we find:

$$Q_3 = W_{c-d} = \int_c^d \left(\frac{nR^* T_C}{V} \right) dV$$ (6.47)

where T_C is the temperature of the tropopause. Because n, R^*, and T_C are constants:

$$Q_3 = W_{c-d} = nR^* T_c \int_c^d \frac{dV}{V}$$ (6.48)

Performing the integral, and substituting mR_d for nR^* (4.50), we obtain:

$$Q_3 = W_{c-d} = mR_d T_C \left[ln \left(V_d \right) - ln \left(V_c \right) \right]$$ (6.49)

which can be rewritten as:

$$Q_3 = W_{c-d} = -mR_d T_C \left[ln \left(V_c \right) - ln \left(V_d \right) \right]$$ (6.50)

or:

$$Q_3 = W_{c-d} = -mR_d T_C \, ln \left(\frac{V_c}{V_d} \right)$$ (6.51)

We'll explain the reason for writing this way shortly.

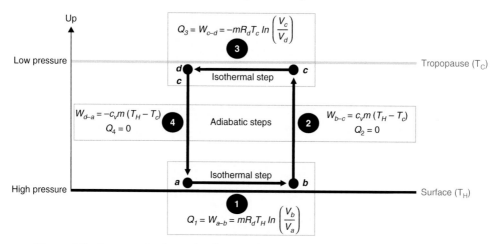

Figure 6.3. Annotated schematic diagram for an atmospheric parcel following the Carnot Cycle. Equations shown are derived in the body of the text. W terms indicate work performed between points in the subscript. Q terms indicate heat transferred in the step indicated by the subscript. (Compare to Figure 6.1.)

Step 4: Adiabatic compression and warming from the tropopause back to Earth's surface. As in Step 2, in this step we assume that the parcel is isolated, so that once again $dq = 0$. The relation shown in (6.43) again applies:

$$dW = -c_v m\, dT$$

which once again must be integrated, this time between points d and a in Figures 6.1 and 6.2. Performing this integral, we find that:

$$\boxed{W_{d-a} = -c_v m\left(T_H - T_C\right)} \tag{6.52}$$

where W_{d-a} is the work performed on the parcel as it descends from the tropopause to Earth's surface, T_H is the temperature of Earth's surface (the high-temperature reservoir), and T_C is the temperature of the base of the stratosphere (the low-temperature reservoir).

With the results of the analysis for Steps 1 through 4, we can create a somewhat more descriptive schematic diagram, shown in Figure 6.3.

Work performed in the Carnot Cycle. According (6.45), the work performed *by* the parcel as it ascends adiabatically from the surface to the tropopause is given by:

$$W_{b-c} = c_v m\left(T_H - T_C\right)$$

while (6.52) describes the work performed *on* the parcel by the environment, as the parcel descends adiabatically from the tropopause back down to Earth's surface:

$$W_{d-a} = -c_v m\left(T_H - T_C\right)$$

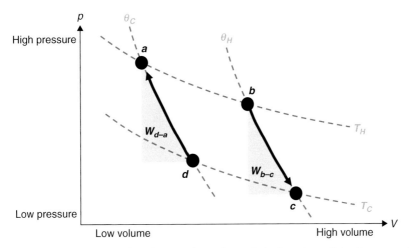

Figure 6.4. p-V diagram showing work performed during the adiabatic steps of a Carnot Cycle. The work performed *by* the parcel *on* the environment ($W > 0$) between points b and c is equal in magnitude but opposite in sign to the work performed *on* the parcel *by* the environment ($W < 0$) between points d and a. (Compare to Figures 3.4 and 6.2.)

Because the two works are equal in magnitude, but opposite in sign, they exactly cancel each other. In other words, *there is no net work performed by the adiabatic steps of the Carnot Cycle.* This means that any *net* work performed by the parcel during the cycle must occur during the isothermal expansion steps and would be the linear sum of the work performed during Step 1 and Step 3.

Mathematically, you can show that the net work is greater than zero by comparing (6.41), which describes the work in the first isothermal step on Earth's surface:

$$W_{a-b} = mR_dT_H \, ln\left(\frac{V_b}{V_a}\right)$$

to (6.51), which describes the work performed in the isothermal step at the tropopause:

$$W_{c-d} = -mR_dT_C \, ln\left(\frac{V_c}{V_d}\right)$$

The net work performed is the linear sum of these two. As long as the product $T_H \, ln(V_b/V_a)$ is greater than $T_C \, ln(V_c/V_d)$, then the net work performed in the cycle is positive, meaning the parcel does more work on the environment in Step 1, than the environment does on the parcel in Step 3.

Graphically, all of this can easily be demonstrated using a few p-V diagrams.[17] Figure 6.4 shows the two components of the work (shaded) performed during the adiabatic steps. As you can see, the two shaded areas are equal, but because they are opposite in sign, they exactly cancel each other. Figure 6.5 shows the work performed

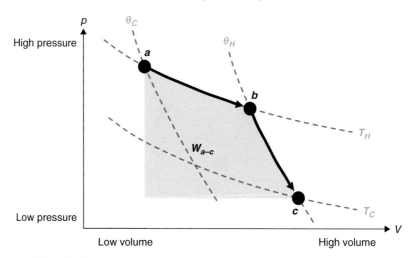

Figure 6.5. p-V diagram showing work performed *by* the parcel during the expansion steps of a Carnot Cycle. Total work (W_{a-c}) during the "forward stroke" of the parcel includes expansion that occurs during the isothermal step on the Earth's surface ($a - b$), and adiabatic step from the Earth's surface to the tropopause ($b - c$). Because the parcel performs work by expanding, the sign of the work is positive. (Compare to Figures 3.4 and 6.2.)

during the "forward stroke" (a term that we would use if this was the piston of an engine rather than an air parcel), when the parcel *expands* first isothermally ($a - b$) and then adiabatically ($b - c$). Figure 6.6 shows the work performed during the "return stroke," when the parcel *contracts* first isothermally ($c - d$) and then adiabatically ($d - a$). For the net work performed during the entire cycle, you must add these two shaded areas together, which is shown in Figure 6.7.

Heat transferred in the Carnot Cycle. By definition, the heat transfers during the two adiabatic steps of the cycle are zero, but there must be a nonzero *net* heat transfer in the cycle, because heat transfer is the energy source driving the parcel's expansion, permitting it to do work. This means that, once again, the net heat transfer must occur during the two isothermal steps.

Before we derive a relationship between these two, please remember the following: the two isothermal heat transfers *can't* add up to zero, because if they did, we wouldn't have any energy left over to perform the work, which we've already shown is nonzero. This is the essence of a steam engine, or an air parcel in this analysis, which behaves like a piston in a steam engine. *Heat is transferred out of the high-temperature reservoir by the working substance. Some of this heat is used to perform work. The remainder of the heat is dumped into the low-temperature reservoir* (see Figure 6.8).

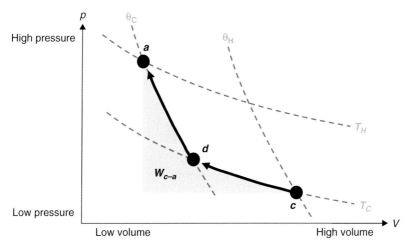

Figure 6.6. p-V diagram showing work performed *on* the parcel during the contraction steps of a Carnot Cycle. Total work (W_{c-a}) during the "reverse stroke" of the parcel includes contraction that occurs during the isothermal step at the tropopause (c – d), and adiabatic step from the tropopause to the Earth's surface (d – a). Because the environment does work on the parcel by forcing it to contract, the sign of the work is negative. (Compare to Figures 3.4 and 6.2.)

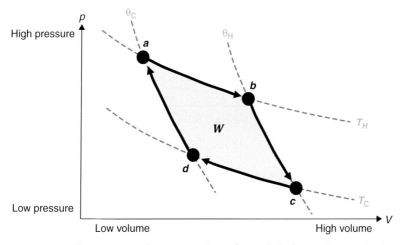

Figure 6.7. p-V diagram showing *net* work performed during a Carnot Cycle. Net work (W; dark shading) performed during all steps of the cycle is the sum of the work performed on the forward stroke (positive) and the work performed on the reverse stroke (negative; lightly shaded). Graphically, you can simply subtract the work shown in Figure 6.6 from the work shown in Figure 6.5. (Compare to Figure 1.7.)

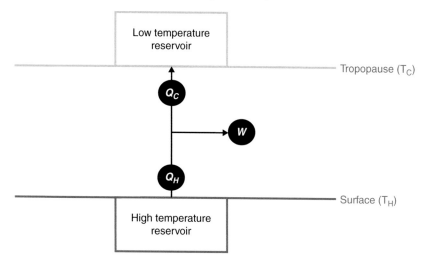

Figure 6.8. Transfer of heat from high-temperature reservoir (Earth's surface) to low-temperature reservoir (tropopause). Q_H is the heat absorbed by the working substance from the high-temperature reservoir, in this case, by a parcel from the surface of the Earth. (Q_H is equivalent to Q_1 in Figure 6.3.) Q_C is the heat deposited by the parcel into the base of the stratosphere, which represents the low-temperature reservoir. (Q_C is equivalent to Q_3 in Figure 6.3.) Net work (W) is performed by the conversion of some of the heat removed from the high-temperature reservoir; therefore, Q_C must be less than Q_H.

We begin this analysis with (5.36), which stated that, for adiabatic processes:

$$T_f = T_i \left(\frac{\alpha_i}{\alpha_f} \right)^{\frac{R_d}{c_v}}$$

where T_f is the final temperature, T_i is the initial temperature, and α_i and α_f are the initial and final specific volumes of the parcel, respectively. R_d[18] and c_v are the individual gas constants for dry air, and the specific heat of dry air at constant volume. Dividing both sides T_i results in:

$$\frac{T_f}{T_i} = \left(\frac{\alpha_i}{\alpha_f} \right)^{\frac{R_d}{c_v}} \tag{6.53}$$

and distributing the exponent into both terms on the RHS yields:

$$\frac{T_f}{T_i} = \frac{\alpha_i^{\frac{R_d}{c_v}}}{\alpha_f^{\frac{R_d}{c_v}}} \tag{6.54}$$

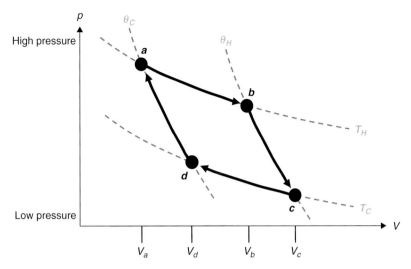

Figure 6.9. p-V diagram for an atmospheric parcel following the Carnot Cycle. Volumes are indicated for comparison. (Compare to Figure 6.2.)

For comparison to Figure 6.9 (which shows the air parcel as it moves through a Carnot Cycle), we can switch to the extensive quantity of volume by first multiplying through by mass, then canceling the extraneous mass terms in both the numerator and denominator of the RHS of (6.54), resulting in the ratio:

$$\frac{T_f}{T_i} = \frac{V_i^{\frac{R_d}{c_v}}}{V_f^{\frac{R_d}{c_v}}} \tag{6.55}$$

With this in mind, we can see that, along the θ_H adiabat, the parcel's *initial* position is on Earth's surface, and its *final* position is at the tropopause, so that:

$$\frac{T_C}{T_H} = \frac{V_b^{\frac{R_d}{c_v}}}{V_c^{\frac{R_d}{c_v}}} \tag{6.56a}$$

By simply inverting both sides of (6.56a), we get:

$$\frac{T_H}{T_C} = \frac{V_c^{\frac{R_d}{c_v}}}{V_b^{\frac{R_d}{c_v}}} \tag{6.56b}$$

Along the θ_C adiabat, the parcel's *initial* position is at the tropopause, and its *final* position is on the surface, so that:

$$\frac{T_H}{T_C} = \frac{V_d^{\frac{R_d}{c_v}}}{V_a^{\frac{R_d}{c_v}}}$$

(6.57)

Comparing (6.56b) and (6.57), we find we have two expressions with the same term on the LHS (T_H/T_C); therefore:

$$\frac{V_c^{\frac{R_d}{c_v}}}{V_b^{\frac{R_d}{c_v}}} = \frac{V_d^{\frac{R_d}{c_v}}}{V_a^{\frac{R_d}{c_v}}}$$

(6.58)

or simply,

$$\frac{V_c}{V_b} = \frac{V_d}{V_a}$$

(6.59)

Cross-multiplying, we obtain:

$$\boxed{\frac{V_c}{V_d} = \frac{V_b}{V_a}}$$

(6.60)

Now let's go back to the two isothermal steps, whose heats were described by (6.41) and (6.51). For the isothermal step on Earth's surface, we had:

$$W_{a-b} = mR_dT_H \, ln\left(\frac{V_b}{V_a}\right)$$

and for the isothermal step at the tropopause, we had:

$$W_{c-d} = -mR_dT_C \, ln\left(\frac{V_c}{V_d}\right)$$

These were equivalent to the heat diabatically absorbed by the parcel at the surface (Q_1, which we can call Q_H), and the heat diabatically dumped into the tropopause (Q_3, which we can call Q_C), that is:

$$Q_H = mR_dT_H \, ln\left(\frac{V_b}{V_a}\right)$$

(6.61)

and:

$$Q_C = -mR_d T_C \, ln\left(\frac{V_c}{V_d}\right) \tag{6.62}$$

The identity is (6.60) states that:

$$\frac{V_c}{V_d} = \frac{V_b}{V_a}$$

therefore:

$$ln\left(\frac{V_c}{V_d}\right) = ln\left(\frac{V_b}{V_a}\right) \tag{6.63}$$

and we can rewrite (6.62) as:

$$Q_C = -mR_d T_C \, ln\left(\frac{V_b}{V_a}\right) \tag{6.64}$$

If *all* of the heat absorbed by the parcel from the Earth's surface were dumped into the tropopause, the ratio Q_C/Q_H would be equal to 1. But, because the parcel performs *work* during the cycle, the First Law requires that some of the heat absorbed from the surface be used up to provide the energy to do the work. Therefore, Q_C must be smaller than Q_H, and the ratio Q_C/Q_H must be less than 1. We can directly compute this by:

$$\frac{Q_C}{Q_H} = \frac{-mR_d T_C \, ln\left(\frac{V_b}{V_a}\right)}{mR_d T_H \, ln\left(\frac{V_b}{V_a}\right)} \tag{6.65}$$

After canceling all the redundant terms, we find that:

$$\frac{Q_C}{Q_H} = \frac{-T_C}{T_H} \tag{6.66}$$

or:

$$\boxed{\left|\frac{Q_C}{Q_H}\right| = \frac{T_C}{T_H}} \tag{6.67}$$

In other words, perhaps unsurprisingly, the ratio of the magnitudes of the two diabatic heat transfers is equal to the ratios of the temperatures of the two environmental heat reservoirs. (Be sure to use the Kelvin scale for both temperatures.)

Efficiency of a Carnot Cycle, or why we can never build a perpetual motion machine. The Carnot Cycle is the most efficient cycle possible, because, when the piston (or parcel) is heated, all of it goes into expansion (performing work). None of the heat is "wasted" on raising the temperature of the parcel. This is obviously an unrealistic model, because when heat is applied to a substance, its temperature will almost certainly increase. In other words, all "real" engines (and parcels) are less efficient that the model described by the Carnot Cycle.

To begin this analysis, let's define efficiency as follows:

$$\boxed{\eta \equiv \frac{W}{Q_H}} \tag{6.68}$$

where η is the efficiency of the engine, scaled between 0 and 1 (unitless), W is the work it performs (J), and Q_H is the heat it absorbs from the high-temperature reservoir (J). If the machine converts *all* the heat it absorbs from the high-temperature reservoir into work, it has an efficiency of 1 (or 100 percent). If it converts *none* of the heat to work, its efficiency is zero.

Next, let's go back to (6.34), which stated that, for a cyclic process:

$$\oint dq = \oint dw$$

when integrated through the complete cycle. So, let's integrate both sides of the equation and convert from intensive to extensive quantities, to obtain:

$$Q_H - Q_C = W \tag{6.69}$$

We have already shown that the RHS of (6.69) is nonzero (see Figure 6.7), so the LHS must also be nonzero. We can then substitute this identity into (6.68), resulting in:

$$\eta = \frac{Q_H - Q_C}{Q_H} \tag{6.70}$$

or:

$$\boxed{\eta = 1 - \frac{Q_C}{Q_H}} \tag{6.71}$$

where we define both heat transfers as positive numbers. We can then appeal to the identity shown in (6.67), namely:

$$\left| \frac{Q_C}{Q_H} \right| = \frac{T_C}{T_H}$$

and substitute the RHS into (6.71), yielding:

$$\boxed{\eta = 1 - \frac{T_C}{T_H}}$$ (6.72)

This relation shows that there are only two ways to "build" a Carnot Engine with 100 percent efficiency:

1. The low-temperature reservoir must have a temperature of absolute zero, which does not occur anywhere in nature, *or*
2. The high-temperature reservoir must have a temperature of infinity, which also does not occur anywhere in nature.

Because a Carnot Engine is more efficient than all real engines, this means that no real engine can operate at 100 percent efficiency. It also precludes the construction of a perpetual motion machine, which must have an efficiency of *at least* 100 percent.

There is another relationship that's often useful. Beginning with (6.71):

$$\eta = 1 - \frac{Q_C}{Q_H}$$

we add the quantity Q_C/Q_H to both sides of the equation:

$$\frac{Q_C}{Q_H} + \eta = 1 - \frac{Q_C}{Q_H} + \frac{Q_C}{Q_H}$$ (6.73)

or:

$$\frac{Q_C}{Q_H} + \eta = 1$$ (6.74)

Next, subtract η from both sides to obtain:

$$\frac{Q_C}{Q_H} + \eta - \eta = 1 - \eta$$ (6.75)

or:

$$\frac{Q_C}{Q_H} = 1 - \eta$$ (6.76)

Finally, we multiply both sides by Q_H and obtain the form:

$$\boxed{Q_C = (1 - \eta)Q_H}$$ (6.77)

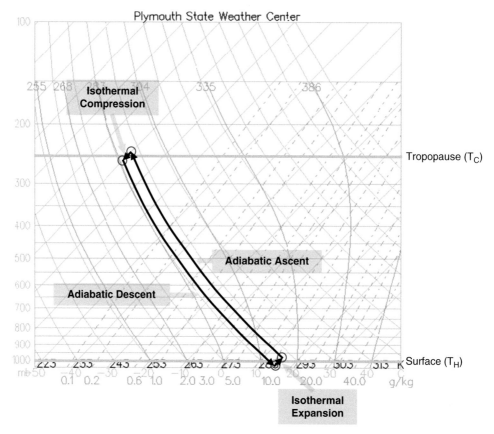

Figure 6.10 Carnot Cycle visualized on a Skew-T Log P diagram. Background image courtesy of Plymouth State University meteorology program (2012).

Example. A 1 kg parcel of dry air follows a Carnot Cycle between the surface of the Earth and the tropopause (see Figure 6.10). On the surface, the parcel diabatically absorbs heat, and expands isothermally by 10 m³. At the tropopause, the parcel diabatically loses heat and contracts isothermally by 10 m³. On the surface:

$T_H = 15\,°\text{C} + 273.16 = 288.16\ \text{K}$
$p = 1000\ \text{hPa} = 10^5\ \text{Pa}$
$\Delta V = 10\ \text{m}^3$

At the tropopause:

$T_C = -80\,°\text{C} + 273.16 = 193.16\ \text{K}$[19]
$p = 250\ \text{hPa} = 2.5 \times 10^4\ \text{Pa}$
$\Delta V = -10\ \text{m}^3$

Compute the efficiency of the parcel's heat transfer process, the total change in entropy of the parcel, the total entropy change of the environment, and the net entropy change of the parcel and the environment.

The efficiency is given by (6.72), which says that:

$$\eta = 1 - \frac{T_C}{T_H} = 1 - \frac{193.16}{288.16} = 0.3297 = 32.97\ \%$$

Because the Carnot Cycle consists of reversible processes, the change in the parcel's entropy is given by (6.2):

$$ds = \frac{dq}{T}$$

which is true for all four steps of the cycle. We know that the two adiabatic steps have $dq = 0$, so only the two isothermal steps (at the surface and at the tropopause) contribute to the parcel's change in entropy. But (6.35) states that, for the isothermal steps:

$$dq = dw = p\,d\alpha$$

so we can rewrite the entropy change on each of these steps as:

$$ds = \frac{p\,d\alpha}{T}$$

On the surface,

$$\Delta S_1 = \frac{p\,\Delta V}{T_H} = \frac{(10^5)(10)}{(288.16)} = +3470.29 \left[\frac{J}{K} \right]$$

with the positive sign indicating an *increase* in the parcel's entropy. At the tropopause:

$$\Delta S_2 = \frac{p\Delta V}{T_H} = -\frac{(2.5 \times 10^4)(-10)}{(193.16)} = -1294.26 \left[\frac{J}{K} \right]$$

with the negative sign indicating a *decrease* in the parcel's entropy. This negative sign is present because the change in volume (ΔV) is negative, meaning that the environment did work on the parcel. The *total* change in parcel's entropy is then given by:

$$\Delta S_{parcel} = \sum_{i=1}^{N} \Delta S_i = \Delta S_1 + \Delta S_2 = +3470.29 - 1294.26 = +2176.03 \left[\frac{J}{K} \right]$$

That is, the overall entropy of the parcel *increases*.

Next, we can compute the entropy change of the environment, using:

$$\Delta Q_H = p\Delta V = (10^5\ \text{Pa})(10\ \text{m}^3) = 10^6\ \text{J}$$

$$\Delta Q_C = (1 - \eta)\Delta Q_H = (1.0000 - 0.3297)(10^6\ \text{J}) = 6.703 \times 10^5\ \text{J}$$

where the value of ΔQ_C was computed using the identity shown in (6.77). Using (6.2), the heat *removed* from the first reservoir results in an environmental entropy change of:

$$\Delta S_1 = \frac{\Delta Q_H}{T_H} = -\frac{10^6}{288.16} = -3470.29 \left[\frac{J}{K} \right]$$

where the negative sign indicates that the environment *lost* heat. Note that the *decrease* in environmental entropy at the surface is of the same magnitude as the *increase* in parcel entropy at the surface. About 67 percent of this heat is then *added* to the second reservoir, resulting in an environmental entropy change of:

$$\Delta S_2 = \frac{\Delta Q_C}{T_C} = +\frac{6.703 \times 10^5}{193.16} = +3470.18 \left[\frac{J}{K} \right]$$

which is positive because the environment *gained* heat. The total change in environmental entropy is then given by:

$$\Delta S_{environment} = \sum_{i=1}^{N} \Delta S_i = \Delta S_1 + \Delta S_2 = -3470.29 + 3470.18 \cong 0 \left[\frac{J}{K} \right]$$

That is, the overall entropy of the *environment* remains constant. We can then compute the *net* entropy change of the parcel and the environment by adding the two totals together (6.32):

$$\Delta S = \sum \left(\Delta S_{parcel} + \Delta S_{environment} \right) = +2176.03 + 0 = +2176.03 \left[\frac{J}{K} \right]$$

and we see that the net entropy increased, which *must* be true, because there were two nonadiabatic steps in the cycle of processes.

6.4. Summary of the Second Law

In this chapter, we defined reversible, irreversible, and impossible processes, and then stated that we would approximate irreversible processes with an infinite series of infinitesimal reversible processes. Next, we defined entropy as the amount of waste heat or disorder in a system, which is written mathematically as:

$$ds \geq \frac{dq}{T}$$

where,
   ds = change in specific entropy [J/(kg K)]
   dq = change in specific heat [J/kg]
   T = temperature [K]

For reversible processes, the equality is applied. In the special case of adiabatic, reversible processes, dq is zero, so ds is zero as well. For irreversible processes, the equality gives you the minimum change in entropy. We also showed that, for adiabatic processes:

$$ds = c_p \frac{d\theta}{\theta}$$

where,

c_p = specific heat coefficient at constant pressure [J/(kg K)]
$d\theta$ = change in potential temperature [K]
θ = potential temperature [K]

which implied that dry adiabats, described in Chapter 5, were also isentropes – that is, lines of equal entropy. For irreversible, diabatic processes:

$$\Delta s \geq c \, ln\left(\frac{T_f}{T_i}\right)$$

where,

Δs = total change in specific entropy [J/(kg K)]
c = specific heat coefficient [J/(kg K)]
T_f = final temperature [K]
T_i = initial temperature [K]

$$\Delta S \geq cm \, ln\left(\frac{T_f}{T_i}\right)$$

where,

ΔS = total change in extensive entropy [J/K]
m = mass [kg]

For diabatic changes resulting in a phase change:

$$\Delta s \geq \frac{l}{T}$$

where,

l = latent heat coefficient for the process in question [J/kg]

$$\Delta S \geq \frac{lm}{T}$$

where,

l = latent heat coefficient for the process in question [J/kg]

A Carnot Cycle is the most efficient cycle an "engine" can follow. In our context as meteorologists, the "engine" consists of the hot surface of Earth, the cold tropopause, and a parcel of air traveling between the two that serves as a "piston." The parcel consists of dry air. The cycle consists of four stages: two diabatic and isothermal (at Earth's surface and the tropopause), and two adiabatic (the vertical journeys in between). For the diabatic, isothermal steps:

$$Q_n = W_{i-f} = mR_dT_n \, ln\left(\frac{V_f}{V_i}\right)$$

where,

Q_n = heat transferred in or out of parcel [J]
W_{i-f} = work performed by parcel in the step [J]
m = mass of parcel [kg]
R_d = individual gas constant for dry air [J/(kg K)]
T_n = temperature of surface or tropopause [K]
V_f = final volume of parcel [m^3]
V_i = initial volume of parcel [m^3]

For the adiabatic steps, $Q_n = 0$, and:

$$W_{i-f} = c_v m\left(T_H - T_C\right)$$

where,

c_v = specific heat of dry air at constant volume [J/(kg K)]
T_H = temperature of the hot reservoir [K]
T_C = temperature of the cold reservoir [K]

The efficiency of an engine is defined as:

$$\eta \equiv \frac{W}{Q_H}$$

where,

η = efficiency of the process [unitless]
W = work performed by the engine [J]
Q_H = heat absorbed by the engine's piston from the high-temperature reservoir [J]

which can be rewritten as:

$$\eta = 1 - \frac{T_C}{T_H}$$

Using efficiency, the relationship between the two heat transfer processes (Q_H and Q_C) can be written:

$$Q_C = \left(1 - \eta\right)Q_H$$

where,

Q_C = heat transferred by the engine's piston to the cold-temperature reservoir [J]

Practice Problems

1. Name and describe three different classes of thermodynamic processes. Be sure to use the word *equilibrium* when describing at least two of them.
2. Write the general expression for an infinitesimal change in entropy with respect to an infinitesimal change in heat at a constant temperature.
3. Show (mathematically) the change in entropy for an adiabatic process.
4. Determine the change in entropy for a change in temperature of a unit mass of liquid water at constant pressure. The temperature of the water increases from $5\,°C$ to $95\,°C$. No evaporation occurs.
5. Determine the change in entropy if 10 kg of water are boiled (isothermally) at $100\,°C$. Note: $l_v = l_{v0} - (2369\,^* T[°C])$; $l_{v0} = 2.5008 \times 10^6$ J kg^{-1}.
6. Determine the change in entropy if 10 kg of water are frozen (isothermally) at $0\,°C$.
7. Determine the efficiency of the heat transfer process if a dry parcel is heated by the Earth's surface at $15\,°C$, and rises adiabatically to the tropopause at $-40\,°C$. Assume the parcel follows the Carnot Cycle.
8. Determine the heat transferred in a Carnot Cycle to the cold reservoir, if the system has an efficiency of 35 percent, and the heat absorbed from the warm reservoir by the working substance is equal to 2.5×10^6 J.
9. Determine the heat absorbed in a Carnot Cycle from the warm reservoir, if the system has an efficiency of 25 percent, and the heat deposited in the cold reservoir by the working substance is equal to 1.0×10^6 J.
10. A 5 kg parcel of dry air follows a Carnot Cycle between the surface of the Earth and the tropopause. On the surface, the parcel diabatically absorbs heat, and expands isothermally by 20 m³. At the tropopause, the parcel diabatically loses heat and contracts isothermally by 20 m³. On the surface:

$T_H = 10\,°C + 273.16 = 283.16$ K

$p = 1013$ hPa $= 1.013 \times 10^5$ Pa

$\Delta V = 20$ m³

At the tropopause:

$T_C = -60\,°C + 273.16 = 213.16$ K

$p = 150$ hPa $= 1.5 \times 10^4$ Pa

$\Delta V = -20$ m³

Compute the efficiency of the parcel's heat transfer process, the total change in entropy of the parcel, the total entropy change of the environment, and the net entropy change of the parcel and the environment.

7

Water Vapor and Phase Transitions

7.1. Introduction

Up to now, we've generally described *isolated* or *closed* systems that are homogenous throughout (as in a small parcel), and have uniform chemical and physical properties, including phase. The presence of water in two or more phases implies *heterogeneous* systems. And if the water is undergoing phase changes, moving back and forth between two or more states, this also implies either *open* systems (exchanging mass and energy with the environment), or *compound* systems like those described in Chapter 1. Figure 7.1 also shows a schematic of a compound system.

In this chapter, we will use the compound system analogy to describe phase changes for water, which includes the derivation of an expression describing the *vapor pressure* above the surface of liquid water (that results from evaporation). The expression is called the *Classius-Clapeyron Equation*.[1] We'll also define several parameters that describe humidity.

7.2. Vapor Pressure

Our first objective is to derive the Classius-Clapeyron relation describing the partial atmospheric pressure resulting from water vapor (or simply *vapor pressure,* given the symbol e) as a function of *dew point* (T_d). We'll define dew point precisely a little later, but for now, accept that it represents the temperature to which a parcel of air must be cooled, at constant pressure, to induce *saturation*. Dew point is a function of the amount of water vapor in the air sample. Saturation is an *equilibrium* condition, so we'll proceed by assuming a state where evaporation and condensation are occurring at the same rate.[2] This means our first step along the way will be to derive the *saturation vapor pressure* (e_s), which is the maximum amount of vapor a sample of air can hold at a given temperature and pressure.

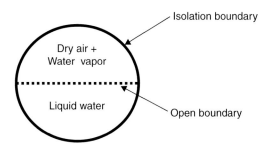

Figure 7.1. Compound system. The lower component consists of liquid water; the upper component of dry air and water vapor. The boundary surrounding the entire system isolates it from the environment. The boundary separating the two components is open, allowing the exchange of mass and energy. Water may evaporate from the surface of the liquid, adding to the vapor in the upper half of the system. Water vapor may also condense out of the air/vapor half of the system, and add to the liquid half of the system below. (Compare to Figure 1.4.)

Deriving the Classius-Clapeyron Equation.[3] Let's begin with the form of the First Law shown in (4.3), which states:

$$dq = du + dw$$

where dq is the heat added, du is the resulting change in internal energy and dw is the work performed. Equation (4.4) expresses work more explicitly:

$$dq = du + pd\alpha$$

where p is the pressure and $d\alpha$ is the change in the parcel's specific volume. Because the pressure we're considering is the saturation *vapor* pressure, we'll let $p = e_s$, or:

$$\boxed{dq = du + e_s d\alpha} \tag{7.1}$$

and because we're considering phase changes, the differential heat added (dq) does not go into changing the temperature (thus is dq_T), but instead results in a phase change, so that:

$$l = dq_T \tag{7.2}$$

which is the same as (4.65). To reduce this differential to a "closed" (finite) solution, and ultimately solve for the vapor pressure, we have to integrate both sides of (7.2), that is:

$$L_{12} = \int_1^2 dq_T \tag{7.3}$$

which would yield the *total* amount of latent heat required to move a substance (in this case, water) from Phase 1 to Phase 2 on the RHS. Note that, because *l* is constant at a given temperature, there's nothing to integrate on the LHS of (7.2). Thus, L_{12} is equal to *l*. Recall (6.2), which states that for a reversible process

$$ds = \frac{dq}{T}$$

or, rearranging,

$$\boxed{Tds = dq} \tag{7.4}$$

which can then be substituted into (7.3), resulting in:

$$L_{12} = \int_1^2 Tds \tag{7.5}$$

Because *T* is a constant during a phase change,

$$L_{12} = T\int_1^2 ds \tag{7.6}$$

which, when integrated, is simply:

$$\boxed{L_{12} = T(s_2 - s_1)} \tag{7.7}$$

The problem now is to evaluate the RHS of (7.7), which is difficult, because we don't know *a priori* the specific entropy of the two phases. But we can solve this by considering a cycle of reversible changes that move water (the substance in question) through two phase changes (one the reverse of the other), returning it to its original state. To illustrate this, we'll adapt the familiar p-V diagram for use with vapor pressure (*e*) and specific volume (α), as shown in Figure 7.2. The cycle we'll consider is shown by the heavy arrows in the figure, and the four *states* we refer to in the following text are shown as heavy black circles lettered *a – d*.

The underlying assumption here is that *e* and α both increase as heat is applied to water in the liquid phase when near the phase transition point. Once the phase shift occurs, transitioning to the water to the vapor phase, *e* begins to decrease again, while α continues to increase. This theoretical *e* as a function of α is shown as the heavy dashed line in the figure and is a fairly accurate description for *evaporation*[4] but is not correct for *melting* (where the first phase is ice, and the second phase is liquid). With this in mind, the derivation that follows is valid for a cycle of evaporation and condensation in equilibrium.

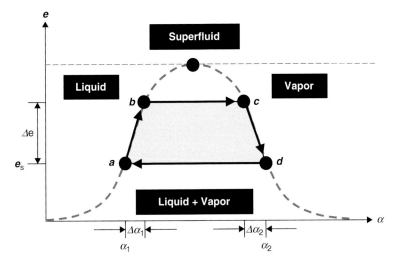

Figure 7.2. p-V (e-α) diagram for water substance moving in a cycle between two phases. Water moves from point *a* through points *b*, *c*, and *d* as it transitions from liquid phase to vapor phase and back to liquid phase. The heavy dashed line indicates the saturation vapor pressure (e_s) and specific volume (α) when the two phases coexist in equilibrium. Inside the bell-shaped region, liquid and vapor coexist in equilibrium. In the region above the top of the bell-shaped curve, water manifests as a *superfluid*, which resembles an ordinary liquid in some ways. (The heavy dot at the top of the bell curve is the *critical point*, which is discussed further in the following text.) The solid arrows indicate the approximation used in the analysis. (Compare to Figure 7.4.)

- In State *a*, the substance is in *equilibrium* in the liquid phase, with state variables[5] equal to e_s, T, and α_1.
- In the first transition, heat is added to the water, and its temperature increases slightly $(+\Delta T)$, causing the substance to move from State *a* to State *b* but remain in the liquid phase. Its vapor pressure and specific volume both increase by a small amount (Δe and $\Delta\alpha_1$, respectively) between State *a* to State *b*, so that its state variables in State *b* are equal to $(e_s + \Delta e)$, $(T + \Delta T)$, and $(\alpha_1 + \Delta\alpha_1)$.
- Next, there is an *isothermal* transition from State *b* to State *c*, as all heat added to the water causes it to undergo the phase change from the liquid to the vapor phase. The vapor pressure remains constant across the phase shift, but the specific volume increases substantially, so that State *c* is described by $(e_s + \Delta e)$, $(T + \Delta T)$, and $(\alpha_2 - \Delta\alpha_2)$.
- In the transition from State *c* to State *d*,[6] there is a slight *increase* in volume ($\Delta\alpha_2$) and *decrease* in temperature $(-\Delta T)$, so that its state variables in State *d* are equal to e_s, T, and α_2.
- In the fourth and final transition, heat is *isothermally* removed from the substance, causing it to move from State *d* back to State *a* and undergo the phase shift from the vapor phase back to the liquid phase. Its final state variables are equal to e_s, T, and α_1.

Returning to (7.1; a form of the First Law), which states:

$$dq = du + e_s\, d\alpha$$

and (7.4; a form of the Second Law), which states:

$$Tds = dq$$

we see that:

$$Tds = du + e_s\, d\alpha \tag{7.8}$$

Integrating both sides through the closed cycle, we obtain:

$$\oint Tds = \oint \left(du + e_s\, d\alpha\right) \tag{7.9}$$

or:

$$\oint Tds = \oint du + \oint e_s\, d\alpha \tag{7.10}$$

but according to (6.33):

$$\oint du = 0$$

so (7.10) reduces to:

$$\boxed{\oint Tds = \oint e_s\, d\alpha} \tag{7.11}$$

Integrating the RHS of (7.11) yields the approximate solution:

$$\boxed{\oint e_s\, d\alpha \cong \Delta e_s\left(\alpha_2 - \alpha_1\right)} \tag{7.12}$$

which, if you examine Figure 7.2, is approximately equal to the shaded area. Because this is a p-V diagram, the area of the integral's approximation is also proportional to the energy involved in the cycle.

To integrate the LHS of (7.11), we use the chain rule, which generically states that:

$$d(xy) = x\,dy + y\,dx \tag{7.13}$$

and an identity for exact differentials,[7] which says:

$$\oint d(xy) = \oint \left(x\,dy + y\,dx\right) = \oint x\,dy + \oint y\,dx = 0 \tag{7.14}$$

Because $d(Ts)$ *is* an exact differential, we can write:

$$\oint d(Ts) = \oint T\,ds + \oint s\,dT = 0 \tag{7.15}$$

or:

$$\oint T\,ds = -\oint s\,dT \tag{7.16}$$

We can evaluate the LHS of (7.11) by computing the value of the RHS of (7.16), which is a little easier. To do this, we break up the *line* integral of $-sdT$ into four *ordinary* integrals that we know how to perform:

$$\boxed{-\oint s\,dT = -\int_a^b s\,dT - \int_b^c s\,dT - \int_c^d s\,dT - \int_d^a s\,dT} \tag{7.17}$$

- From State *a* to State *b*:

$$-\int_a^b s\,dT \cong -s_1 \Delta T \tag{7.18}$$

where s_1 is the mean specific entropy in Phase 1 and ΔT is the temperature *increase* resulting from the heat added in the first transition.

- From State *b* to State *c*:

$$-\int_b^c s\,dT = 0 \tag{7.19}$$

because the transition from State *b* to State *c* is isothermal.

- From State *c* to State *d*:

$$-\int_c^d s\,dT \cong -s_2 \left(-\Delta T\right) = s_2 \Delta T \tag{7.20}$$

where s_2 is the mean specific entropy in Phase 2 and $-\Delta T$ is the temperature *decrease* associated with the transition from State *c* to State *d*.

- From State *d* back to State *a*:

$$-\int_d^a s\,dT = 0 \tag{7.21}$$

because this transition is also isothermal.

Combining all four, we see that (7.17) can be rewritten by:

$$-\oint s\,dT \cong -s_1 \Delta T - 0 + s_2 \Delta T - 0 \tag{7.22}$$

or simply:

$$-\oint s\,dT \cong (s_2 - s_1)\Delta T \tag{7.23}$$

Because both sides of (7.23) are also equal to the LHS of (7.11), we see that:

$$\oint T\,ds \cong (s_2 - s_1)\Delta T \tag{7.24}$$

But from (7.11) and (7.12), we also know that:

$$\underbrace{\oint T\,ds = \overbrace{\oint e_s\,d\alpha}^{(7.12)} \cong \Delta e_s\,(\alpha_2 - \alpha_1)}_{(7.11)} \tag{7.25}$$

Combining (7.24) and (7.25), we obtain:

$$(s_2 - s_1)\Delta T \cong \Delta e_s\,(\alpha_2 - \alpha_1) \tag{7.26}$$

Reducing the two finite differences to infinitesimal differences, we have:

$$(s_2 - s_1)dT \cong de_s\,(\alpha_2 - \alpha_1) \tag{7.27}$$

and rearranging yields:

$$\boxed{(s_2 - s_1) \cong (\alpha_2 - \alpha_1)\frac{de_s}{dT}} \tag{7.28}$$

which you should compare to (7.7), which said that:

$$L_{12} = T(s_2 - s_1)$$

The objective was to find another way to write the entropy change on the RHS of this expression, and with (7.28), we have it. Using the RHS of (7.28) in its place, we obtain:

$$L_{12} = T\left[(\alpha_2 - \alpha_1)\frac{de_s}{dT}\right] \tag{7.29}$$

Dividing through by temperature and the difference between the two specific volumes yields:

$$\boxed{\frac{de_s}{dT} = \frac{L_{12}}{T(\alpha_2 - \alpha_1)}} \tag{7.30}$$

which is the Classius-Clapeyron Equation, and describes the variation of saturation water vapor pressure (de_s) with respect to temperature (dT). It shows that changes in the saturation vapor pressure are a function of the latent heat required to move water between two adjacent phases (L_{12}), the temperature of the phase change (T), and the specific volumes (α_1 and α_2) of the two adjacent phases.

Integrating the Classius-Clapeyron Equation. The next objective is to obtain a closed solution describing the saturation vapor pressure as a function of several independent variables. Along the way, we also want to replace the specific volume terms in (7.30) with something more routinely used in meteorology.

To begin, we'll make two simplifying assumptions. The first is that the specific volume of vapor (α_2 in 7.30) is much greater than the specific volume of liquid water (α_1 in 7.30), so that:

$$\left(\alpha_2 - \alpha_1\right) \cong \alpha_2 \tag{7.31}$$

The second simplifying assumption is that the system is never far from equilibrium, so that the Ideal Gas Law applies. When written for vapor, we can use the form shown in (2.80):

$$e\alpha_v = R_v T$$

where R_v is the individual gas constant for water vapor. In the saturation/equilibrium context, the relationship becomes:

$$e_s \alpha_2 = R_v T \tag{7.32}$$

Solving for the specific volume of the vapor yields:

$$\alpha_2 = \frac{R_v T}{e_s} \tag{7.33}$$

but (7.30) has the inverse of this, that is,

$$\frac{1}{\alpha_2} = \frac{e_s}{R_v T} \tag{7.34}$$

Beginning with (7.30),

$$\frac{de_s}{dT} = \frac{L_{12}}{T\left(\alpha_2 - \alpha_1\right)}$$

and inserting the first assumption, we have:

$$\frac{de_s}{dT} \cong \frac{L_{12}}{T\alpha_2} = \frac{L_{12}}{T\alpha_v} \tag{7.35}$$

Substituting the identity in (7.34), we obtain:

$$\frac{de_s}{dT} \cong \frac{L_{12}}{T}\frac{e_s}{R_v T} = \frac{L_{12}e_s}{R_v T^2} \qquad (7.36a)$$

or simply:

$$\boxed{\frac{de_s}{dT} \cong \frac{l_v e_s}{R_v T^2}} \qquad (7.36b)$$

where we've substituted l_v (the latent heat of vaporization) for L_{12}. Rearranging to collect like terms results in:

$$\frac{de_s}{e_s} = \frac{l_v dT}{R_v T^2} \qquad (7.37a)$$

(where we have assumed an equality) or:

$$\frac{de_s}{e_s} = \left(\frac{l_v}{R_v}\right)\frac{dT}{T^2} \qquad (7.37b)$$

which we know how to integrate. The parenthetical terms on the RHS are approximately constant. The integral of the LHS of (7.37b) is the natural log of e_s, and the solution to the integral on the RHS of (7.37b) is of the form:

$$\int_i^f \frac{dx}{x^2} = -\left[\frac{1}{x_f} - \frac{1}{x_i}\right] \qquad (7.38)$$

Setting up the integral on both sides of (7.37b), we have:

$$\int_i^f \frac{de_s}{e_s} = \int_i^f \left(\frac{l_v}{R_v}\right)\frac{dT}{T^2} \qquad (7.39)$$

and using the assumption that l_v and R_v are approximately constant over a short range of temperatures, we can simplify the integral on the RHS by:

$$\int_i^f \frac{de_s}{e_s} = \frac{l_v}{R_v}\int_i^f \frac{dT}{T^2} \qquad (7.40)$$

Next, we apply the identities for solutions to both of these integrals:

$$ln\left[(e_s)_f\right] - ln\left[(e_s)_i\right] = -\frac{l_v}{R_v}\left[\frac{1}{T_f} - \frac{1}{T_i}\right] \tag{7.41}$$

This can be simplified by applying a logarithm identity on the LHS and distributing the negative sign to the left of the constants on the RHS, obtaining:

$$ln\left[\frac{(e_s)_f}{(e_s)_i}\right] = \frac{l_v}{R_v}\left[\frac{1}{T_i} - \frac{1}{T_f}\right] \tag{7.42}$$

We can interpret the initial (*i*) condition as a reference point. At this reference point, the temperature is 273.16 K (0 °C), where the known saturation vapor pressure is 611.12 Pa. The former we'll call T_0, and the latter we'll call e_0. Substituting these two identities into (7.42) and dropping the *f* subscript from the other temperature and pressure results in:

$$ln\left(\frac{e_s}{e_0}\right) = \frac{l_v}{R_v}\left[\frac{1}{T_0} - \frac{1}{T}\right] \tag{7.43}$$

Next, we apply the inverse of the natural log function, namely the exponentiation function ($\exp(x)$ or e^x), to both sides of (7.43), resulting in:

$$\frac{e_s}{e_0} = \exp\left[\frac{l_v}{R_v}\left(\frac{1}{T_0} - \frac{1}{T}\right)\right] \tag{7.44}$$

Multiplying through by the reference vapor pressure (e_0) yields a useful, closed form of the Classius-Clapeyron Equation:

$$\boxed{e_s^l = e_0\exp\left[\frac{l_v}{R_v}\left(\frac{1}{T_0} - \frac{1}{T}\right)\right]} \tag{7.45}$$

which says that the saturation vapor pressure e_s above liquid (hence the superscript *l*) at temperature *T* is a function of constants e_0, l_v, R_v, and T_0, and the temperature itself. Both *T* and T_0 must be in Kelvins, but l_v and R_v can be in any units, provided they're the *same type of* units. If e_0 is in Pascals, then e_s will be in Pascals. To compute the *in situ* vapor pressure – that is, the amount of water *actually* in the air sample – simply replace the temperature with the dew point:

$$\boxed{e = e_0\exp\left[\frac{l_v}{R_v}\left(\frac{1}{T_0} - \frac{1}{T_d}\right)\right]} \tag{7.46}$$

Both (7.45) and (7.46) have been shown, by comparison to experimental data obtained in a laboratory, to produce vapor pressure values correct to within a few percent in the range of temperatures between $-30\,^{\circ}\mathrm{C}$ and $+30\,^{\circ}\mathrm{C}$. Its accuracy can be improved somewhat by replacing the constant l_v with a temperature-dependent value computed from the relation shown in (4.66); that is:

$$l_v = l_{v0} - \left(2369 T_c\right)$$

where l_{v0} is the reference latent heat, valid at $0\,^{\circ}\mathrm{C}$, and equal to 2.5008×10^6 [J/kg], and T_c is the temperature in Celsius degrees. Alternatively, you can use a temperature-appropriate value of l_v from Table 4.2.

7.3. Solid-Vapor and Solid-Liquid Equilibrium Conditions

In addition to the liquid-vapor equilibrium state described by (7.45), there are two more than must be described. The first of these is the solid-vapor equilibrium state, which is associated with sublimation and deposition. Fortunately for us, many of the same assumptions we use to move from (7.30) to (7.45) also apply, except in this case, the latent heat under consideration is the latent heat of sublimation (l_s). We can still use the two main simplifying assumptions mentioned previously, that is, that the volume of a given mass of water vapor is much greater than the volume of the same mass of water ice, and, because we assume equilibrium between ice and vapor, the Ideal Gas Law applies. With these points in mind, we can write a similar closed solution for the saturation vapor pressure (e_s) above ice by:

$$e_s^i = e_0 \exp\left[\frac{l_s}{R_v}\left(\frac{1}{T_0} - \frac{1}{T} \right) \right] \tag{7.47}$$

which is another exponential curve. The superscript i indicates that the value is valid above *ice*.

The solid-liquid (i.e., melting-freezing) equilibrium condition begins with the form of the Classius-Clapeyron Equation shown in (7.30):

$$\frac{de_s}{dT} = \frac{L_{12}}{T\left(\alpha_2 - \alpha_1\right)}$$

and makes the following substitutions:

- p for e_s, where p indicates the pressure at which melting and freezing occurs,
- l_f for L_{12}, where l_f is the latent heat of fusion ($\cong 3.337 \times 10^5$ J/kg),[8] and
- $\Delta\alpha$ for $\alpha_2 - \alpha_1$, where $\Delta\alpha$ is the difference between the specific volume of liquid water (α_2) and ice (α_1). Because liquid water is denser than ice,[9] $\alpha_2 < \alpha_1$, and $\Delta\alpha$ is a *negative number*.

It can be computed by noting that, at a temperature of 273.16 K, $\alpha_{liquid} = 1.000 \times 10^{-3}$ m³/kg, and $\alpha_{ice} = 1.091 \times 10^{-3}$ m³/kg, so that $\Delta\alpha = -0.091 \times 10^{-3}$ m³/kg.

With these substitutions, we have:

$$\boxed{\frac{dp}{dT} = \frac{l_f}{T \Delta\alpha}} \tag{7.48}$$

Rearranging (7.48), we obtain:

$$dp = \left(\frac{l_f}{\Delta\alpha}\right)\frac{dT}{T} \tag{7.49}$$

where the parenthetical term on the RHS can be taken as approximately constant over a short range of temperatures. Next we integrate both sides from the reference point (p_0, T_0) to the melting point (p_m, T_m):

$$\int_0^m dp = \left(\frac{l_f}{\Delta\alpha}\right)\int_0^m \frac{dT}{T} \tag{7.50}$$

resulting in:

$$p_m - p_0 = \left(\frac{l_f}{\Delta\alpha}\right)\left[ln\left(T_m\right) - ln\left(T_0\right)\right] \tag{7.51}$$

or:

$$p_m = \left(\frac{l_f}{\Delta\alpha}\right)ln\left(\frac{T_m}{T_0}\right) + p_0 \tag{7.52}$$

which states that, if $T_m = T_0$, then p_m (the melting pressure) is equal to p_0 (the reference pressure). Returning to the form shown in (7.51), we can also obtain:

$$\boxed{p_m = \left(\frac{l_f}{\Delta\alpha}\right)ln\left(T_m\right) - \left(\frac{l_f}{\Delta\alpha}\right)ln\left(T_0\right) + p_0} \tag{7.53}$$

or:

$$p_m = \beta_1 ln\left(T_m\right) + \beta_0 \tag{7.54}$$

where p_m is the melting pressure, $\beta_1 = (l_f/\Delta\alpha)$ is the slope of a linear equation, $ln(T_m)$ is the natural log of the melting temperature (T_m), and $\beta_0 = -(l_f/\Delta\alpha)ln(T_0) + p_0$ is the intercept. Because $\Delta\alpha$ is a negative number, making the slope of the linear equation negative, (7.54) indicates that, if the pressure *increases*, the melting temperature *decreases*.

The numerical value of the slope (β_1) is -3.6670×10^9 Pa, and the numerical value of the intercept (β_0) is 2.0572×10^{10} Pa. A few sample calculations show that a very large change in melting pressure is associated with a very small change in melting temperature. Put another way, the melting temperature is nearly constant over the ordinary range of pressures in Earth's troposphere.

7.4. Phase Diagram of Evaporation, Sublimation, and Melting for Water

We'll adapt the thermodynamic diagrams used throughout this book to show the phases of water as functions of temperature and pressure. By computing all possible values of e_s (using 7.45 and 7.47) and p_m (using 7.54) in the temperature range between absolute zero and several hundred Kelvins above zero, we can generate curves for the liquid-vapor, the solid-vapor, and the solid-liquid equilibrium states. These are shown in Figure 7.3.

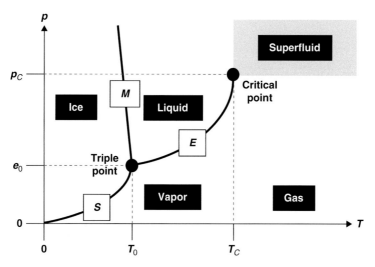

Figure 7.3. Phase diagram for water substance. Heavy lines labeled S, M, and E are the sublimation, melting, and evaporation lines, computed using (7.47), (7.54), and (7.45), respectively. All three lines meet at the reference point, corresponding to $p = 611.12$ Pa, and $T = 273.16$ K, meaning that all three phases can coexist in equilibrium at this temperature and pressure. The three ordinary states of water (ice, liquid, and vapor) occur at relatively low pressures and temperatures. At very high temperatures, but relatively low pressures, water vapor is sometimes called a "gas," although this distinction is somewhat subjective. At very high temperatures and very high pressures (above the critical point corresponding to p_c and T_c), the distinction between liquid and gas disappears, and the substance enters a *superfluid* (or simply *fluid*) state.

Some important points illustrated by Figure 7.3.

- At a temperature of absolute zero, the pressure and specific volume go to zero, consistent with the gas laws of Gay-Lussac (which assumed an ideal gas, that is, one in which the vapor molecules have zero volume).
- The *triple point of water* is the temperature and pressure where all three phases (ice, liquid, and vapor) are in equilibrium. This corresponds to a temperature of 273.16 [K] and a pressure of 611.12 [Pa]. For ice, the triple point specific volume is 1.091×10^{-3} [m^3/kg]; for liquid water, it is 1.000×10^{-3} [m^3/kg]; for vapor, it is 206 [m^3/kg].
- The evaporation curve begins at the triple point and curves exponentially up to the right, terminating at the *critical point of water*. The critical point corresponds to a temperature of 647 [K], a pressure of 2.22×10^7 [Pa] (218 atmospheres), and a specific volume of 3.07×10^{-3} [m^3/kg].
- Along the evaporation curve, liquid water and vapor are in equilibrium, so that $e_s = e$ for vapor over liquid (7.45).
- *Boiling* is a form of rapid evaporation that occurs when the total atmospheric pressure (approximately 1,013 [hPa] at sea level) is equal to the saturation vapor pressure above liquid water.[10] This corresponds to a temperature of 373.16 [K]. If the total atmospheric pressure *decreases*, the boiling temperature also *decreases*.
- At temperatures and pressures beyond the critical point, the distinction between vapor (or gaseous water) and liquid water disappears, and water manifests as a *superfluid* (or simply *fluid*). Superfluids look like ordinary liquids but have no viscosity (internal resistance to movement).[11]
- The sublimation curve begins at the triple point and arcs downward toward lower temperatures and pressures. Along the sublimation curve, ice and vapor are in equilibrium, so that $e_s = e$ for vapor over ice (7.47). If the atmospheric pressure *decreases*, the sublimation temperature also *decreases*.
- The melting curve begins at the triple point and slopes sharply upward toward higher pressures and lower temperatures. At ordinary pressures in Earth's troposphere, the melting temperature is almost constant. Along the melting curve, liquid and vapor are in equilibrium, so that (7.54) applies. If the pressure *increases*, the melting temperature *decreases*.

The sum of all states that water can manifest can be specified by computing all possible values of its state variables, allowing the construction of a *three-dimensional* p-V-T diagram showing the thermodynamic surface of water. Figure 7.4 shows the result of such a calculation.

7.5. Additional Functions for Expressing Vapor Pressure above Liquid and Ice

In addition to the Classius-Clapeyron Equation, several other functions have been derived to compute vapor pressure. Some have been developed for greater ease of use; others, for greater accuracy when compared to laboratory data. In the interest of brevity, we'll just list a few them here without their associated derivations.[12]

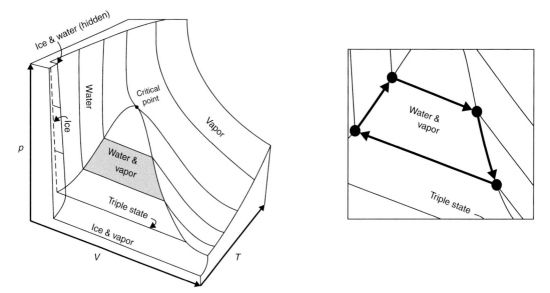

Figure 7.4. p-V-T diagram. The thermodynamic surface of water seen in three dimensions. This is adapted from Hess (1959), with axes relabeled p, V, T from the original e, α, T. (Hess means *liquid water* where it says "water.") If you rotate the model on the left clockwise about 45 degrees, and vertically about 45 degrees, it will closely resemble the p-T phase diagram shown in Figure 7.3. Where isotherms are "horizontal" (parallel to the V axis), two or more phases are in equilibrium. The region of the thermodynamic surface used for the development of the Classius-Clapeyron Equation (7.30) (represented in Figure 7.2) is highlighted in light gray in the panel on the left and indicated with heavy arrows in the expanded panel on the right.

Advanced form of the Classius-Clapeyron Equation. If you're going to compute vapor pressure (or saturation vapor pressure) with a calculator, use the simple Classius-Clapeyron Equation shown in (7.46), or one of the simple functions shown in (7.56) and (7.57). If you're going to program a computer to compute e for you, use an advanced form of Classius-Clapeyron, which states:

$$e = e_0 \exp\left[\frac{l_{v0}}{R_v}\left(\frac{1}{T_0} - \frac{1}{T_d}\right) - \left(\frac{c - c_p^{\text{vapor}}}{R_v}\right) ln\left(\frac{T_d}{T_0}\right) + T_0\left(\frac{1}{T_0} - \frac{1}{T_d}\right)\left(\frac{c - c_p^{\text{vapor}}}{R_v}\right) \right] \quad (7.55)[13]$$

where e_0 is the reference pressure (611.12 Pa), l_{v0} is the latent heat of vaporization at 0 °C (2.5008 × 10^6 J/kg),[14] R_v is the individual gas constant for water vapor (461.2 J/(kg K)),[15] T_0 is the reference temperature (273.16 K), T_d is the *in situ* dew point [K], c is the specific heat of liquid water at 0 °C (4215 J/(kg K)),[16] and c_p^{vapor} is the specific heat of water vapor at constant pressure (1844.8 J/(kg K)).[17] Note that, if you use (7.55), there is no need to correct l_v for temperature. And, as with the simpler Classius-Clapeyron Equation, if you want to compute the saturation (or equilibrium) vapor pressure (e_s), rather than the *in situ* vapor pressure (e), simply substitute the *in situ* temperature for the dew point.

Bolton (1980) showed that the following relation, which describes the vapor pressure above liquid water, is correct to within less than 1 percent in the temperature range between −30 °C and +35 °C:

$$ e = e_0 \exp\left[\frac{17.67 T_d}{T_d + 243.5} \right] \tag{7.56} $$

In Bolton's function, e_0 is equal to 611.2 [Pa] and the dew point temperature (T_d) is in degrees C.[18]

Buck (1981) showed that, for vapor pressure above liquid water at pressures greater than 800 hPa, the following relationship was sufficiently precise for most purposes between −30 °C and +50 °C:

$$ e = e_0 \exp\left[\frac{17.502 T_d}{T_d + 240.97} \right] \tag{7.57a} $$

where e_0 is equal to 613.65 [Pa]. For the vapor pressure above ice:

$$ e = e_0 \exp\left[\frac{22.452 T_d}{T_d + 272.55} \right] \tag{7.57b} $$

where e_0 is equal to 613.59 [Pa]. In both cases, the dew point temperature (T_d) is in degrees C.[19]

Wexler (1976, 1977) developed empirical expressions for vapor pressure above liquid water and ice, based on laboratory data, and which are similar to the *virial expansion* Equation of State shown in Chapter 2.[20] The general form on these equations is given by:

$$ e = \exp\left[\frac{c_1}{T_d^2} + \frac{c_2}{T_d} + c_3 + c_4 T_d + c_5 T_d^2 + c_6 T_d^3 + c_7 T_d^4 + c_8 \ln T_d \right] \tag{7.58} $$

where the vapor pressure (e) has units of Pascals, and the dew point temperature (T_d) is in Kelvins.[21] As with the other forms, you can substitute the *in situ* temperature T for T_d to compute the saturation vapor pressure (e_s). The coefficients c_i, relevant to either liquid water or ice, are given in Table 7.1, and Table 7.2 shows values of e_s between +40 °C and −40 °C, computed from Wexler's (1976, 1977) equations.

Table 7.1. *Coefficients for Equation (7.58), for vapor pressure (e) above liquid water and ice.*[i]

Coefficient	For e above Liquid Water	For e above Ice
c_1	−2991.272	0.0
c_2	−6017.0128	−5865.3696
c_3	18.87643854	22.241033
c_4	−0.028354721	0.013749042
c_5	−0.17838301 × 10^{-4}	−0.34031774 × 10^{-4}
c_6	−0.84150417 × 10^{-9}	0.26967687 × 10^{-7}
c_7	0.44412543 × 10^{-12}	0.0
c_8	2.858487	0.6918651

[i] Taken from Table 5.1 in Brock and Richardson (2001); and Buck (1981), Equations (5a) and (5b).

Table 7.2. *Saturation vapor pressure above liquid water and ice.*[i]

Compare column 2 to values computed using (7.45), and column 3 to values computed using (7.47).

Temperature [°C]	e_s above Liquid [Pa]	e_s above Ice [Pa]
40	7381.27	
35	5626.45	
30	4245.20	
25	3168.74	
20	2338.54	
15	1705.32	
10	1227.94	
5	872.47	
0	611.21	611.15
−5	421.84	401.78
−10	286.57	259.92
−15	191.44	165.32
−20	125.63	103.28
−25	80.90	63.30
−30	51.06	38.02
−35	31.54	22.36
−40	19.05	12.85

[i] Taken from Table 2.1 in Rogers and Yau (1989), which is based on Wexler (1976) and Wexler (1977). Values of e_s above liquid water at temperatures below 0 °C are extrapolated from Wexler (1976).

7.6. Humidity

Once you have determined vapor pressure, you can also compute several variations on "humidity" to describe the moisture content of air.

Absolute humidity (ρ_v) is simply the density of water vapor, determined by solving the Ideal Gas Law for vapor. Beginning with (2.66), which states:

$$e = \rho_v R_v T$$

we find that:

$$\boxed{\rho_v \equiv \frac{e}{R_v T}} \tag{7.59}$$

where ρ_v is the absolute humidity [kg/m³], e is the vapor pressure computed by one of the means described in the preceding text [Pa], R_v is the individual gas constant for water vapor (461.2 J/(kg K)), and T is the temperature [K]. For the saturation value, substitute e_s for e.

Mixing ratio (w) is defined as the ratio of the mass of water vapor to the mass of *dry* in a sample of air, that is:

$$\boxed{w \equiv \frac{m_v}{m_d}} \tag{7.60}$$

which has units of kg$_{vapor}$/kg$_{dry\ air}$, or, in other words, is unitless. Because $\rho \equiv m/V$ (the basic definition of density), we know that $m = \rho V$, so:

$$w = \frac{\rho_v V}{\rho_d V} = \frac{\rho_v}{\rho_d} \tag{7.61}$$

Because density is not usually reported in weather observations, we need another way to write both the numerator and denominator of (7.61). We can appeal to the Ideal Gas Law for both. We've already obtained another expression for the numerator in (7.61) – that is, the absolute humidity. For the denominator, the Ideal Gas Law states:

$$p - e = \rho_d R_d T \tag{7.62}$$

The LHS of (7.62) is the total atmospheric pressure (p) minus the vapor pressure (e). The difference is taken because we're interested in only describing the pressure attributed to the dry air component of the total air sample. On the RHS, the subscript d indicates dry air values. With this in mind, we can rewrite ρ_d as:

$$\rho_d = \frac{p - e}{R_d T} \tag{7.63}$$

Making these substitutions into (7.61) yields:

$$w = \frac{\rho_v}{\rho_d} = \frac{\left(\dfrac{e}{R_v T}\right)}{\left(\dfrac{p-e}{R_d T}\right)} \tag{7.64}$$

or:

$$w = \frac{\left(\dfrac{e}{R_v}\right)}{\left(\dfrac{p-e}{R_d}\right)} \tag{7.65}$$

or:

$$w = \frac{e}{p-e}\left(\frac{R_d}{R_v}\right) \tag{7.66}$$

We've seen the parenthetical term on the RHS of (7.66) before. In fact, (2.69) defines this ratio as epsilon (ε), that is:

$$\varepsilon \equiv \frac{R_d}{R_v} = 0.622$$

so that (7.66) can be rewritten more simply as:

$$\boxed{w = \frac{e\varepsilon}{p-e}} \tag{7.67}$$

One important simplifying assumption can be made by noting that $p \gg e$, so that $p - e \cong p$, or:

$$w \cong \frac{e\varepsilon}{p} \tag{7.68}$$

which we will use again later. To compute the *saturation* mixing ratio (w_s), substitute e_s for e in (7.67) or (7.68).

While mixing ratio is generally unitless (it *must* be if used in calculations), it is sometimes multiplied by 1000, giving it units of $g_{vapor}/kg_{dry\ air}$. See Appendix 6 for a discussion of how saturation mixing ratio varies with temperature and pressure.

Integrated Precipitable Water (W and IPW) is defined as "the total atmospheric water vapor contained in a vertical column of unit cross-sectional area [i.e., a column of air above a 1 m² area of the Earth's surface] extending between any two specified levels [usually from the surface to the 'top' of the atmosphere], commonly expressed in terms of the height to which that water substance would stand if completely condensed and collected in a vessel of the same unit cross section."[22] It refers to the theoretical maximum amount of precipitation the atmosphere is capable of producing, and is computed as the vertical integral of w in the pressure domain:

$$\boxed{W \equiv -\frac{1}{g}\int_i^f w\,dp}$$

(7.69)

where W is integrated precipitable water, g is the gravitational acceleration near Earth's surface (a positive number equal to about 9.81 m/s²), w is the mixing ratio [$kg_{vapor}/kg_{dry\,air}$], and dp is an infinitesimal change in pressure [Pa]. Because we integrate from p_i (high pressure near Earth's surface) to p_f (low pressure aloft), we need the additional negative sign outside the integral to make W come out a positive number.

In practice, we determine the *mean* mixing ratio through some shallow layer of the atmosphere (usually from radiosonde data) and compute a series of finite sums by:

$$W_{lyr} = -\frac{\bar{w}}{g}\left(p_f - p_i\right)$$

(7.70)

where $p_f < p_i$. The subscript *lyr* indicates W through the shallow vertical layer. A unit analysis indicates:

$$[W_{lyr}] = \frac{[w]}{[g]}[p] = \frac{\left(\dfrac{kg_{vapor}}{kg_{dry\,air}}\right)}{\left(\dfrac{m}{s^2}\right)}(Pa)$$

(7.71)

The mixing ratio (w) units cancel and $Pa \equiv N/(m^2)$.[23] With these two points in mind, (7.71) becomes:

$$[W] = \left(\frac{s^2}{m}\right)\left(\frac{N}{m^2}\right)$$

(7.72)

But, $N \equiv (kg\,m)/(s^2)$,[24] so we have:

$$[W] = \left(\frac{s^2}{m}\right)\left(\frac{1}{m^2}\right)\left(\frac{kg\,m}{s^2}\right)$$

(7.73)

After canceling extraneous units, we're left with:

$$[W] = \left(\frac{kg}{m^2}\right) \tag{7.74}$$

Or, in English, the *natural* units of Integrated Precipitable Water are kilograms of liquid precipitation per unit area of the Earth's surface. We usually measure the *depth* of precipitation, not its mass, so the result in (7.74) isn't practical. But we can convert W to a linear depth by dividing by the density (ρ) of liquid water (almost exactly 1000 kg/m³ at sea level, in the normal range of temperatures at the Earth's surface), that is:

$$\frac{[W]}{[\rho_{\text{liquid}}]} = \left(\frac{kg}{m^2}\right)\left(\frac{m^3}{kg}\right) = m \tag{7.75}$$

which is better than the previous result, but still not really useful. If we reported (or predicted) rainfall in meters, we'd end up with many small amounts to the right of the decimal point. *Millimeters* are an even better unit, which we can obtain if we use units of $g_{\text{vapor}}/kg_{\text{dry air}}$ for the mixing ratio, rather than the unitless $kg_{\text{vapor}}/kg_{\text{dry air}}$ (it effectively adds another factor of 1000, resulting in mm at the end rather than m). With this in mind, we'll define *IPW* as the Integrated Precipitable Water in millimeters:

$$IPW \equiv -\frac{\overline{w}}{g\rho_{\text{liquid}}}\left(p_f - p_i\right) \tag{7.76}$$

where the mean mixing ratio in the layer ($\overline{w}$) has units of $g_{\text{vapor}}/kg_{\text{dry air}}$.

Specific humidity (q) is defined as the ratio of the mass of water vapor to the mass of *moist* air in a sample of air, that is:

$$q \equiv \frac{m_v}{m_m} \tag{7.77}$$

which is similar to mixing ratio. It is left as a practice problem at the end of this chapter to show that:

$$q = \frac{e\varepsilon}{p - (1-\varepsilon)e} \tag{7.78}$$

Because $p \gg e$, and, therefore, $p \gg (1-\varepsilon)e$, we know that $p - (1-\varepsilon)e \cong p$, and we can make the simplifying assumption that:

$$q \cong \frac{e\varepsilon}{p} \tag{7.79}$$

which you should compare to (7.68). The natural units of q are $kg_{vapor}/kg_{moist\ air}$ (which you must use when performing physical calculations), but we often express it in $g_{vapor}/kg_{moist\ air}$. To compute the *saturation* specific humidity (q_s), substitute e_s for e in (7.78) and (7.79).

Relative Humidity (r and RH) is the ratio of the amount of water vapor in a sample of air to the total amount that would be present of the sample was saturated. Like w and q, it is unitless and ranges in value from 0 to 1. Its definition is given by:

$$\boxed{r \equiv \frac{e}{e_s}} \tag{7.80}$$

where e and e_s are the vapor pressure and saturation vapor pressure, respectively, but it can also be estimated by:

$$r \cong \frac{w}{w_s} \cong \frac{q}{q_s} \tag{7.81}$$

where w and w_s are the mixing ratio and saturation mixing ratio, respectively, and q and q_s are the specific humidity and saturation specific humidity, respectively. The approximation symbol on the RHS takes advantage of the identities shown in (7.68) and (7.79). Use any of these forms when performing physical calculations.

Relative humidity is the most common form of humidity discussed in meteorology, especially when speaking with the general public. But this audience is accustomed to thinking of humidity as a *percentage*; that is, scaled between 0 and 100 percent. The forms shown in the preceding text can be converted to the commonly expected scale by:

$$\boxed{RH \equiv r \times 100\%} \tag{7.82}$$

Relative humidity can also be computed directly from psychrometric data, which is discussed in greater detail in Chapter 8.

7.7. Summary of Water Vapor and Phase Transitions

In this chapter we derived several expressions for describing the amount of water vapor in a parcel of air and the total amount the parcel could hold if it were saturated. We began by deriving the Classius-Clapeyron Equation, and then we integrated it to obtain a closed form that describes the *in situ* vapor pressure as a function of dew point. From there, we continued on to describe functions for sublimation and melting as well. Alternative expressions were given for estimating vapor pressure. The open form of the Classius-Clapeyron Equation is:

$$\frac{de_s}{dT} = \frac{L_{12}}{T\left(\alpha_2 - \alpha_1\right)}$$

where:

    de_s/dT = variation in saturation vapor pressure with respect to temperature [Pa/K]

    T = *in situ* temperature [K]

    L_{12} = latent heat for transition from phase 1 to phase 2 [J/kg]

    α_2 = specific volume of substance in phase 2 [m³/kg]

    α_1 = specific volume of substance in phase 1 [m³/kg]

Its closed form, describing the equilibrium point between evaporation and condensation is:

$$e_s = e_0 \exp\left[\frac{l_v}{R_v}\left(\frac{1}{T_0} - \frac{1}{T}\right)\right]$$

where:

    e_s = saturation vapor pressure above liquid water [Pa]

    e_0 = reference pressure (611.12 Pa)

    l_v = latent heat of vaporization ($\cong 2.5008 \times 10^6$ J/kg)[25]

    R_v = individual gas constant for water vapor (461.2 J/(kg K))

    T_0 = reference temperature (273.16 K)

The "melting line" is described by:

$$p_m = \left(\frac{l_f}{\Delta\alpha}\right)ln\left(T_m\right) - \left(\frac{l_f}{\Delta\alpha}\right)ln\left(T_0\right) + p_0$$

where:

    p_m = melting pressure of water [Pa]

    l_f = latent heat of fusion ($\cong 3.337 \times 10^5$ J/kg)

    $\Delta\alpha$ = difference between specific volumes of ice and liquid water (-0.091×10^{-3} m³/kg)

    T_m = melting temperature of water [K]

    T_0 = reference temperature (273.16 K)

    p_0 = reference pressure (611.12 Pa)

Another form of the Classius-Clapeyron Equation can also be used to describe vapor pressure above liquid water:

$$e_s = e_0 \exp\left[\frac{l_{v0}}{R_v}\left(\frac{1}{T_0} - \frac{1}{T}\right) - \left(\frac{c - c_p^{vapor}}{R_v}\right)ln\left(\frac{T}{T_0}\right) + T_0\left(\frac{1}{T_0} - \frac{1}{T}\right)\left(\frac{c - c_p^{vapor}}{R_v}\right)\right]$$

where:

lv_0 = latent heat of vaporization at $0\,°C$ (2.5008×10^6 J/kg)

c = specific heat of liquid water at $0\,°C$ (4215 J/(kg K))

c_p^{vapor} = specific heat of water vapor at constant pressure (1844.8 J/(kg K))

as can Bolton's expression:

$$e_s = e_0 \exp\left[\frac{17.67T}{T + 243.5}\right]$$

where:

e_0 = reference pressure (611.2 Pa)

T = *in situ* temperature [°C]

and several others.

After completing vapor pressure, we derived several additional parameters describing water vapor content. These included absolute humidity, mixing ratio, integrated precipitable water, specific humidity, and relative humidity.

$$\rho_v \equiv \frac{e}{R_v T}$$

where,

ρ_v = absolute humidity [kg/m^3]

e = *in situ* vapor pressure [Pa]

$$w_s = \frac{e_s \varepsilon}{p - e_s}$$

where:

w_s = saturation mixing ratio [kg$_{\text{vapor}}$/kg$_{\text{dry air}}$]

ε = ratio of molar masses of vapor and dry air (0.622)

$$w_s \cong \frac{e_s \varepsilon}{p}$$

$$W \equiv -\frac{1}{g}\int_i^f w\,dp$$

where:

W = integrated precipitable water [kg/m^2]

g = gravitational acceleration (9.81 m/s^2)

$$\text{IPW} \equiv -\frac{\overline{w}}{g\rho_{\text{liquid}}}\left(p_f - p_i\right)$$

where:

  IPW = integrated precipitable water [mm]

  $\bar{w}$ = mean mixing ratio through column [g_{vapor}/$kg_{dry\ air}$]

  ρ_{liquid} = density of liquid water (1000 kg/m^3)

  p_f = pressure at top of column [Pa]

  p_i = pressure at bottom of column [Pa]

$$q_s \equiv \frac{e_s \varepsilon}{p-\left(1-\varepsilon\right)e_s}$$

where:

  q_s = saturation specific humidity [kg_{vapor}/$kg_{moist\ air}$]

$$q_s \cong \frac{e_s \varepsilon}{p}$$

$$r \equiv \frac{e}{e_s} \cong \frac{w}{w_s} \cong \frac{q}{q_s}$$

where:

  r = relative humidity [unitless]

$$RH \equiv r \times 100\%$$

where:

  RH = relative humidity [%]

Practice Problems

1. Define the critical point of water.
2. Define the triple point of water.
3. Compute the equilibrium (saturation) vapor pressure above a pool of liquid water if the water is at 10 °C.
   - Use the integrated form of the Classius-Clapeyron Equation shown in (7.46). Be sure to include any corrections needed to the latent heat coefficient.
   - Use the advanced form of the Classius-Clapeyron Equation shown in (7.55).
   - Compare your answers obtained by the two different methods.
4. Repeat the preceding for a pool of water at 60 °C.
5. Using the Classius-Clapeyron Equation, derive an expression for calculating RH (in units of %) directly from temperature and dew point (both in °C). Remember that l_v is a function of temperature.
6. Compute the *in situ* vapor pressure for a parcel of air with a temperature of 15 °C and a relative humidity of 65 percent.
7. Define boiling in terms of vapor pressure and total atmospheric pressure, then compute the boiling temperature for total pressure 800 hPa.

8. Calculate the absolute humidity (ρ_v) if $e = 10$ hPa and $T = 40\,^\circ$C.
9. Calculate the mixing ratio (w) and saturation mixing ratio (w_s) if $T = 35\,^\circ$C and $T_d = 10\,^\circ$C. Assume $p = 1$ atmosphere (1,013.25 hPa). Use temperature-appropriate values of l_v.
10. Beginning with the definition of the specific humidity (q):

$$q \equiv \frac{m_v}{m_m}$$

Show that q may be written:

$$q = \frac{e\varepsilon}{p - \left(1 - \varepsilon\right)e}$$

8

Moisture Considerations: Effects
on Temperature

8.1. Introduction

When water is added to a parcel of air, it changes the behavior of the parcel in important ways. One effect is to reduce the density of the air in the parcel, and another is to change the rate at which it cools when lifted aloft. In this chapter, we'll work with several hypothetical temperatures, and examine our method of describing the cooling rate of a saturated parcel when it is lifted aloft. We'll also see how these parameters are all interrelated.

8.2. Hypothetical Temperatures

In meteorology we use a series of temperatures derived from surface data, radiosonde data, or both. These temperatures describe physical parameters such as the true density of a sample of air and the temperature at which condensation will occur. There are several variations on the latter.

Virtual temperature (T_v) is the temperature of a parcel of dry air with the same density as a parcel of moist air. The molecular masses of vapor and dry air are given by (2.64) and Table 2.1, shown in Chapter 2:

$$M_d = 28.97 \left[\frac{kg}{kmol} \right]$$

$$M_v = 18.02 \left[\frac{kg}{kmol} \right]$$

Comparing these two, you can see that adding a molecule of water vapor to an open parcel of air causes its density to decrease, if we assume the addition of the lighter vapor molecule results in the expulsion of one of the heaver molecules of the dry air.[1] This means that the Ideal Gas Law, in forms such as (2.37):

$$p = \rho R_d T$$

(where R_d is substituted for R_i) would *not* yield the correct value for density if solved for ρ. The water vapor in the parcel reduces its actual density, so the value of ρ computed by (2.37) would to be too high. This would, in turn, result in errors computed for other quantities, such as *geopotential height* (which we'll discuss later).

We correct for this problem by modifying the temperature. Taking advantage of the fact that another way to reduce density is to *raise the temperature*, we can go back to our assumption of a perfectly dry air sample (and continue to use R_d in the Ideal Gas Law) if we create a hypothetical temperature, slightly warmer than the true *in situ* temperature, that incorporates information about the amount of vapor in the air. That's the *virtual temperature*.

We begin with Dalton's Law (2.57), which (in this context) states that the total atmospheric pressure (p) is the sum of the partial pressures resulting from the dry air (p_d) and the water vapor (e) in the column of air above a point on the Earth's surface:

$$p = p_d + e \tag{8.1}$$

Using the Ideal Gas Law, we can rewrite this as:

$$p = p_d + e = \rho_d R_d T + \rho_v R_v T \tag{8.2}$$

or simply:

$$p = \rho_d R_d T + \rho_v R_v T \tag{8.3}$$

Combining this with the definitions of density, mixing ratio, and ε, we eventually arrive at:

$$p = \rho R_d T \left(\frac{1 + \dfrac{w}{\varepsilon}}{1 + w} \right) \tag{8.4}$$

which looks like the Ideal Gas Law with an extra term. The steps required to take you from (8.3) to (8.4) constitute one of the practice problems at the end of this chapter. *This parenthetical term is the correction for the water vapor content of the parcel.* We'll combine the temperature T with the vapor correction term to define the *virtual temperature* as:

$$\boxed{T_v \equiv T \left(\frac{1 + \dfrac{w}{\varepsilon}}{1 + w} \right)} \tag{8.5}$$

where T must be entered in Kelvins, and w must be in its unitless form. It can also be approximated by the functions:

$$T_v \cong T\left(1 + 0.61w\right) \tag{8.6a}$$

where T is in Kelvins and w is in its unitless form,[2] and:

$$T_v \cong T + \frac{w}{6} \tag{8.6b}$$

where T may have units of either °C or Kelvins, and w is in g/kg.[3] The form shown in (8.6b) is often used with manual analysis of a radiosonde sounding plotted on a Skew-T log p diagram. A third alternative listed in the U.S. Standard Atmosphere[4] is:

$$T_v \cong \frac{T}{1 - 0.0379\dfrac{e}{p}} \tag{8.6c}$$

where T is in Kelvins, e is the vapor pressure, and p is the total atmospheric pressure. The latter two can be in any units, as long as they are the same units. In the middle latitudes, T_v is rarely more than a Celsius degree warmer than T. In the tropics, where water vapor mixing ratios can be quite high, T_v may be warmer than T by a few degrees.

Using virtual temperature, the Ideal Gas Law for moist air is given by:

$$\boxed{p = \rho R_d T_v} \tag{8.7}$$

where p is the total atmospheric pressure [Pa], ρ is the density of the moist air parcel [kg/m³], R_d is the individual gas constant for dry air (286.8 J/(kg K)), and T_v is the virtual temperature [K].

Example. Compute the virtual temperature using (8.5), (8.6a), and (8.6b), given an *in situ* temperature (T) of 275.00 K and water vapor mixing ratio (w) of 0.005kg$_{\text{vapor}}$/kg$_{\text{air}}$ (5 g$_{\text{vapor}}$/kg$_{\text{air}}$), and compare the results.

By the analytical solution in (8.5), we have:

$$T_v = T\left(\frac{1 + \dfrac{w}{\varepsilon}}{1 + w}\right) = 275.00\left(\frac{1 + \dfrac{0.005}{0.622}}{1 + 0.005}\right) = 275.8315K \cong 275.83\,K$$

Using the approximation in (8.6a) we obtain:

$$T_v \cong T\left(1 + 0.61w\right) = 275.00\left[1 + 0.61\left(0.005\right)\right] = 275.8388K \cong 275.84\,K$$

and the approximation in (8.6b) yields:

$$T_v \cong T + \frac{w}{6} = 275.00 + \frac{5}{6} = 275.8333K \cong 275.83\,K$$

and, we see that all three functions produce very similar results.

Wet-bulb temperature [T_w] is the temperature to which an air parcel (or some other object) may be cooled by forcing water to evaporate at constant pressure. The *sensible* heat of the air (measured by its temperature) is converted to the *latent* heat needed to move water from the liquid to the vapor phase. A perfectly dry air parcel would be able to absorb a great deal of vapor without becoming saturated, therefore its wet-bulb temperature would be much lower than its *in situ* temperature (also known as its *dry-bulb temperature*).

Dry-and wet-bulb temperatures are directly measured using a *sling psychrometer*, which is a device consisting of two thermometers, one of which is covered by a cotton sock (Figure 8.1). The user wets the sock with some clean, distilled water, and twirls the two thermometers holding the handle. Periodically both temperatures are checked. When the wet-bulb temperature stops falling, the air around the sling psychrometer has become saturated, and no further evaporative cooling of the thermometer can occur. The dry and wet bulbs can then be combined to compute several atmospheric parameters.

Wet-bulb temperature can be computed by:

$$T_w = T - \frac{l_v}{c_p}\left[\frac{\varepsilon}{p} A \exp\left(\frac{-B}{T_w} \right) - w \right] \tag{8.8}$$

where T_w is the wet-bulb temperature [K], T is the *in situ* temperature [K], l_v is the latent heat of vaporization ($\cong 2.5008 \times 10^6$ J/kg),[5] c_p is the specific heat of dry air at constant pressure ($\cong 1003.8$ J/(kg K)),[6] p is the *in situ* pressure [Pa], and w is the mixing ratio [$kg_{vapor}/kg_{dry\ air}$]. A in (8.8) is the *pressure constant* and is equal to 2.5×10^{11} [Pa]. B is the *temperature constant*, equal to 5.4×10^3 [K].[7]

The relationship shown for the wet-bulb temperature is a *transcendental equation*, that is, it can't be solved by ordinary algebraic means. T_w is on both sides of the equation and can't be isolated. There are several techniques available for solving this class of equation, such as *Newton's Method* (also known as the *Newton–Raphson Method*), which finds the *roots* of the equation. Roots are values of the unknown (T_w, in this case) that make the equation true. Most high-end scientific calculators also come equipped with root finders.

Psychrometric data can be used to estimate relative humidity (*RH*) from tables. This tabular method involves computing the *wet-bulb depression*, which is simply the difference between the dry- and wet-bulb temperatures from the psychrometer.

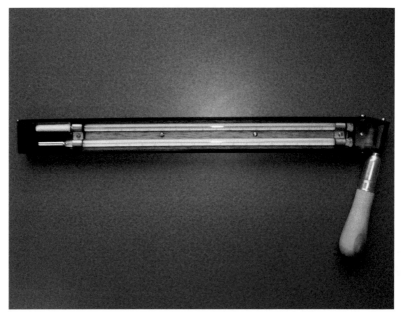

Figure 8.1. Sling psychrometer. Handheld device used to measure the *in situ* vapor content of the air, by forcing evaporation from the wet-bulb thermometer (upper tube, with cotton sock), which is cooled by the removal of heat to supply the liquid-vapor phase change. This one is held by hand and twirled manually, but various schemes have been developed to create similar automated devices, such as using a small fan to cool the wet bulb. (Image from Wikipedia, 2012.)

Table 8.1 shows relative humidity as a function of T and T_w for the range of ordinary atmospheric pressures at sea level.

Wet-bulb potential temperature (θ_w) is the *in situ* wet-bulb temperature taken *moist adiabatically* to 1000 hPa.[8] We'll discuss the Moist Adiabatic Lapse Rate (Γ_m) in the following text. Wet-bulb potential temperature is used in a number of severe weather applications.

Dew point temperature (T_d) is the temperature to which an air parcel must be cooled at constant pressure, *without* adding vapor (w is constant), to induce saturation.

(a) Dew point can be computed analytically from the wet- and dry-bulb temperatures (collectively known as *psychrometric data*) by considering the following analysis. We begin with the Classius-Clapeyron identity for *in situ* vapor pressure above liquid water (7.46), which states that:

$$e = e_0 \exp\left[\frac{l_v}{R_v} \left(\frac{1}{T_0} - \frac{1}{T_d} \right) \right]$$

Table 8.1. *Relative humidity as a function of* in situ *temperature and wet-bulb depression.*[i] Wet-bulb depression is the difference between the psychrometer's dry- and wet-bulb temperatures. All relative humidity values shown are percentages, and rounded to the nearest whole value.

Temp [°C]	Wet-Bulb Depression [°C]															
	0	1	2	3	4	5	6	7	8	9	10	11	12	13	14	15
30	100	93	86	79	72	66	61	55	49	44	39	34	29	25	20	16
28	100	93	86	78	71	65	59	53	47	42	36	31	26	21	17	12
26	100	92	85	77	70	64	57	51	45	39	34	28	23	18	13	9
24	100	92	84	76	69	62	55	49	42	36	30	25	20	14	9	4
22	100	92	83	75	68	60	53	46	40	33	27	21	15	10	4	0
20	100	91	82	74	66	58	51	44	36	30	23	17	11	5	0	
18	100	91	81	72	64	56	48	40	33	26	19	12	6	0		
16	100	90	80	71	62	54	45	37	29	21	14	7	1			
14	100	89	79	69	60	50	41	33	25	16	8	1				
12	100	88	78	67	57	48	38	28	19	10	2					
10	100	88	76	65	54	43	33	24	13	4						
8	100	87	74	62	51	39	28	17	6							
6	100	86	72	59	46	35	22	10	0							
4	100	85	70	56	42	27	14									
2	100	83	67	51	36	20	6									
0	100	81	63	45	28	11										
−2	100	79	58	37	20	1										
−4	100	77	54	32	11											
−6	100	73	48	20	0											
−8	100	71	41	13												
−10	100	66	33	0												
−12	100	61	23													
−14	100	55	11													
−16	100	48	0													
−18	100	40														
−20	100	28														

[i] State University of New York (1994); Lide (1997).

Our first objective is to solve this for dew point. Dividing through by e_0 and taking the natural log of both sides yields:

$$ln\left(\frac{e}{e_0}\right) = \frac{l_v}{R_v}\left(\frac{1}{T_0} - \frac{1}{T_d}\right) \tag{8.9}$$

We then multiply through by l_v/R_v, and subtract $1/T_0$ from both sides:

$$-\frac{1}{T_0} + \frac{R_v}{l_v}ln\left(\frac{e}{e_0}\right) = -\frac{1}{T_d} \tag{8.10}$$

Multiplying by -1 and inverting results in:

$$T_d = \frac{1}{\dfrac{1}{T_0} - \dfrac{R_v}{l_v}ln\left(\dfrac{e}{e_0}\right)} \tag{8.11}$$

The second step begins with the approximation for water vapor mixing ratio given by (7.68):

$$w \cong \frac{e\varepsilon}{p}$$

which we solve for e:

$$\boxed{e \cong \frac{pw}{\varepsilon}} \tag{8.12}$$

and substitute the result into (8.11), yielding:

$$T_d = \frac{1}{\dfrac{1}{T_0} - \dfrac{R_v}{l_v}ln\left(\dfrac{pw}{e_0\varepsilon}\right)} \tag{8.13}$$

Next we solve the expression for wet-bulb temperature given in (8.8) for w:

$$T_w = T - \frac{l_v}{c_p}\left[\frac{\varepsilon}{p}A\exp\left(\frac{-B}{T_w}\right) - w\right]$$

First, distribute the constant to the left of the square brackets on the RHS:

$$T_w = T - \frac{l_v}{c_p}\frac{\varepsilon}{p}A\exp\left(\frac{-B}{T_w}\right) + \frac{l_v}{c_p}w \tag{8.14}$$

Next, subtract T_w and $(l_v/c_p)w$ from both sides:

$$-\frac{l_v}{c_p}w = T - \frac{l_v}{c_p}\frac{\varepsilon}{p}A\exp\left(\frac{-B}{T_w}\right) - T_w \tag{8.15}$$

and multiply through by $-c_p/l_v$:

$$w = -\frac{c_p}{l_v}\left[T - \frac{l_v}{c_p}\frac{\varepsilon}{p}A\exp\left(\frac{-B}{T_w}\right) - T_w\right] \tag{8.16}$$

and distribute the negative sign into the square brackets on the RHS:

$$w = \frac{c_p}{l_v}\left[T_w - T + \frac{l_v}{c_p}\frac{\varepsilon}{p}A\exp\left(\frac{-B}{T_w}\right)\right] \tag{8.17}$$

We now substitute this expression for w into (8.13):

$$T_d = \frac{1}{\dfrac{1}{T_0} - \dfrac{R_v}{l_v}\ln\left(\left[\dfrac{p}{e_0\varepsilon}\right]\dfrac{c_p}{l_v}\left[T_w - T + \dfrac{l_v}{c_p}\dfrac{\varepsilon}{p}A\exp\left(\dfrac{-B}{T_w}\right)\right]\right)} \tag{8.18}$$

and combine a few constants into one term:

$$T_d = \frac{1}{\dfrac{1}{T_0} - \dfrac{R_v}{l_v}\ln\left(\dfrac{pc_p}{e_0\varepsilon l_v}\left[T_w - T + \dfrac{l_v}{c_p}\dfrac{\varepsilon}{p}A\exp\left(\dfrac{-B}{T_w}\right)\right]\right)} \tag{8.19}$$

We can then distribute and simplify the compound constants inside the natural log argument on the RHS:

$$T_d = \frac{1}{\dfrac{1}{T_0} - \dfrac{R_v}{l_v}\ln\left[\left(\dfrac{pc_p}{e_0\varepsilon l_v}\right)T_w - \left(\dfrac{pc_p}{e_0\varepsilon l_v}\right)T + \dfrac{A}{e_0}\exp\left(\dfrac{-B}{T_w}\right)\right]} \tag{8.20}$$

or:

$$\boxed{T_d = \frac{1}{\dfrac{1}{T_0} - \dfrac{R_v}{l_v}\ln\left[\left(\dfrac{pc_p}{e_0\varepsilon l_v}\right)(T_w - T) + \dfrac{A}{e_0}\exp\left(\dfrac{-B}{T_w}\right)\right]}} \tag{8.21}$$

Similar expressions can be derived for the *frost point temperature*, which is the saturation point above ice rather than liquid water.

(b) Dew point can also be computed from psychrometric data using tabular data in the Smithsonian Meteorological Tables (SMTs)[9] or tables *derived* from the SMTs. Table 8.2 provides for dew point estimation at ordinary sea-level pressures.

Table 8.2. *Dew point as a function of* in situ *temperature and wet-bulb depression.*[i]

Wet-bulb depression is the difference between the psychrometer's dry- and wet-bulb temperatures. All dew point values shown are in degrees Celsius, and rounded to the nearest whole degree.

Temp [°C]	Wet-Bulb Depression [°C]															
	0	1	2	3	4	5	6	7	8	9	10	11	12	13	14	15
30	30	29	27	26	24	23	21	19	18	16	14	12	10	8	5	1
28	28	27	25	24	22	21	19	17	16	14	11	9	7	4	1	−3
26	26	25	23	22	20	18	17	15	13	11	9	6	3	0	−4	−9
24	24	23	21	20	18	16	14	12	10	8	6	2	−1	05	−10	−18
22	22	21	19	17	16	14	12	10	8	5	3	−1	−5	−10	−19	
20	20	19	17	15	14	12	10	7	4	2	−2	−5	−10	−19		
18	18	16	15	13	11	9	7	4	2	−2	−5	−10	−19			
16	16	14	13	11	9	7	4	1	−1	−6	−10	−17				
14	14	12	11	9	6	4	1	−2	−5	−10	−17					
12	12	10	8	6	4	1	−2	−5	−9	−16						
10	10	8	6	4	1	−2	−5	−9	−14	−28						
8	8	6	3	1	−2	−5	−9	−14								
6	6	4	1	−1	−4	−7	−13	−21								
4	4	1	−1	−4	−7	−11	−19									
2	2	−1	−3	−6	−11	−17										
0	0	−3	−6	−9	−15	−24										
−2	−2	−5	−8	−13	−20											
−4	−4	−7	−12	−17	−29											
−6	−6	−10	−14	−22												
−8	−8	−12	−18	−29												
−10	−10	−14	−22													
−12	−12	−18	−18													
−14	−14	−21	−36													
−16	−16	−24														
−18	−18	−28														
−20	−20	−33														

[i] State University of New York (1994); Lide (1997).

(c) Psychrometric data can be used to estimate dew point graphically with a thermodynamic diagram or a simpler graph. The graphical relationship between temperature, wet bulb, and dew point on a Skew-T is discussed further in the following text and shown in Figure 8.4. A simpler diagram, such as that shown in Figure 8.2, can be derived by contouring tabular data similar to that shown in Table 8.2.

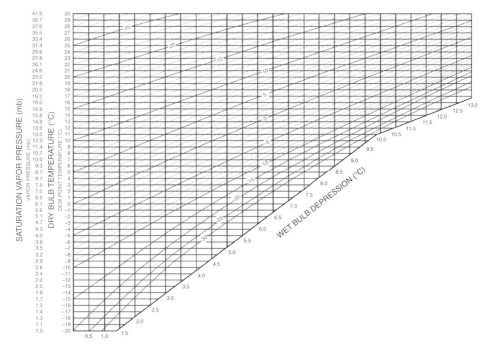

Figure 8.2. Psychrometric diagram. The data shown in Table 8.2 can be contoured to create a graphical diagram for computing dew point from psychrometric data. First the dry-bulb temperature is identified on the vertical scale to the left (red). Next, the wet-bulb depression is identified on the vertical scale below (blue). The user locates the intersection of the two, and then traces the point from the intersection along the diagonal lines back to the scale on the left. Dew point is then read off the vertical temperature scale. The vapor pressure (associated with the dew point) and saturation vapor pressure (associated with the temperature) can also be read off the outer vertical scale on the left and used to compute relative humidity from Equations (7.80) and (7.82). Image borrowed from Aviles (2012), who created this version of the psychrometric diagram for use in the laboratory exercises taught by the meteorology program at Plymouth State University.

(d) If the mixing ratio is already known by other means, the dew point can be directly computed by:

$$T_d = \frac{B}{ln\left(\dfrac{A\varepsilon}{wp}\right)} \qquad (8.22)$$

where B is the temperature constant (5.4×10^3 K), A is the pressure constant (2.5×10^{11} Pa), ε is the usual ratio of the molecular masses of vapor and water, and p is the pressure [Pa].[10] (8.22) also clearly illustrates the direct (but not 1:1) relationship between dew point and pressure: *for a fixed mixing ratio, if pressure decreases, the dew point does as well.*

(e) The *vertical variation* in dew point is similar to the concept of a temperature lapse rate.[11] Beginning with the form of the Classius-Clapeyron Equation shown in (7.36b), and assuming (1) a saturated atmosphere, (2) that the hydrostatic relation (5.77) and the Ideal Gas Law (several forms discussed in Chapter 2) are both reasonable approximations, and (3) that $T_d \cong T$ on the absolute scale,[12] the variation of dew point with respect to vertical distance z can be estimated by:

$$\frac{dT_d}{dz} \cong -\frac{g}{l_v}\frac{R_v}{R_d}T_d \qquad (8.23)$$

Using $g = 9.81$ m/s², $T_d = 273.16$ K, $l_v \cong 2.5008 \times 10^6$ J/kg, and the known values of R_v and $R_{d,}$[13] the RHS of (8.23) has a value of −0.0017 K/m, or −1.7 K/km. Adopting the same convention used for the Dry Adiabatic Lapse Rate (5.82), we define a *positive* "dew point lapse rate" as a *decrease* in dew point with *increasing* altitude.

As the dew point temperature *decreases*, (8.23) shows that its lapse rate also *decreases*, and does so linearly. This means that, when looking at a thermodynamic diagram (such as a Skew-T), at higher dew point temperatures, the mixing ratio lines – which are used to track dew point variation with height – cross the isotherms at greater angles. As the dew point temperature decreases, the mixing ratio lines cross the isotherms at progressively shallower angles. Prove this to yourself by looking at Figure 1.12 (in Chapter 1), or Figures 8.3 and 8.4.

(f) Once the dew point near Earth's surface is known, it can be combined with *in situ* temperature to estimate the bases of newly forming cumuliform clouds, using a cloud-base height diagram or a Skew-T. (The Skew-T method is discussed in the following text.) These cloud bases are assumed to form at the *Lifted Condensation Level* (LCL), which is the altitude to which a parcel must be lifted in order for its temperature to fall (adiabatically) to its dew point. Figure 8.3 reproduces a generic cloud-base height diagram.

Equivalent temperature (T_e)[14] is the temperature a parcel of moist air would have if all the vapor in the parcel were condensed at constant pressure. The latent heat released by the condensation of the water vapor would raise the sensible temperature of the air. Analytically, it is given by:

$$T_e = T\exp\left[\frac{l_v w_s}{c_p T}\right] \qquad (8.24)$$

where T is the *in situ* temperature [K], l_v is the latent heat of vaporization ($\cong 2.5008 \times 10^6$ J/kg), w_s is the saturation mixing ratio [kg$_{vapor}$/kg$_{dry\ air}$], and c_p is the specific heat of dry air at constant pressure ($\cong 1003.8$ J/(kg K)). It can also be approximated by the linear form:

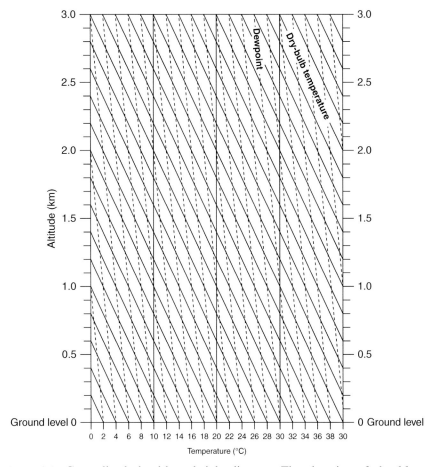

Figure 8.3. Generalized cloud-base height diagram. The elevation of cloud-bases (in km above local ground level) for newly forming cumulus can be estimated by finding the intersection of the isotherms corresponding to the *in situ* temperature and dew point temperature for an air parcel near Earth's surface. Temperatures (dry bulb) are solid diagonal lines from lower right to upper left. Dew point temperatures are dashed diagonal lines. Notice that both temperatures and dew points decrease with height: temperature decreases at the Dry Adiabatic Lapse Rate, and the dew point decreases at the "dew point lapse rate" described by (8.23). This diagram is widely available on the Internet, and is in the public domain, but this copy was borrowed from Kluge (2012). An older version of the diagram was called the Convective Cloud-Base Height Diagram, which was published by the U.S. National Weather Service in the Office of the Federal Coordinator for Meteorology (OFCM) (1982), as well as earlier publications. For more, see Craven et al. (2002).

$$\boxed{T_e \cong T + \frac{l_v w}{c_p}}$$ (8.25)

where w is the *in situ* water vapor mixing ratio [$kg_{vapor}/kg_{dry\ air}$]. A simple unit analysis will convince you that the second term on the RHS has units of Kelvins.

Equivalent potential temperature (θ_e)[15] is the temperature a parcel of moist air would have if all the vapor in the parcel were condensed at constant pressure *and* the parcel is taken dry adiabatically to 1000 hPa. It is given by:

$$\theta_e = \theta \exp\left[\frac{l_v w_s}{c_p T}\right] \qquad (8.26)$$

where θ is the potential temperature, computed by *Poisson's Equation* (5.27):

$$\theta = T_i \left(\frac{1000}{p_i}\right)^{\kappa}$$

where T_i is the *in situ* (initial) temperature [K], p_i is the initial pressure [Pa], and κ is R_d/c_p [unitless]. Equivalent potential temperature is used in a number of severe weather applications.

Isentropic condensation temperature (T_{ic})[16] is the temperature at which saturation is reached when a parcel of moist air is lifted dry adiabatically, with its vapor content (w) held constant. Put another way, this is the temperature of a lifted parcel at the LCL. An analytical approximation, which is also another transcendental equation, is given by:

$$T_{ic} \cong \frac{B}{\ln\left[\frac{A\varepsilon}{wp_i}\left(\frac{T_i}{T_{ic}}\right)^{\frac{1}{\kappa}}\right]} \qquad (8.27)$$

where T_i and p_i are the starting (initial) temperature and pressure of the parcel, respectively [units K and Pa], A is the pressure constant (2.5×10^{11} Pa), B is the temperature constant (5.4×10^3 K), ε is the ratio of the molecular masses of vapor and water, w is the water vapor mixing ratio of the parcel [$kg_{vapor}/kg_{dry\ air}$], and κ is R_d/c_p for dry air ($\cong 0.286$).[17]

Visualizing the relationship between these temperatures using a thermodynamic diagram. The relationship between wet-bulb, dew point, and isentropic condensation temperature; the LCL; the Dry Adiabatic Lapse Rate; *and* the Moist Adiabatic Lapse Rate is summed up as follows. If a parcel beginning with a temperature T and mixing ratio w (giving it an initial dew point T_d) is lifted dry adiabatically, its temperature will eventually decrease to its dew point (which also varies as the pressure falls with height), and the parcel will become saturated. This occurs at the LCL, and the temperature at which this occurs is the isentropic condensation temperature. If the T_{ic} is then followed moist adiabatically back to the point of the parcel's origin, the resulting temperature is the parcel's initial wet-bulb temperature. Figure 8.4 illustrates this relationship with a Skew-T Log P diagram.

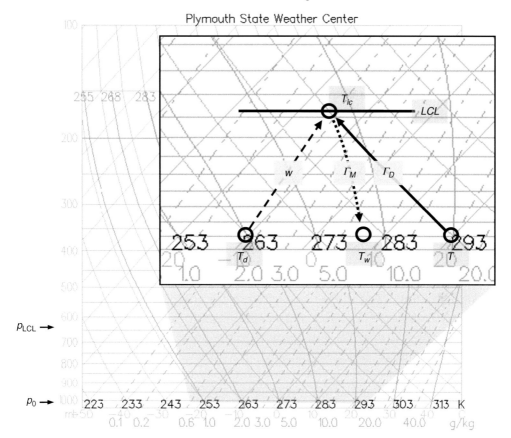

Figure 8.4. Illustration of the relationship between several thermodynamic variables using a Skew-T Log P diagram. A parcel with initial temperature T (20 °C) and dew point T_d (−10 °C) originates at p_0 (1000 hPa). Then an external force lifts the parcel aloft. Its temperature cools at the Dry Adiabatic Lapse Rate (Γ_D), and its dew point follows a line of constant water vapor mixing ratio (w). (Note that T_d decreases as z increases, even though w remains constant.) The parcel temperature continues to fall dry adiabatically until reaching the isentropic condensation temperature (T_{ic}; about −15 °C) at the LCL (near 640 hPa), where $T = T_d$. Visible cloud bases appear at the LCL as vapor condenses into liquid cloud droplets. If the T_{ic} is followed moist adiabatically (Γ_M) back to the parcel's point of origin (1000 hPa), one can compute the parcel's wet-bulb temperature (T_w; about 7 °C). (Background image courtesy of Plymouth State University meteorology program, 2012.)

8.3. The Moist Adiabatic Lapse Rate

The Moist Adiabatic lapse Rate (Γ_M) is the only remaining isopleth to describe in the relationships shown in Figure 8.4. The scenario under examination is as follows. An unsaturated parcel near the Earth's surface is lifted aloft until reaching the LCL. Lifting continues, carrying the parcel above the LCL, forcing condensation to begin inside the saturated parcel. As water vapor condenses into liquid droplets, *latent* heat is released, which adds to the parcel's *sensible* heat. The internal heating of the parcel has the effect of slowing the rate of cooling resulting from the parcel's vertical motion. This is why saturated parcels cool more slowly than unsaturated parcels when they're lifted aloft.

If we assume that the water droplets produced by the condensation of the parcel's vapor remain inside, then the process can be reversed, because the droplets can be converted back into vapor, reabsorbing the heat they surrendered when they condensed. This is a *saturated* or *moist adiabatic process*, and it corresponds to the notion of a compound parcel (that's isolated from the surrounding environment) discussed at the beginning of the previous chapter (Figure 7.1).

However, if we assume that the droplets grow to precipitation-sized particles and fall out of the parcel, then the process is *not* reversible, because the droplets cannot then evaporate and reabsorb the heat they surrendered to the parcel when they condensed. This is a *pseudoadiabatic process*, and it corresponds to the open-parcel model mentioned at the beginning of the chapter. Some realistic estimates indicate that differences between the parcel's heat content resulting from moist adiabatic processes and pseudoadiabatic processes is on the order of a few percent,[18] so our first simplifying assumption will be that the more realistic pseudoadiabatic process can be closely approximated by the simpler moist adiabatic process.

The full derivation of the Moist Adiabatic Lapse Rate is given in Appendix 7; so here we will only list the simplifying assumptions and show the result. It's important that you understand the assumptions (i.e., the weaknesses in the model) before you use this result. They are:

- We assume an isolated, compound parcel, with water vapor coexisting in equilibrium with liquid water. The two halves of the compound parcel are open to each other, and consist of (1) liquid water and (2) water vapor mixed with dry air, but are isolated from the surrounding environment.
- We assume that, as vapor condenses, the resulting liquid water drops remain inside the compound parcel, so that when they evaporate again, they can take up exactly the same amount of latent heat that they release when they condense. This makes the processes reversible and allows the heat to be conserved. That is, *the Moist Adiabatic Lapse Rate is a model of heat conservation.*
- We assume that the combined mass of the parcel's vapor and dry air is equal to one kilogram.
- We assume that all latent heat released during condensation is taken up by the parcel's *dry air*, and none of it goes into the parcel's remaining load of water vapor.
- We assume that the total atmospheric pressure (p) is much greater than the saturation vapor pressure (e_s), so that $p - e_s \cong p$.
- We assume that $w_s = e_s \varepsilon / p$, and equivalently, that $e_s = p w_s / \varepsilon$.
- Finally, we assume that temperature varies only as a function of elevation.

With these in mind, the Moist Adiabatic Lapse Rate is given by:

$$\Gamma_M = \Gamma_D \frac{\left(\dfrac{l_v w_s}{R_d T} + 1\right)}{\left(\dfrac{l_v^2 \varepsilon w_s}{c_p R_d T^2} + 1\right)} \tag{8.28}$$

where Γ_M is the Moist Adiabatic Lapse Rate [K/m], Γ_D is the Dry Adiabatic Lapse Rate (9.77 K/km), l_v is the latent heat of vaporization ($\cong 2.5008 \times 10^6$ J/kg), w_s is the *in situ* saturation mixing ratio [$\text{kg}_{\text{vapor}}/\text{kg}_{\text{dry air}}$], R_d is the individual gas constant for dry air (286.8 J/(kg K)), T is the in situ temperature (K), ε is the ratio of the molar masses of water vapor and dry air (0.622), and c_p is the specific heat of dry air at constant pressure (1003.8 J/(kg K)).

Comparison of the moist and dry adiabatic lapse rates. We saw in Chapter 5 that Γ_D has a fixed value equal to 9.77 K/km. We should expect that Γ_M would be less than this, since the latent heat released during condensation will slow the rate of cooling as the parcel is lifted. We can test this assumption empirically.

Let's assume a reasonable temperature for a parcel near the surface of the Earth: $10\,^\circ\text{C}$ (283.16 K). This means that, using (7.45), the saturation vapor pressure is:

$$e_s = e_0 \exp\left[\frac{l_v}{R_v}\left(\frac{1}{T_0} - \frac{1}{T}\right)\right] = (611.12)\exp\left[\left(\frac{2.4774 \times 10^6}{461.2}\right)\left(\frac{1}{273.16} - \frac{1}{283.16}\right)\right]$$
$$= 1223.87\, Pa$$

where the value of l_v was taken from Table 4.2, and R_v was taken from (2.68). Assuming a surface pressure of 1000 hPa (10^5 Pa), we then use (7.67) to compute the saturation mixing ratio:

$$w_s = \frac{e_s \varepsilon}{p - e_s} = \frac{(1223.87 \times 0.622)}{(10^5 - 1223.87)} \cong 0.0077$$

The numerator of the ratio in (8.28) is, therefore:

$$\left(\frac{l_v w_s}{R_d T} + 1\right) = \left(\frac{2.4774 \times 10^6 \times 0.0077}{286.8 \times 283.16} + 1\right) \cong 1.235$$

where R_d was taken from (2.65). The denominator of (8.28) is:

$$\left(\frac{l_v^2 \varepsilon w_s}{c_p R_d T^2} + 1\right) = \left[\frac{\left(2.4774 \times 10^6\right)^2 \times 0.622 \times 0.0077}{1003.8 \times 286.8 \times \left(283.16\right)^2} + 1\right] \cong 2.273$$

where the value of c_p was taken from Table 4.1, assuming a temperature near $0\,^\circ\mathrm{C}$. Combining these results, (8.28) has a numerical value of:

$$\Gamma_M = \Gamma_D \frac{\left(\dfrac{l_v w_s}{R_d T} + 1\right)}{\left(\dfrac{l_v^2 \varepsilon w_s}{c_p R_d T^2} + 1\right)} \cong \left(9.77\left[\frac{K}{km}\right]\right)\left(\frac{1.235}{2.273}\right) \cong 5.31\left[\frac{K}{km}\right]$$

meaning that saturated parcels (with typical values of T and p) near the Earth's surface cool at about half the rate as unsaturated parcels cool when lifted vertically.

Let's repeat this for a high-altitude parcel, assuming a pressure of 500 hPa (5×10^4 Pa) and a typical mid-latitude 500 hPa temperature of $-20\,^\circ\mathrm{C}$ (253.16 K). In this case, the saturation vapor pressure is:

$$e_s = e_0 \exp\left[\frac{l_v}{R_v}\left(\frac{1}{T_0} - \frac{1}{T}\right)\right] = (611.12)\exp\left[\left(\frac{2.5495 \times 10^6}{461.2}\right)\left(\frac{1}{273.16} - \frac{1}{253.16}\right)\right]$$
$$= 123.54\, Pa$$

Where the value of l_v was taken from Table 4.2. This yields a saturation mixing ratio of:

$$w_s = \frac{e_s \varepsilon}{p - e_s} = \frac{(123.54 \times 0.622)}{(5 \times 10^4 - 123.54)} \cong 0.0015$$

The numerator of the ratio in (8.28) is:

$$\left(\frac{l_v w_s}{R_d T} + 1\right) = \left(\frac{2.5495 \times 10^6 \times 0.0015}{286.8 \times 253.16} + 1\right) \cong 1.053$$

and the denominator of (8.28) is:

$$\left(\frac{l_v^2 \varepsilon w_s}{c_p R_d T^2} + 1\right) = \left[\frac{(2.5495 \times 10^6)^2 \times 0.622 \times 0.0015}{1003.0 \times 286.8 \times (253.16)^2} + 1\right] \cong 1.329$$

where the value of c_p was taken from Table 4.1, assuming a temperature near 250 K. Combining these results, (8.28) has a numerical value of:

$$\Gamma_M = \Gamma_D \frac{\left(\dfrac{l_v w_s}{R_d T} + 1\right)}{\left(\dfrac{l_v^2 \varepsilon w_s}{c_p R_d T^2} + 1\right)} \cong \left(9.77\left[\frac{K}{km}\right]\right)\left(\frac{1.053}{1.329}\right) \cong 7.74\left[\frac{K}{km}\right]$$

These two results illustrate how, unlike the dry rate, the Moist Adiabatic Lapse Rate is highly dependent on the current condition of the parcel. *As the temperature decreases, the moist rate speeds up.* Additional analysis would show that, at low temperatures (below about −40 °C), the moist rate essentially parallels the dry rate. This is because at temperatures below −40 °C, the saturation vapor pressure (and, therefore, the saturation mixing ratio) approaches zero (see Table 7.2). If $w_s \cong 0$, then the numerator and the denominator of (8.28) are both $\cong 1$, which makes $\Gamma_M \cong \Gamma_D$. Table 8.3 shows sample values of Γ_M computed from this function, over a wide range of temperatures and pressures.

Illustration of moist and dry adiabatic lifting on a thermodynamic diagram. Figure 8.5 illustrates the vertical ascent of a parcel originating near the Earth's surface. (This is the same parcel used in Figure 8.4.) Beginning with a temperature of 20 °C and dew point of −10 °C, the parcel initially ascends dry adiabatically. (The dew point tracks upward parallel to a mixing ratio line.) At a pressure of about 640 hPa (the LCL), the temperature and dew point intersect at the isentropic condensation temperature. This indicates the parcel is saturated. If additional lift is supplied, the parcel ascends moist adiabatically, which, while initially slower than the dry rate, becomes parallel to the dry rate above a pressure of about 350 hPa. In general, Figure 8.5 also demonstrates that at low temperatures (below about −40 °C), the Moist Adiabatic Lapse Rate is approximately equal to the Dry Adiabatic Lapse Rate.

8.4 Summary of Moisture Considerations

In this chapter, we derived several theoretical temperatures that were used in different contexts. These include virtual temperature, wet-bulb temperature, dew point, equivalent temperature, equivalent potential temperature, and the isentropic condensation temperature.

$$T_v \equiv T \left(\frac{1 + \dfrac{w}{\varepsilon}}{1 + w} \right)$$

where:

T_v = virtual temperature [K]
T = *in situ* temperature [K]
w = *in situ* mixing ratio [kg$_{vapor}$/kg $_{dry\ air}$]
ε = ratio of molar masses of water vapor and dry air (0.622)

$$T_v \cong T \left(1 + 0.61 w\right)$$

where:

T_v = virtual temperature [K]
T = *in situ* temperature [K]

Table 8.3. *Moist Adiabatic Lapse Rate as a function of in situ temperature and pressure.*

Values shown are in K/km, and were computed using (4.66) for the latent heat of vaporization (l_v), the Classius-Clapeyron Equation (7.46) for vapor pressure, and (7.67) for water vapor mixing ratio (w). R_d was set to 286.6 J/(kg K), the ratio epsilon (ε) was set to 0.622, and the Dry Adiabatic Lapse Rate (Γ_D) was set to 9.77 K/km. Using these parameters, the Moist Adiabatic Lapse Rate (Γ_M) was then computed using (8.28). Some regions of the table (such as temperatures of $+40\,^{\circ}$C at 300 hPa) are clearly unrealistic in the context of Earth's atmosphere.

Pressure [hPa]	Temperature [°C]																
	−40	−35	−30	−25	−20	−15	−10	−5	0	+5	+10	+15	+20	+25	+30	+35	+40
300	9.02	8.63	8.13	7.51	6.81	6.08	5.36	4.70	4.13	3.66	3.29	3.00	2.78	2.61	2.50	2.41	2.36
350	9.12	8.78	8.32	7.76	7.10	6.39	5.67	5.00	4.40	3.90	3.49	3.16	2.92	2.73	2.59	2.49	2.42
400	9.19	8.89	8.48	7.96	7.34	6.65	5.95	5.27	4.65	4.12	3.68	3.33	3.05	2.84	2.68	2.57	2.48
450	9.25	8.98	8.60	8.12	7.53	6.88	6.19	5.51	4.88	4.32	3.86	3.48	3.18	2.95	2.77	2.64	2.55
500	9.30	9.05	8.70	8.25	7.70	7.07	6.40	5.72	5.08	4.52	4.03	3.63	3.30	3.05	2.86	2.71	2.61
550	9.34	9.11	8.79	8.37	7.85	7.24	6.58	5.92	5.28	4.69	4.19	3.77	3.42	3.15	2.94	2.78	2.67
600	9.38	9.16	8.86	8.46	7.97	7.39	6.75	6.09	5.45	4.86	4.34	3.90	3.54	3.25	3.03	2.85	2.72
650	9.41	9.20	8.92	8.55	8.08	7.52	6.90	6.25	5.61	5.01	4.48	4.03	3.65	3.35	3.11	2.92	2.78
700	9.43	9.24	8.98	8.63	8.18	7.64	7.04	6.40	5.76	5.16	4.62	4.15	3.76	3.44	3.19	2.99	2.84
750	9.45	9.28	9.03	8.69	8.26	7.75	7.16	6.54	5.90	5.30	4.75	4.27	3.86	3.53	3.26	3.05	2.89
800	9.47	9.30	9.07	8.75	8.34	7.85	7.28	6.66	6.03	5.43	4.87	4.38	3.96	3.62	3.34	3.12	2.95
850	9.49	9.33	9.11	8.80	8.41	7.93	7.38	6.78	6.16	5.55	4.99	4.49	4.06	3.70	3.41	3.18	3.00
900	9.50	9.35	9.14	8.85	8.48	8.01	7.48	6.88	6.27	5.66	5.10	4.59	4.15	3.78	3.48	3.24	3.05
950	9.52	9.37	9.17	8.89	8.53	8.09	7.56	6.98	6.38	5.77	5.20	4.69	4.24	3.87	3.55	3.30	3.11
1000	9.53	9.39	9.20	8.93	8.59	8.16	7.65	7.08	6.48	5.88	5.31	4.79	4.33	3.94	3.62	3.36	3.16

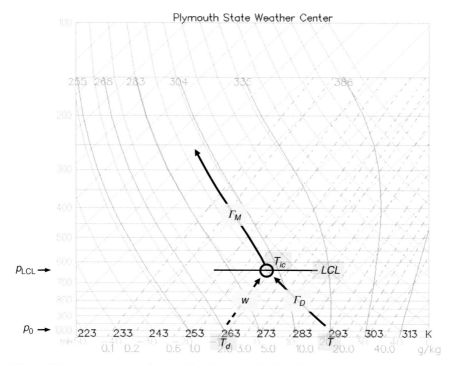

Figure 8.5. An unsaturated parcel begins near the Earth's surface, and is lifted aloft until reaching the LCL. Below the LCL, the temperature cools at the Dry Adiabatic Lapse Rate (Γ_D); above the LCL, the temperature cools at the Moist Adiabatic Lapse Rate (Γ_M). (Background image courtesy of Plymouth State University meteorology program, 2012.)

$$T_v \cong T + \frac{w}{6}$$

where:

T_v = virtual temperature [°C]
T = *in situ* temperature [°C]
w = *in situ* mixing ratio [g_{vapor}/kg $_{dry\ air}$]

$$T_v \cong \frac{T}{1 - 0.0379\dfrac{e}{p}}$$

where:

T_v = virtual temperature [K]
T = *in situ* temperature [K]
e = vapor pressure
p = total pressure

$$T_w = T - \frac{l_v}{c_p}\left[\frac{\varepsilon}{p}A\exp\left(\frac{-B}{T_w}\right) - w\right]$$

where:

T_w = wet-bulb temperature [K]

l_v = latent heat of vaporization (2.5008×10^6 J/kg)

c_p = specific heat of dry air at constant pressure (1003.8 J/(kg K))

A = pressure constant (2.5×10^{11} Pa)

B = temperature constant (5.4×10^3 K)

w = *in situ* mixing ratio [$kg_{vapor}/kg_{dry\ air}$]

$$T_d = \frac{1}{\dfrac{1}{T_0} - \dfrac{R_v}{l_v}\ln\left[\left(\dfrac{pc_p}{e_0\varepsilon l_v}\right)(T_w - T) + \dfrac{A}{e_0}\exp\left(\dfrac{-B}{T_w}\right)\right]}$$

where:

T_d = dew point temperature [K]

c_p = specific heat of dry air at constant pressure ($\cong 1003.8$ J/(kg K))

e_0 = reference pressure (611.12 Pa)

$$T_d = \frac{B}{ln\left(\dfrac{A\varepsilon}{wp}\right)}$$

$$T_e = T\exp\left[\frac{l_v w_s}{c_p T}\right]$$

where:

T_e = equivalent temperature [K]

w_s = saturation mixing ratio [$kg_{vapor}/kg_{dry\ air}$]

$$T_e \cong T + \frac{l_v w}{c_p}$$

$$\theta_e = \theta\exp\left[\frac{l_v w_s}{c_p T}\right]$$

where:

θ_e = equivalent potential temperature [K]

θ = potential temperature [K]

$$T_{ic} \cong \cfrac{B}{ln\left[\cfrac{A\varepsilon}{wp_i}\left(\cfrac{T_i}{T_{ic}}\right)^{\frac{1}{\kappa}}\right]}$$

where:

T_{ic} = isentropic condensation temperature [K]
T_i = initial temperature [K]
p_i = initial pressure [Pa]
$\kappa = R_d/c_p$ for dry air ($\cong 0.286$)

Assuming a fixed water vapor mixing ratio and a parcel lifting vertically, the dew point decreases at its own "dew point lapse rate" approximated by:

$$\frac{dT_d}{dz} \cong -\frac{g}{l_v}\frac{R_v}{R_d}T_d$$

where:

R_v = individual gas constant for water vapor (461.2 J/(Kg K))
R_d = individual gas constant for dry air (286.8 J/(kg K))

The last parameter derived was the Moist Adiabatic Lapse Rate, which is a suitable substitute for the more physically accurate Pseudoadiabatic Lapse Rate:

$$\Gamma_M = \Gamma_D \cfrac{\left(\cfrac{l_v w_s}{R_d T}+1\right)}{\left(\cfrac{l_v^2 \varepsilon w_s}{c_p R_d T^2}+1\right)}$$

where:

Γ_M = moist adiabatic lapse rate [K/m]
Γ_D = dry adiabatic lapse rate [K/m]

Practice Problems

1. Using the definitions of density, mixing ratio, and ε, show that:

$$p = \rho_d R_d T + \rho_v R_v T$$

can be written as:

$$p = \rho R_d T \left(\cfrac{1+\cfrac{w}{\varepsilon}}{1+w}\right)$$

2. Calculate the mixing ratio (w) and saturation mixing ratio (w_s) if $T = 35\,°C$ and $T_d = 10\,°C$. Assume $p = 1$ atmosphere (1013.25 hPa). Use temperature-appropriate values of l_v.

3. Calculate the virtual temperature for No. 2.

4. Calculate the density of the moist air in No. 2 using (8.7). Calculate the density of the equivalent dry air parcel using (2.37), and compare to the result for moist air.

5. Calculate the equivalent temperature for No. 2 using the linear form.

6. Using a Skew-T, calculate the LCL, isentropic condensation temperature (T_{ic}), and wet-bulb temperature (T_w) for No. 2.

7. List the independent variables and constants that define the Moist Adiabatic Lapse Rate.

8. If the dew point and temperature are both $0\,°C$, and the pressure is 850 hPa, at what rate [K/km] will the saturated parcel cool over a short vertical distance?

9. Record the *station* pressure using a barometer:

   $p = $ ——————— hPa = ——————— Pa

10. Use a sling psychrometer and your calculator to determine the following. Show all work.

    a. Wet- and dry-bulb temperatures.

      $T = $ ——————— °F = ——————— °C = ——————— K

      $T_w = $ ——————— °F = ——————— °C = ——————— K

    b. Dew point.

      $T_d = $ ——————— °F = ——————— °C = ——————— K

    c. Vapor pressure and saturation vapor pressure.

      $e = $ ——————— hPa = ——————— pa

      $e_s = $ ——————— hPa = ——————— pa

    d. Mixing ratio and saturation mixing ratio.

      $w = $ ——————— kg/kg × 1000 = ——————— g/kg

      $w_s = $ ——————— kg/kg × 1000 = ——————— g/kg

    e. Virtual temperature.

      $T_v = $ ——————— °F = ——————— °C = ——————— K

    f. Relative humidity.

      $RH = $ ——————— %

11. Starting with the station pressure, and the dry-bulb and dew point temperatures determined in the preceding text, use a Skew-T to calculate the following. A Skew-T tutorial can be found at http://vortex.plymouth.edu/~stmiller/stmiller_content/Tutorials/skew_t.html

    a. Wet-bulb temperature.

      $T_w = $ ——————— °F = ——————— °C = ——————— K

    b. Vapor pressure and saturation vapor pressure.

      $e = $ ——————— hPa = ——————— pa

      $e_s = $ ——————— hPa = ——————— pa

    c. Mixing ratio and saturation mixing ratio.

      $w = $ ——————— kg/kg × 1000 = ——————— g/kg

      $w_s = $ ——————— kg/kg × 1000 = ——————— g/kg

    d. Virtual temperature.

      $T_v = $ ——————— °F = ——————— °C = ——————— K

12. Evaluate the differences in your results using the two methods. For example, did you get the same virtual temperature by both methods? How large was the difference? How would you use the two methods in different circumstances?

9

Atmospheric Statics

9.1. Introduction

In this chapter we'll take a closer look at how gravity varies over Earth's surface, and how this variability affects the atmosphere. We'll also examine some techniques for forecasting sensible weather such as precipitation type (i.e., whether a storm will produce rain, snow, or something else), and for computing useful information about the vertical pressure profile.

9.2. Effective Gravity

Recall (5.77), the hydrostatic relation, which stated that:

$$\frac{dp}{dz} = -\rho g$$

where dp/dz is the vertical variation in pressure [Pa/m], ρ is the density of air [kg/m³], and g is the gravitational acceleration ($\cong 9.81$ m/s²). The negative sign implies that, as z increases, p decreases. Up until now, we've accepted g as being the *Newtonian* gravitational acceleration, defined in (1.8) as:

$$\vec{g} = -\left(G \frac{m_e}{r^2} \right) \hat{r}$$

where G is the Universal Gravitational Constant (6.673×10^{-11} Nm²/kg²), m_e is the mass of Planet Earth (5.988×10^{24} kg), and r is the distance above the center of Earth [m]. $\hat{r}$ is a unit vector pointing from the center of the Earth to some test mass (giving the vector quantity $\vec{g}$ a direction), and the negative sign on the RHS defines "down" as toward the Earth's center of mass. On the surface of Earth, little r is the radius of Earth (R), so we can write the scalar version of (1.8) as:

$$g^* = \frac{Gm_e}{R^2} \tag{9.1}$$

where the star (*) symbol next to little g means that the expression describes *Newtonian* gravity, and R is Earth's radius.

While the atmosphere is shallow compared to the radius of the Earth, its depth can't be completely ignored. We can generalize (9.1) a little further by adding a term for the height above Earth's surface. In other words, let's redefine the radius term in the denominator of the expression as:

$$r = R + z \qquad (9.2)$$

where z is the height above the surface. Substituting this into (9.1) yields:

$$g^* = \frac{Gm_e}{(R+z)^2} \qquad (9.3)$$

The relation in (9.3) is a small improvement over (9.1), but it still ignores two important details: (1) The radius of Earth (R) is *not constant*, but in fact increases from pole to equator (i.e., it is dependent on latitude φ, so should be written as R_φ), and, (2) Earth is rotating on its axis, creating a centrifugal force (proportional to an acceleration) that slightly offsets gravity, especially near the equator. We can attack the first by noting the following two facts:

• Earth's *polar* radius (R_p) is 6356.91 km, and
• The *equatorial* radius (R_e) is 6378.39 km, which is about 0.338 percent greater than the polar radius.

Thus, there is a small increase in Earth's radius (ΔR), from the pole to the equator, equal to 21.48 km (Figure 9.1). This equatorial bulge is caused by the rotation of Earth and is much more pronounced in planets that rotate more rapidly, such as Jupiter.[1]

We can write a general expression for Earth's radius as a function of latitude, by assuming that the change in radius is a *smooth* function of latitude. We know that the cosine of an angle (the latitude, in this case) goes to *zero* if the angle is 90° (the North Pole), and goes to *one* if the angle is 0° (the equator). Thus, for any latitude (φ), the radius can be approximated by:

$$R_\varphi = R_p + \Delta R \cos(\varphi) \qquad (9.4)$$

which converges to the polar radius if the latitude is 90°, and the equatorial radius if the latitude is zero degrees. Substituting this into (9.3), we obtain:

$$g^* = \frac{Gm_e}{(R_\varphi + z)^2} \qquad (9.5)$$

or, more explicitly:

$$\boxed{g^* = \frac{Gm_e}{(R_p + \Delta R \cos(\varphi) + z)^2}} \qquad (9.6)$$

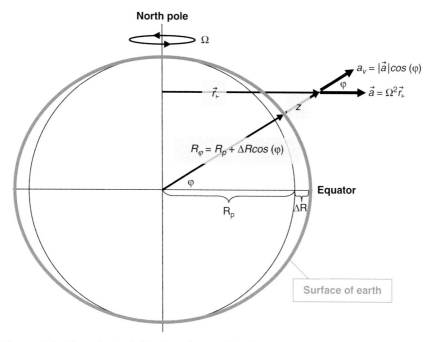

Figure 9.1. Nonspherical Earth and centrifugal acceleration. Earth is an *oblate spheroid*, with an equatorial radius about 0.3 percent larger than the polar radius (exaggerated representation in gray). The radius as a function of latitude (R_φ) can be estimated by adding the polar radius (R_p) to a small, latitude-dependent correction ($\Delta R cos(\varphi)$). The total centrifugal acceleration is given by $\Omega^2 \vec{r}_\llcorner$, where Ω is Earth's angular velocity (360 degrees in slightly less than 24 hours, or about 7.292×10^{-5} [s^{-1}]), and $\vec{r}_\llcorner$ is a radius vector perpendicular to the rotational axis of the planet, pointing to an object at latitude φ and altitude z above surface. The local vertical component of centrifugal acceleration is also dependent on latitude, equal to the full magnitude of $\Omega^2 \vec{r}_\llcorner$ at the equator, and going to zero at the pole.

The expression for gravity in (9.6) is an improvement on simple Newtonian gravity, but we still have to account for the offsetting effect of the outward centrifugal acceleration. In general, centrifugal acceleration[2] is defined by:

$$\vec{a} \equiv \omega^2 \vec{r} \qquad (9.7)$$

where ω is the angular velocity of a rotating object [s^{-1}] and $\vec{r}$ is a vector pointing from the center (focus) of rotation to the point under consideration [m].[3] If we bring this into the Earth context, we find that:

$$\vec{a} = \Omega^2 \vec{r}_\llcorner \qquad (9.8)$$

where Ω is the angular velocity of the Earth. If the period (T) of Earth's rotation is $\cong$ 24 hours[4] (86,164 seconds), then $\Omega \cong 2\pi/86,164 = 7.292 \times 10^{-5}$ [s^{-1}].

In *this* case, $\vec{r}_{\perp}$ is a vector perpendicular to Earth's *axis*, pointing to a location on or above the surface (Figure 9.1). Its magnitude is equal to Earth's radius (plus some arbitrary distance above the surface) at the equator and decreases to zero as the latitude increases from zero to 90°. With this in mind, we can write the scalar magnitude of $\vec{r}_{\perp}$ as:

$$|\vec{r}_{\perp}| = \left(R_{\varphi} + z\right)cos\left(\varphi\right) \tag{9.9}$$

where R_{φ} is the correct Earth radius for latitude φ (computed by 9.4) [m], and z is the elevation above the surface [m]. Note that the resulting acceleration vector ($\vec{a}$) is positive, meaning that it points "outward" from the axis. If $\vec{r}_v$ has a magnitude of zero, then $\vec{a}$ also has a magnitude equal to zero.

We're interested in the *local vertical component* (upward) of the centrifugal acceleration (a_v), because Newtonian gravity is a vector pointing in the local vertical direction defined as "down." At the equator, a_v is parallel to $\vec{a}$, while at the pole, they are perpendicular. In other words, like R_{φ}, it's a function of the latitude. At the equator, a_v has a magnitude equal to the full magnitude of $\vec{a}$, while at the pole (because a_v is perpendicular to $\vec{a}$, *and* $\vec{a}$ also has a magnitude of zero) a_v has a magnitude of zero. With this in mind, we can write:

$$a_v = |\vec{a}|cos\left(\varphi\right) \tag{9.10}$$

where $|\vec{a}|$ is the magnitude of the centrifugal acceleration vector, computed by taking the scalar value of (9.8). Substituting the explicit expression for the magnitude of $\vec{a}$ (9.8) into (9.10), we obtain:

$$a_v = \Omega^2 |\vec{r}_{\perp}|cos\left(\varphi\right) \tag{9.11}$$

and substituting (9.9) into (9.11), we obtain:

$$a_v = \Omega^2 \left[\left(R_{\varphi} + z\right)cos\left(\varphi\right)\right]cos\left(\varphi\right) \tag{9.12}$$

or:

$$a_v = \Omega^2 \left(R_{\varphi} + z\right)cos^2\left(\varphi\right) \tag{9.13}$$

Finally, substituting the identity for R_{φ} given in (9.4) into (9.13), we see that:

$$\boxed{a_v = \Omega^2 \left(R_p + \Delta R\,cos\left(\varphi\right) + z\right)cos^2\left(\varphi\right)} \tag{9.14}$$

By subtracting the local *outward-pointing* vertical component of centrifugal acceleration given by (9.14) from the local *inward-pointing* Newtonian gravity given by (9.6), we can define the magnitude of the *local effective gravity* (g), that is:

Table 9.1. *Local effective gravity (g) for a range of latitudes and elevations.*

All values of g are given in m/s^2, and were computed using Equation (9.16). The troposphere extends from 0 to about 10 km in the middle latitudes – corresponding to the first five data columns in the table (lightly shaded). The outermost layer of the atmosphere, called the *exosphere* (beginning at the top of the thermosphere) is generally considered to begin at an altitude of about 500 km.

Latitude [°]	Elevation [km]									
	0	1	2	5	10	20	50	100	200	500
90	9.89	9.88	9.88	9.87	9.86	9.83	9.73	9.58	9.29	8.50
80	9.88	9.87	9.87	9.86	9.84	9.81	9.72	9.57	9.28	8.49
70	9.86	9.86	9.86	9.85	9.83	9.80	9.71	9.56	9.27	8.48
60	9.85	9.84	9.84	9.83	9.82	9.78	9.69	9.54	9.25	8.46
50	9.83	9.83	9.83	9.82	9.80	9.77	9.68	9.53	9.24	8.45
40	9.82	9.81	9.81	9.80	9.79	9.76	9.66	9.52	9.23	8.44
30	9.81	9.80	9.80	9.79	9.77	9.74	9.65	9.50	9.22	8.43
20	9.80	9.79	9.79	9.78	9.76	9.73	9.64	9.49	9.21	8.42
10	9.79	9.79	9.78	9.77	9.76	9.73	9.64	9.49	9.20	8.41
0	9.79	9.79	9.78	9.77	9.76	9.73	9.64	9.49	9.20	8.41

$$g \equiv g^* - a_v \tag{9.15}$$

or, more explicitly:

$$\boxed{g = \frac{Gm_e}{\left(R_p + \Delta R\cos(\varphi) + z\right)^2} - \Omega^2 \left(R_p + \Delta R\cos(\varphi) + z\right)\cos^2(\varphi)} \tag{9.16}[5]$$

For most purposes in practical meteorology, the approximation given by:

$$\boxed{g \cong g_\varphi \left(1 - 3.14 \times 10^{-7} z\right)} \tag{9.17}$$

is sufficient. In (9.17), g_φ is the local effective gravity at sea level [m/s^2], and z is the altitude ASL [m].[6]

Table 9.1 shows values of g computed using (9.16) and rounded to the nearest 0.01 m/s^2. From equator to pole, at sea level (at an elevation of zero km), gravity increases from 9.79 to 9.89 m/s^2, which is approximately 1 percent. A smaller variation, on the order of a third of a percent, applies when comparing values of g between sea level and the tropopause (10 km), at any of the latitudes shown. In the middle latitudes, between the surface and 10 km, the frequently used approximation $g = 9.81$ m/s^2 is reasonable.

9.3. Geopotential and Geopotential Height

Geopotential surfaces in the atmosphere are theoretical two-dimensional surfaces above the surface where geopotential is constant. That sounds like a tautology (*it is*), but we'll explain it more clearly in a moment.

Geopotential. The mathematical definition of geopotential is given by:

$$\boxed{d\Phi \equiv g\,dz} \tag{9.18}$$

where $d\Phi$ is differential geopotential, g is effective gravity [m/s^2], and dz is a small difference is height [m][7]. A unit analysis of $d\Phi$ shows:

$$[d\Phi] = \left(\frac{m}{s^2}\right)(m) = \frac{m^2}{s^2} \tag{9.19}$$

But, checking back with Table 1.4, we find that:

$$J \equiv kg\frac{m^2}{s^2} \tag{9.20}$$

so that:

$$\frac{J}{kg} = \frac{m^2}{s^2} \tag{9.21}$$

which means the units of geopotential are equivalent to energy per unit mass. Because the force that's performing work in this context is effective gravity (g), the energy per unit mass described by geopotential is the potential energy of position, or *altitude*. This potential energy is converted into the kinetic energy of motion if the object is dropped.

The final velocity (v_f) [m/s] of the dropped object (assuming no wind resistance) can be computed by the energy equivalence relation:

$$\frac{1}{2}mv_f^2 = mgz_i \tag{9.22}$$

where m is the mass of the object [kg], g is the effective gravity acting on it [m/s^2], and z_i is the (initial) altitude of the object before it's dropped [m]. The LHS of (9.22) is kinetic energy, and the RHS is the potential energy of position in a gravitational field. Canceling the mass on both sides of (9.22) and solving for final velocity results in:

$$v_f = \sqrt{2gz_i} \tag{9.23}$$

We have seen that g varies by about 1 percent from pole to equator, which means that the final velocity of an object dropped from a fixed altitude would vary as a function of latitude. A straightforward example will illustrate this very nicely. Let's assume we are at latitude 45° and elevation 10 km. To compute v_f using (9.23), we need a representative value of g, which can be estimated using Table 9.1. To begin, let's estimate the value of g at the surface by taking the arithmetic mean of g at 40 and 50°:

$$g_{sfc}^{45°} \cong \frac{\left(g_{sfc}^{40°} + g_{sfc}^{50°}\right)}{2} = \frac{(9.82 + 9.83)}{2} = 9.825 \frac{m}{s^2} \tag{9.24a}$$

Similarly, we can estimate the mean value of g at 10 km by:

$$g_{10km}^{45°} \cong \frac{\left(g_{10km}^{40°} + g_{10km}^{50°}\right)}{2} = \frac{(9.79 + 9.80)}{2} = 9.795 \frac{m}{s^2} \tag{9.24b}$$

Next, let's use these two estimates to approximate the mean effective gravity from the surface to 10 km:

$$\boxed{\overline{g}_{sfc-10km}^{45°} \cong \frac{\left(g_{sfc}^{45°} + g_{10km}^{45°}\right)}{2} = \frac{(9.825 + 9.795)}{2} = 9.8105 \frac{m}{s^2}\left(\cong 9.81 \frac{m}{s^2}\right)} \tag{9.24c}$$

We can now plug this result into (9.23) and obtain:

$$v_{final}^{45°} = \sqrt{2gz_i} = \sqrt{2(9.8105)(10,000)} = 442.96 \frac{m}{s} \tag{9.25}$$

where the initial elevation of the object (z_i) is in meters. Now, using the surface and 10 km values for g at 0° and 90°, we find that the vertical mean values of effective gravity at these latitudes are approximately:

$$\overline{g}_{sfc-10km}^{0°} \cong \frac{\left(g_{sfc}^{0°} + g_{10km}^{0°}\right)}{2} = \frac{(9.79 + 9.76)}{2} = 9.775 \frac{m}{s^2} \tag{9.26a}$$

and:

$$\overline{g}_{sfc-10km}^{90°} \cong \frac{\left(g_{sfc}^{90°} + g_{10km}^{90°}\right)}{2} = \frac{(9.89 + 9.86)}{2} = 9.875 \frac{m}{s^2} \tag{9.26b}$$

Plugging these into (9.23), we find that the final velocities for objects dropped from 10 km at the equator (0°) and the pole (90°) are:

$$v_{final}^{0°} = \sqrt{2gz_i} = \sqrt{2(9.775)(10,000)} = 442.15 \frac{m}{s} \tag{9.27a}$$

and:

$$v_{final}^{90°} = \sqrt{2gz} = \sqrt{2(9.875)(10,000)} = 444.41 \frac{m}{s} \tag{9.27b}$$

which you should compare to (9.25). This result shows that the object dropped from a fixed height above the equator has a *smaller* final velocity than the object dropped from the same height at 45° (because gravity is *weaker*), and the object dropped from the same height above the pole has a *greater* final velocity than the object dropped in the middle latitudes (because gravity is *stronger*).

Returning to the tautology stated at the beginning of this section, we can now clarify matters a little: *Geopotential surfaces are those hypothetical surfaces above Earth's surface where a dropped object would achieve the same final velocity, regardless of latitude.* Because gravity is stronger at higher latitudes, geopotential surfaces are lower above the poles than they are at the equator. How much does the height of a geopotential surface vary from equator to pole? We can answer this question by returning to the same example we've already used.

Let's start with the object dropped from 10 km at 45° that had a final velocity of 442.96 m/s. If we invert (9.23) to solve for z_i, we find that:

$$z_i = \frac{v_f^2}{2g} \tag{9.28}$$

Using v_f for 45° and vertical mean effective gravity for 0 and 90°, we obtain the initial heights (z_i) for the two latitudes:

$$z_i^{0°} = \frac{(442.96)^2}{2(9.775)} = 10,036.32\,m \tag{9.29a}$$

at the equator, and:

$$z_i^{90°} = \frac{(442.96)^2}{2(9.875)} = 9934.68\,m \tag{9.29b}$$

at the pole. The geopotential surface with an altitude of 10,000 meters at 45° is about 65 meters *lower* at the pole, and about 35 meters *higher* at the equator (Figure 9.2).

Geopotential height. An object dropped from the same geopotential surface would achieve the same final velocity (i.e., it would begin with the same *geopotential*) regardless of its latitude. The elevation of this geopotential surface is called its *geopotential height*, which we define by:

$$\boxed{\zeta \equiv \frac{1}{g_0} \int_0^z g\,dz} \tag{9.30}$$

where g is the local value of effective gravity (which must be vertically averaged between elevation 0 and z) [m/s^2], g_0 is a fixed reference value of effective gravity,

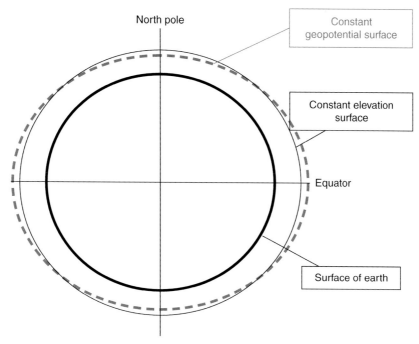

North pole

Constant geopotential surface

Constant elevation surface

Equator

Surface of earth

Figure 9.2. Height of a geopotential surface. Because effective gravity (g) is stronger at the pole than at the equator, an object dropped from a fixed elevation at the pole would achieve a greater final velocity at the pole. A surface of constant geopotential is the latitude-varying altitude from which a dropped object would achieve the *same* final velocity at all latitudes. Geopotential surfaces slope upward from pole to equator.

equal to 9.8 m/s², and dz (linear elevation) is the variable of integration [m]. ζ is the geopotential height and has units of geopotential meters [gpm], which are squeezed (shortened) as you go toward the pole, or stretched (elongated) as you go toward the equator. The height of the geopotential surface shown in Figure 9.2 has a *varying* linear height (lower at the pole and higher at the equator) but a *fixed* geopotential height.

How big is a geopotential meter? We can answer this by going back to the example we used in the preceding text, and focus on 45° latitude. If we integrate (9.30) from the surface to $z = 10$ km, we can use our estimated $\bar{g}$ from (9.24c), which was 9.8105 m/s². Using this, (9.30) becomes:

$$\zeta^{45°}\left[gpm\right] = \frac{\left(9.8105\left[\frac{m}{s^s}\right]\right)}{\left(9.8\left[\frac{m}{s^s}\right]\right)}(10,000)\left[m\right] = 10,010.7\left[gpm\right] \tag{9.31}$$

or 1.000 geopotential meter is equal to 0.9989 linear meter, if integrated between the surface and 10 km above sea level (ASL). At the 90°, $\bar{g}$ is 9.875 m/s² between sea level and 10 km ASL (9.26b), so:

$$\zeta^{90°}[gpm] = \frac{\left(9.875\left[\dfrac{m}{s^s}\right]\right)}{\left(9.8\left[\dfrac{m}{s^s}\right]\right)}(10,000)[m] = 10,076.5[gpm] \tag{9.32a}$$

meaning that 1.000 gpm = 0.9924 m, (a little *less* than at 45°). At the equator, $\bar{g}$ is 9.775 m/s² between sea level and 10 km ASL (9.26a), so:

$$\zeta^{0°}[gpm] = \frac{\left(9.775\left[\dfrac{m}{s^s}\right]\right)}{\left(9.8\left[\dfrac{m}{s^s}\right]\right)}(10,000)[m] = 9974.5[gpm] \tag{9.32b}$$

meaning that 1.000 gpm = 1.0026 m (a little *more* than at 45°). Both solutions show that, to within a few millimeters, 1 gpm is almost exactly equal to 1 linear meter, when considered through the depth of the troposphere (roughly 10 km deep).

Geopotential meters are the physically meaningful unit when considering most aspects of meteorology, but because geopotential meters are almost the same as linear meters, to within an accuracy of less than 1 percent, linear meters are often substituted for gpm in the equations. As long as we're concerning ourselves primarily with the troposphere (which contains about 85 percent of the mass of the atmosphere[8]), this doesn't introduce errors larger than any of our other simplifying assumptions.

9.4. The Hypsometric Equation and "Thickness"

Our next objective is to derive an expression that relates the vertical distance between two adjacent pressure levels to the mean temperature of the layer between them. One application of this is determining sea-level pressure from *in situ* (station) pressure. Another way this is used is to make a first guess about the phase of forecast precipitation.

Theoretical basis of the Hypsometric Equation. By combining the improved form of the Ideal Gas Law involving virtual temperature (8.7), which says:

$$p = \rho R_d T_v$$

with the hydrostatic relation (5.77), which states:

$$\frac{dp}{dz} = -\rho g$$

the definition of geopotential height (9.30) can be rewritten:

$$\Delta\zeta = -\frac{R_d}{g_0}\int_{p_1}^{p_2}T_v\frac{dp}{p} \tag{9.33}$$

The details of these substitutions are left for the student to show in one of the practice problems at the end of this chapter. The approximate solution to (9.33) is given by:

$$\Delta\zeta \cong -\frac{R_d\overline{T}_v}{g_0}ln\left(\frac{p_2}{p_1}\right) \tag{9.34}$$

which is the Hypsometric Equation. In (9.34), $\overline{T}_v$ is the mean virtual temperature in the layer between p_1 (higher pressure, lower elevation) to p_2 (lower pressure, higher elevation). Any pressure units for p_1 and p_2 are permissible, as long as they're the same. To eliminate the negative sign, the ratio inside the natural log argument on the RHS can be flipped, that is:

$$\boxed{\Delta\zeta \cong \frac{R_d\overline{T}_v}{g_0}ln\left(\frac{p_1}{p_2}\right)} \tag{9.35}$$

resulting in a cleaner form of the Hypsometric Equation. By defining the scaling height (H) of the atmosphere as:

$$\boxed{H \equiv \frac{R_d\overline{T}_v}{g_0}} \tag{9.36}$$

we can rewrite the Hypsometric Equation in an even more compact form:

$$\boxed{\Delta\zeta \cong Hln\left(\frac{p_1}{p_2}\right)} \tag{9.37}$$

The difference ($\Delta\zeta$) on the LHS of (9.33) through (9.35), and (9.37), is also known as the *thickness* of the layer from p_1 to p_2, and has units of geopotential meters [gpm]. (Because we have shown that one gpm is almost exactly the same as one linear meter, you can assume that all thickness equations produce answers in simple linear meters without any meaningful loss of accuracy.) Very cold or very dry air has a relatively low virtual temperature, and therefore, a low thickness. Warm, moist air has a higher virtual temperature, and therefore, a higher thickness.

Sensitivity of thickness to water vapor content of the air. Let's examine the form of the Hypsometric Equation given by (9.35) and estimate the sensitivity of thickness

($\Delta\zeta$) to the amount of water vapor in the air. The question we're really asking here is whether or not it's necessary to use the mean virtual temperature. Will using the simpler mean *temperature* give a reasonable result?

Let's assume that we are dealing with a layer between 1000 hPa and 850 hPa, and that it has a mean temperature of 0 °C and a mean dew point of −5 °C. First, we can compute the mean vapor pressure in the layer using the Classius-Clapeyron Equation (7.46):

$$e = e_0 \exp\left[\frac{l_v}{R_v}\left(\frac{1}{T_0} - \frac{1}{T_d}\right)\right]$$

where l_v is the latent heat of vaporization at 0 °C (2.5008×10^6 J/kg),[9] R_v is the individual gas constant for water vapor (461.2 J/(kg K)),[10] e_0 and T_0 are the reference pressure and reference temperature equal to 611.12 Pa and 273.16 K, respectively, and T_d is the *in situ* dew point [K]. Plugging in our mean dew point, we obtain:

$$e = (611.12)\exp\left[\frac{(2.5008\times10^6)}{461.2}\left(\frac{1}{273.16} - \frac{1}{268.16}\right)\right] = 422.07\,Pa$$

Next we'll compute the water vapor mixing ratio using (7.67), that is:

$$w = \frac{e\varepsilon}{p - e}$$

where ε is the ratio of the molecular mass of water vapor to the molecular mass of dry air (0.622) and p is the total atmospheric pressure. The vapor pressure (e) and p can be in any units, as long as they're both in the *same* units. In the current context, we'll assume the total pressure is equal to the mean pressure through the layer (925 hPa or 92,500 Pa), so that:

$$w = \frac{(422.07\times0.622)}{92500 - 422.07} = 0.0029\,\frac{kg_{vapor}}{kg_{dry\,air}}$$

We'll compute the virtual temperature using (8.5):

$$\bar{T}_v = \bar{T}\left(\frac{1+\dfrac{w}{\varepsilon}}{1+w}\right)$$

Plugging in the value of w computed in the preceding text and the mean temperature through the layer:

$$\bar{T}_v = (273.16) \left(\frac{1 + \dfrac{0.0029}{0.622}}{1 + 0.0029} \right) = 273.64 \, K$$

With this, we can now compute the thickness of the layer using (9.35):

$$\Delta \zeta = \frac{(286.8)(273.64)}{(9.8)} \ln \left(\frac{1000}{850} \right) = 1301.5 \, gpm$$

Now let's see what happens if we just use the mean temperature through the layer, rather than the mean virtual temperature, that is substitute $\bar{T}$ for $\bar{T}_v$ in (9.35):

$$\Delta \zeta = \frac{(286.8)(273.16)}{(9.8)} \ln \left(\frac{1000}{850} \right) = 1299.2 \, gpm$$

The second result differs from the first by about 0.18 percent, which means that going to all the extra trouble to compute the virtual temperature when determining the thickness, rather than just using the temperature, doesn't really result in much practical advantage. If you're going to use a computer program to compute thickness, it makes sense to do it the hard way using virtual temperature. (Your computer won't mind the extra work, and you'll get a slightly better result.) But if you're computing thickness with a small calculator, it's just about as good (and a great deal faster) to simply use the temperature.

Variations on the scaling height. In (9.36), we defined H as the scaling height of the atmosphere. The result just demonstrated indicates that the scaling height is almost completely independent of the water vapor content of the atmosphere, at least in the Midlatitudes. But what about other parts of the world, or even on other worlds?

For Earth, R_d and g_0 are constants, so the only variable we'll consider for now is the mean temperature through a column of air. If we assume a mean surface-to-500 hPa temperature of $0 \, °C$ (273.16 K),[11] the scaling height is:

$$H = \frac{(286.8)(273.16)}{(9.8)} = 7994.1 \, gpm$$

or about 8 km. If a typical wintertime high-latitude mean temperature is $-20 \, °C$ (253.16), we have:

$$H = \frac{(286.8)(253.16)}{(9.8)} = 7408.8 \, gpm$$

or about 7.5 km. In the tropics, a representative mean temperature is +20 °C (293.16), so we have:

$$H = \frac{(286.8)(293.16)}{(9.8)} = 8579.4 \, gpm$$

which is roughly 8.5 km. From this, we see than the scaling height of Earth's atmosphere is roughly 8 km, plus or minus about 500 meters. Given a fixed pair of pressure levels (such as 1000 and 500 hPa), this result indicates lower thickness at the poles and higher thickness at the equator. *In other words, the atmosphere extends vertically at the equator to a greater height than it does at the poles.*

Now let's look at Venus, which in some ways is nearly the geological twin of Earth. *Climatologically, it's quite different.* Venus suffers from a runaway greenhouse, and has a mean surface temperature of about 730 K and a surface pressure of more than 92 *bars* (or 92,000 hPa). The radius of Venus is 6.052×10^3 km, which is only about 300 km less than Earth's polar radius of 6.357×10^3 km.[12] The mass of Venus is 4.869×10^{24} kg, which is about 20 percent less than Earth's mass of 5.988×10^{24} kg.[13]

One of the important differences between Venus and Earth is that Venus rotates on its axis once every 243 Earth days – that is, its "day" is 243 times longer than an Earth day. This implies two important things:

- The radius of the planet shows much less latitude dependence than does the Earth's radius, so it can be assumed constant, and
- There is essentially no centrifugal force offsetting gravity, because the angular velocity of the planet on its axis is so small.

Another important difference is the composition of Venus' atmosphere, which, unlike Earth's, is more than 95 percent carbon dioxide (see Table 9.2). Before we can compute the scaling height of its atmosphere, the difference in Venus' mass and radius means that we have to recompute the local gravity, and the difference in its atmospheric composition means that we have to redetermine the individual gas constant.

We'll start with the individual gas constant. In Chapter 2, it was stated that the mean molecular mass of a gas ensemble ($\bar{M}$) is given by (2.62):

$$\bar{M} = \frac{\sum_{i=1}^{N} m_i}{\sum_{i=1}^{N} \left(\frac{m_i}{M_i} \right)}$$

where m_i are the masses of the individual gases in the ensemble [kg] and M_i are the molecular masses of the individual gases [kg/kmol]. If we assume a 100 kg sample

Table 9.2. *Chemical composition of the atmosphere of Venus. Compare to Table 2.1, which shows similar data for Earth's atmosphere.*

Substance	Chemical Symbol	Percentage of "Air" by Mass	Molar Mass (M_i) [kg/kmol]
Carbon Dioxide	CO_2	96.500	44.01
Nitrogen	N_2	3.472	28.01
Sulfur Dioxide	SO_2	0.015	64.06
Argon	Ar	0.007	39.95
Water Vapor	H_2O	0.002	18.02
Carbon Monoxide	CO	0.002	28.01
Helium	He	0.001	4.00

of Venusian atmosphere, we can use the percentages shown in the third column of Table 9.2 as our m_i values. Adapting (2.62) for our use here results in:

$$M_{Ve} = \frac{\sum_{i=1}^{7} m_i}{\sum_{i=1}^{7}\left(\dfrac{m_i}{M_i}\right)} = \frac{m_{CO_2} + m_{N_2} + m_{SO_2} + m_{Ar} + m_{H_2O} + m_{CO} + m_{He}}{\left(\dfrac{m_{CO_2}}{M_{CO_2}}\right) + \left(\dfrac{m_{N_2}}{M_{N_2}}\right) + \left(\dfrac{m_{SO_2}}{M_{SO_2}}\right) + \left(\dfrac{m_{Ar}}{M_{Ar}}\right) + \left(\dfrac{m_{H_2O}}{M_{H_2O}}\right) + \left(\dfrac{m_{CO}}{M_{CO}}\right) + \left(\dfrac{m_{He}}{M_{He}}\right)}$$

(9.38)

where the subscript *Ve* on the LHS indicates the Venusian atmosphere. Plugging the values from Table 9.2 into (9.38) results in a mean molecular mass of 43.15 kg/kmol, just slightly less than that of CO_2. *Sanity check:* Given that carbon dioxide is more than 95 percent of Venus' atmosphere, the mean molecular mass of the ensemble *should* be very close to that of CO_2. From here we can compute the individual gas constant using (2.40), which, when adapted for the current purpose, states:

$$R_i = \frac{R^*}{M_{Ve}} = \frac{8310 \left[\dfrac{J}{kmol\,K}\right]}{43.15 \left[\dfrac{kg}{kmol}\right]} = 192.6 \left[\frac{J}{kg\,K}\right]$$

(9.39)

where R^* is the Universal Gas Constant. Next we'll compute g_{Ve}, the mean gravitational acceleration for Venus. For this, we'll use the scalar value of (1.8), which, when adapted for this context, states:

$$g_{Ve} = G \frac{m_{Ve}}{\left(R_{Ve}\right)^2}$$

(9.40)

where G is the Universal Gravitational Constant (6.673×10^{-11} Nm²/kg²), m_{Ve} is the mass of the planet (4.869×10^{24} kg), and R_{Ve} is its radius (6.052×10^6 m). Plugging these into (9.40) yields:

$$g_{Ve} = \left(6.673 \times 10^{-11}\right) \frac{\left(4.869 \times 10^{24}\right)}{\left(6.052 \times 10^6\right)^2} = 8.87 \frac{m}{s^2} \qquad (9.41)^{14}$$

which you should compare to Table 9.1 – the gravity on the surface of Venus is similar to the gravity a few hundred kilometers above Earth's surface. Finally, we can compute the scaling height for Venus (H_{Ve}) using (9.36):

$$H_{Ve} = \frac{R_i \bar{T}}{g_{Ve}} = \frac{(192.6)(730)}{(8.87)} = 15,851 \, \text{"}g\text{"}pm \qquad (9.42)$$

or about 16 km. (The g in "g"pm is in quotes because the word "geo" refers to Earth, not Venus.) Compare this to the results we computed for Earth at the poles and at the equator. This result indicates that the Venusian atmosphere has about *twice* the vertical thickness of Earth's, if computed for a common pair of pressure levels.

Thickness and precipitation type. Back to Earth. The winter season can produce all phases of stratiform precipitation. If we assume most Midlatitude winter stratiform precipitation begins as growing ice crystals in the middle troposphere, then we can also see that the phase of the precipitation that finally reaches the ground is a function of what the falling particle experiences on the way down.

If we start with ice crystals at 700 hPa, then the vertical virtual temperature profile between Earth's surface (near 1,000 hPa) and 700 hPa can be used to predict the phase of the precipitation reaching the ground. Note the *assumption* that precipitation particles develop near 700 hPa, which may not be correct, but we'll set that aside for now.[15] There are several ways to put this idea into operation. One takes a fast, simple approach, using thickness as a proxy for the mean temperature in the layer between two adjacent pressure levels. Another uses a more sophisticated approach, by computing and then comparing the positive and negative temperature areas (equivalent to energy[16]) on a thermodynamic diagram.[17] Yet another makes explicit use of the microphysics of precipitation. We'll take a close look at the first method and leave the rest for more advanced courses.

The classic scenario for generating multiple phases of wintertime precipitation involves the approach of a warm front. Initially, when the warm front is several hundred kilometers to the south of a station,[18] the advection of moisture and warm air (in the form of a maritime tropical, or mT air mass) in the middle troposphere results in the formation of an increasingly dense nimbostratus cloud, which precipitates continuous light snow. The intensity of the precipitation increases, and the base of the nimbostratus clouds get progressively lower, as the warm front moves closer to the station location. At some point, so much warm air invades the mid-levels that partial or complete melting of the falling snow crystals occurs, but, because so much cold air remains firmly entrenched *near the surface* (in the underlying continental polar, or cP air mass), the precipitation particles refreeze into a pellet of bulk ice, creating either ice pellets or snow pellets (collectively known as sleet), before reaching the ground.

As the front comes even closer, the melting region in the middle troposphere gets deeper and warmer, while the remaining cold air adjacent to the surface grows shallower. Surface temperatures may increase to only a degree or two below the freezing point. As this continues, the falling precipitation particles undergo complete melting in the middle levels but do *not* have time to refreeze when falling into the shallow cold air near the surface. Instead, they reach the ground as supercooled liquid water droplets that freeze on contact with objects on the surface, creating glaze ice. This is freezing rain. Finally, as the front comes within a few kilometers of the station, the influx of warm air reaches the surface, and the precipitation transitions to ordinary rain.

This sequence is illustrated in Figure 9.3, which assumes stations near sea level. Looking at the figure, you'll notice that the temperature profile for snow and ice pellets (panel b) is almost identical to the temperature profile for freezing rain (panel c). Whether the melted precipitation particle has a chance to refreeze into a ball of ice before reaching the ground is very sensitive to the depth, and temperature, of the shallow cold air layer near the surface, as well as the mass of the particle, and the temperature it reaches in the melting layer. This is further confounded by complex terrain. If there are many hills and valleys, the shallow cold air layer may be hundreds of meters deeper in some locations (the valleys) than in others (the hilltops), often horizontally separated by only a few kilometers, regardless of the location of the surface warm front. For this reason, weather forecasters working in regions of complex terrain (such as the New England states in the northeastern United States) will often use phrases such as "wintry mix" in their generalized public zone forecasts.

The thickness method of forecasting type usually considers two layers of the troposphere: 1000–850 hPa and 850–700 hPa are most often used. The former represents the layer of air near the surface (where refreezing may occur), and the latter represents the mid-level layer (where melting may occur). Recalling that a low mean layer temperature implies a low thickness (9.35), we can write the conclusions shown in Table 9.3 about the four precipitation types in question. Notice that the conditions implied for pellets and freezing precipitation are identical.

Elliot (1988) takes this a step further and attempts to account for many variations in the vertical profiles that produce these four precipitation types. For example, for *snow* to occur, Elliott (1988) states the temperature at 850 hPa must be $0\,°C$ or colder, and the freezing level must be no higher than 500 meters AGL. If the freezing level is higher than 500 meters AGL, mixed precipitation in the form of rain and snow is more likely than simple snow. However, if conditions in the low levels are very dry, then

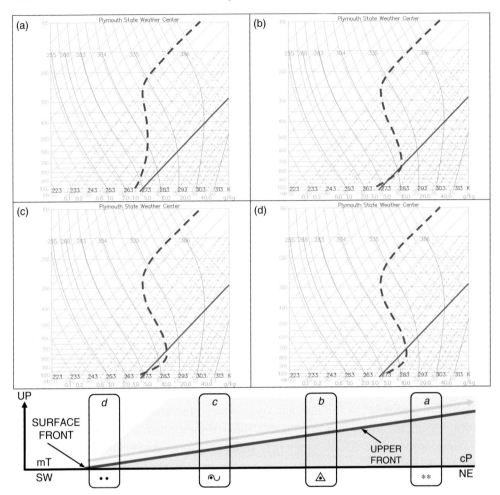

Figure 9.3. Precipitation types occurring with the approach of a warm front. Upper four panels show hypothetical temperature profiles at four stations northeast of a surface warm front. Station *a* is farthest from the front; station *d* is closest. The blue diagonal line highlights 0 °C isotherm; dashed red line shows typical temperature profile at each station. An increasingly pronounced mid-level warm region becomes evident at stations nearer the surface warm front. The lower panel shows vertical schematic with locations of all four surface stations. The heavy red line shows the *upper front* – the boundary between approaching maritime tropical (mT) air mass in the southwest, and slowly receding continental polar (cP) air mass in the northeast. The heavy green arrow indicates warm air overrunning; the light green background indicates region of nimbostratus cloud resulting from the overrunning. Rounded boxes in the lower panel indicate typical locations of the four soundings shown in upper panels, along with precipitation types resulting from each profile. Station *a*: Snow; Station *b*: Ice pellets; Station *c*: Freezing rain; Station *d*: Rain (Precipitation symbols from National Weather Service, 2012. Skew-T image courtesy of Plymouth State University meteorology program, 2012).

Table 9.3. *Mean layer temperature and thickness associated with winter precipitation type.*

Thickness $(\Delta\zeta)$ values were computed using (9.35), with T substituted for T_v.

Precipitation Type	1000–850 hPa	850–700 hPa
Snow	$T < 0\,^\circ C$; $\Delta\zeta < 1299$ gpm	$T < 0\,^\circ C$; $\Delta\zeta < 1552$ gpm
Ice and snow pellets	$T < 0\,^\circ C$; $\Delta\zeta < 1299$ gpm	$T > 0\,^\circ C$; $\Delta\zeta > 1552$ gpm
Freezing rain	$T < 0\,^\circ C$; $\Delta\zeta < 1299$ gpm	$T > 0\,^\circ C$; $\Delta\zeta > 1552$ gpm
Ordinary rain and drizzle	$T > 0\,^\circ C$; $\Delta\zeta > 1299$ gpm	$T > 0\,^\circ C$; $\Delta\zeta > 1552$ gpm

evaporative cooling[19] may prevent the falling snowflakes from melting, producing snow at ground level, even if the freezing level is above 500 meters AGL. Of particular interest is *freezing rain*, which is extremely hazardous to aviation, transportation in general, and all kinds of infrastructure. According to Elliot (1988), for freezing precipitation to occur:

• The low-level freezing layer should be about 1500 feet (460 meters) deep, corresponding to approximately 46 hPa.
• The mid-level melting layer should be about 5000 feet (1525 meters) deep, corresponding to about 150 hPa.

Elliot (1988) then summarizes the forecasting of precipitation type using a two-dimensional diagram (see Figure 9.4), where 1000–850 hPa $\Delta\zeta$ is on the horizontal axis and 850–700 hPa $\Delta\zeta$ is on the vertical axis (Elliot refers to ice and snow pellets using the term *sleet*, usually reserved for conversations with nonspecialists). Some points about the diagram are as follows:

• The provisional forecasts obtained from the diagram are valid for locations in North America, east of 90° west longitude.
• There is an implicit assumption in Elliot's (1988) method, that the surface station experiencing the precipitation is at or very near sea level (near 1000 hPa).
• In general, conditions in the *lower right* corner of the diagram (high $\Delta\zeta_{1000-850}$, low $\Delta\zeta_{850-700}$) are the most unstable.[20] That is, they are associated with very warm conditions near Earth's surface, and very cold conditions aloft. Unstable conditions are conducive to showery precipitation produced by cumuliform clouds. Conditions in the *upper left* (low $\Delta\zeta_{1000-850}$, high $\Delta\zeta_{850-700}$) are the most stable and are likely to be associated with continuous precipitation produced by stratiform clouds.
• Rain occurs on the RHS side of the diagram, associated with higher values of $\Delta\zeta_{1000-850}$ and $\Delta\zeta_{850-700}$. It is separated from frozen and freezing forms of precipitation by a line that slopes from the upper left to the lower right. The slope of this demarcation line is about -2.0 $[\text{gpm}_{850-700}/\text{gpm}_{1000-850}]$. If $\Delta\zeta_{1000-850}$ is high, then $\Delta\zeta_{850-700}$ can be relatively low, and the atmosphere will still produce ordinary rain. This implies warm conditions near the surface

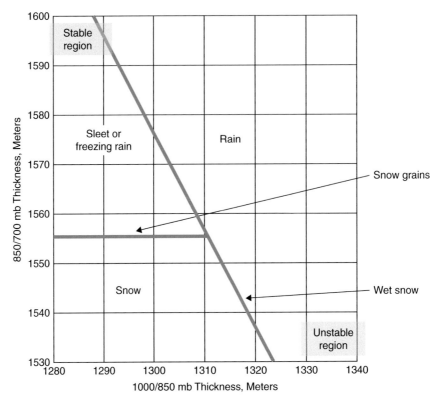

Figure 9.4. Precipitation type as a function of low- and mid-level thickness. Precipitation occurring with thickness conditions in the lower right is likely to be showery, produced by cumuliform clouds. Precipitation occurring with thickness conditions in the upper left is likely to be continuous, produced by stratiform clouds. The former is typical of post cold-frontal conditions, with a warm surface, and cold air advection cooling the middle troposphere. The latter is typical of an approaching warm front, with a cool surface, and warm air advection warming the middle troposphere. Provisional forecasts obtained by this method are valid for stations near sea level in North America, east of 90° west longitude. Adapted from Elliot (1988).

and a strong vertical lapse rate (resulting in cooling aloft), which is a scenario likely to produce showery precipitation. (Picture an afternoon following the passage of a cold front in the spring or fall months.) The atmosphere will also produce ordinary rain if $\Delta\zeta_{1000-850}$ is low and $\Delta\zeta_{850-700}$ is relatively high. This implies cold conditions near the surface and a weak or negative vertical lapse rate (resulting in little cooling, or even warming aloft), which is a scenario likely to produce continuous precipitation. (This is the warm frontal scenario described in Figure 9.3.)

• Note that ordinary rain can occur even with $\Delta\zeta_{1000-850}$ of less than 1299 gpm. Figure 9.4 shows ordinary rain occurring if the low-level thickness is as low as 1288 gpm, which, by inverting (9.40), corresponds to a mean layer temperature of $-2.4\,°C$ ($\cong 27.8\,°F$). This can be interpreted in two ways: (1) Even though the *mean layer temperature* is below the freezing point, the vertical lapse rate in this layer is such that, at the surface, the temperature just reaches 0 °C (32 °F), or (2) even though the *air temperature* is slightly below freezing, *objects* on the ground have temperatures slightly above freezing, so that glaze ice does not form.

- The $\Delta\zeta_{1000-850}$ value separating rain from other forms of precipitation varies between 1288 and 1323 and has a mean value of about 1305 gpm. The latter value, with appropriate error bars to indicate the range, is often used by weather forecasters as an aid in predicting precipitation type.

- Snow only occurs with $\Delta\zeta_{850-700}$ values below about 1555 gpm. By inverting (9.40), we see that this corresponds to a mean mid-level temperature of about 0.5 °C (32.9 °F). Snowflakes generated above the mid-level melting layer will *not* experience significant melting, provided the mean layer temperature between 850 and 700 hPa remains below 0.5 °C. If $\Delta\zeta_{850-700}$ is *greater* than 1555 gpm, and $\Delta\zeta_{1000-850}$ is in the correct range, partial or complete melting of the snowflakes will occur in the mid-level, and pellets or freezing precipitation will reach the surface instead. The intermediate form called *snow grains* is likely to occur if the mid-level thickness value is near the demarcation. Snow grains occur when the original snowflake experiences a very slight amount of melting before falling into the shallow cold layer nearer Earth's surface. The crystalline structure of the flake is mostly preserved, but it is initially coated with a small amount of liquid water, which then refreezes into an outer glaze layer.

- Snow can occur with $\Delta\zeta_{1000-850}$ values as high as 1323 gpm, which, using (9.40), corresponds to a mean low-level temperature of 5.0 °C (41 °F). This only occurs with very low mid-level thickness (mean temperature) values. This scenario is characterized by a warm surface layer and cold mid-level temperatures – that is, a strong vertical positive temperature lapse rate, resulting in an unstable atmosphere (as described previously). Therefore, these are snow showers (rather than continuous snow), and the snowflakes produced in the cumuliform clouds only experience above-freezing temperatures for the final few seconds of their descent, and do not have time to melt before reaching the ground. This effect is aided by very dry conditions in the low levels, resulting in evaporative cooling of the falling snowflakes, which may prevent them from melting.

- Along the rain-snow demarcation, as the $\Delta\zeta_{1000-850}$ value *decreases*, the $\Delta\zeta_{850-700}$ *increases*. If snow is occurring at the maximum mid-level temperature of 0.5 °C (corresponding to the $\Delta\zeta_{850-700}$ demarcation value of 1555 gpm), the low-level mean temperature can be as high as 2.3 °C (36 °F, corresponding to a $\Delta\zeta_{1000-850}$ value of 1310 gpm), which is *also* above freezing. In this case, while the falling snowflake experiences an above-freezing environment from 700 all the way down to 1000 hPa, it does not completely melt. The precipitation reaching the ground would be "wet snow."

- Sleet (ice pellets or snow pellets) and freezing rain (and freezing drizzle)[21] can only occur when the $\Delta\zeta_{850-700}$ exceeds 1555 gpm, corresponding to a mean layer temperature of more than 0.5 °C. The corresponding $\Delta\zeta_{1000-850}$ value is 2.3 °C (36 °F), meaning that freezing precipitation can still occur with low-level *air* temperatures above freezing. Objects on the ground can still have temperatures below freezing, and glaze ice can form if liquid precipitation falls on them.

- On the line demarking rain from sleet and freezing rain, as the $\Delta\zeta_{1000-850}$ value (mean low-level layer temperature) *decreases*, the $\Delta\zeta_{850-700}$ value (mean mid-level layer temperature) *increases*. In other words, as the amount of heat gained in the mid-level by the falling precipitation particle *increases*, the temperature of the low-level layer must *decrease* (thus improving its ability to recool the falling particles), in order for sleet or freezing rain to occur (very dry conditions in the low levels adds evaporative cooling, which further improves the ability to recool the falling precipitation particles). If the 1000–850 hPa layer isn't cold enough ($\Delta\zeta_{1000-850}$ value is too high), ordinary rain will occur instead.

Andrew Foley (my graduate student) and I created a program, using Matlab,[22] that plots observed precipitation type as a function low- and mid-level thickness. The objective was to see how well the Elliot (1988) method works for stations near sea level, as well as stations at high elevation, and to see if alternative choices of layers might work better. The precipitation type observations come from hourly Meteorological Aerodrome Report (METAR) data, and the thickness values were computed using Radiosonde Observation (RAOB). The latter was done by simply differencing the heights reported for various pressure values.

As an initial test, we looked at data for Portland, Maine. This station was chosen because (1) it is very near sea level (about 23 meters ASL), (2) the terrain around Portland is a coastal plain with rolling hills of relatively low relief (so uneven trapping of cold air in valleys is not significant), (3) RAOBs are available from the Gray, Maine National Weather Service Weather Forecast Office, which is only about 25 km away, and (4) Portland is in New England, which experiences every imaginable kind of precipitation.

We compared ten years of precipitation-type observations to their corresponding thickness values. The ten-year period begins January 1, 1999, at 0000 UTC, and ends December 31, 2008, at 2300 UTC. To make the comparison between METAR reports, which are usually recorded once per hour, and RAOBs, which are recorded once every twelve hours, we allowed a one-hour window around the cardinal hours associated with the RAOBs. This means that precipitation types recorded at 2300, 0000, and 0100 UTC were compared to the 0000 UTC RAOBs, and that precipitation types recorded at 1100, 1200, and 1300 UTC were compared to the 1200 UTC RAOBs. There are additional complications that were ignored. Among these are the fact that the RAOBs time-stamped at "0000 UTC" actually begin at 2300 UTC the previous day, and continue until about 0100 UTC. Thus the low- and mid-level data are all recorded *before* 0000 UTC. Similar arguments can be made for the "1200 UTC" RAOB.

The ten-year period represents a total of 87,672 hours, of which 1094 hours are missing from the hourly METAR reports. In the remaining reports, there are 199 hours of freezing rain or freezing drizzle, 55 hours of ice pellets or snow pellets, 6104 hours of ordinary rain or drizzle, and 2241 hours of snow or snow grains. The mean values and standard deviations of the low- and mid-level thickness for each precipitation category were computed from the RAOBs, and plotted on a diagram similar to Elliot's (1988). The initial results were not very encouraging. Even a one standard deviation ($\pm 1\sigma$) box[23] (i.e., centered on the mean value of the thickness associated with the precipitation type, with borders drawn at a distance of one standard deviation from the mean) showed considerable overlap between the four different types of precipitation. It occurred to us that the overlap may only be *apparent*; that is, there may be a third parameter (another dimension in the parametric space of the diagram) that would separate the $\pm 1\sigma$ boxes associated with each precipitation type. After some experimentation, we found that a useful third dimension was the 700 hPa wind direction (see Figure 9.5).

Figure 9.5 shows that Elliot's (1988) method *can* be adapted to Portland, as long as one is willing to accept a degree of uncertainty (manifested here as overlapping $\pm 1\sigma$ boxes of the different precipitation types). The upper left panel shows the expected distribution of rain and snow, with the former centered near *higher* values of $\Delta\zeta_{1000-850}$ and $\Delta\zeta_{850-700}$, and the latter centered near *lower* values of both. Freezing precipitation and pellets have somewhat higher values of both thicknesses than do snow and snow grains, although there appears to be quite a bit of overlap between the former and the latter. (Freezing rain and pellets are still impossible to separate from *each other* using this method.) But by adding the 700 hPa wind direction, we can see that freezing rain (199 hours in ten years) only occurs with wind directions between 150 and 260°, and the even-rarer ice and snow pellets (55 hours in ten years) occurs in the narrower range of wind directions between 210 and 240°. Sorting by wind direction also seems to pull apart the rain and snow boxes over most of the azimuth circle. To be a useful forecast tool, a similar climatological study would have to be accomplished for each station in the Area of Responsibility (AOR), which would be used to determine the critical values of thickness for each precipitation type, and the mid-level wind directions (or values of some other third variable) associated with freezing precipitation and pellets.

9.5. Pressure Calculations

The Hypsometric Equation can also be inverted and used to compute the pressure at a desired elevation. This (with *many* caveats and compromises) is the theoretical basis for our calculation of sea-level pressure, a key component of the global weather observing program. It is also used in some facets of satellite meteorology: if the sea-level pressure is known, and a vertical profile of virtual temperature can be derived (this is what the sounders[24] flying on the satellites do), then the pressure at any elevation can be estimated. (The virtual temperature profile is used to compute the vertical thickness profile.)

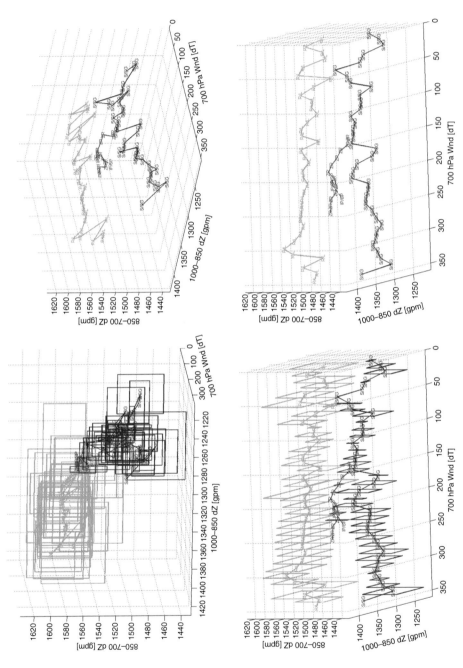

Figure 9.5. Precipitation type as a function of low- and mid-level thickness and 700 hPa wind direction for Portland, Maine (KPWM). Green and blue boxes lines correspond to the ±1σ boxes (left two panels) and mean thickness values (right two panels) for rain/drizzle and snow/snow grains, respectively. The solid red line indicates ±1σ box (left) and mean values of thickness (right) for freezing rain. Dashed red line indicates ±1σ box (left) and mean values of thickness (right) for ice pellets and snow pellets.

Beginning with the form of the Hypsometric Equation shown in (9.37), that is:

$$\Delta \zeta \cong H \ln\left(\frac{p_1}{p_2}\right)$$

where $p_1 > p_2$, we assume an equality and write the LHS explicitly:

$$\zeta_2 - \zeta_1 = H \ln\left(\frac{p_1}{p_2}\right) \tag{9.43}$$

where ζ_2 and ζ_1 refer to the geopotential heights of p_2 and p_1, respectively [gpm]. In this case, let's assume we know the two geopotential heights and one of the pressures (p_1). H is easily computed from the mean virtual temperature profile ($\overline{T_v}$), gravity (g_0), and the individual gas constant for dry air (R_d). p_2 is the unknown. Next, we divide through by H and exponentiate both sides:

$$\exp\left(\frac{\zeta_2 - \zeta_1}{H}\right) = \frac{p_1}{p_2} \tag{9.44}$$

and then multiply through by p_2:

$$p_2 \exp\left(\frac{\zeta_2 - \zeta_1}{H}\right) = p_1 \tag{9.45}$$

Next, we divide through by the exponential term on the LHS:

$$p_2 = \frac{p_1}{\exp\left(\dfrac{\zeta_2 - \zeta_1}{H}\right)} \tag{9.46}$$

Recall the identity that states:

$$\frac{1}{e^x} = e^{-x} \tag{9.47}$$

which we now apply to (9.46), resulting in:

$$\boxed{p_2 = p_1 \exp\left[-\left(\frac{\zeta_2 - \zeta_1}{H}\right)\right]} \tag{9.48}$$

where, $p_1 > p_2$, and therefore, $\zeta_1 < \zeta_2$. If we define p_1 as the sea-level pressure, and ζ_1 as sea level, then we can select an elevation ζ_2 and compute the *in situ* pressure p_2 at that altitude.

Example. At sea level, the pressure is 1010 hPa. The average temperature in the lowest 3000 meters of the atmosphere is 0 °C (273.16 K). What is the pressure at 2500 meters ASL?

In this case, p_1 is the sea-level pressure (1010 hPa), and ζ_1 is sea level (0 meters). ζ_2 is the elevation for which we want the pressure (2500 m). From (9.41), the scaling height is given by:

$$H = \frac{(286.8)(273.16)}{(9.8)} = 7994.1\, gpm$$

Plugging these into (9.53), we find that the pressure at 2,500 meters is:

$$p_2 = (1010)\exp\left[-\left(\frac{2500}{7994}\right)\right] = 738.8$$

which has units of hPa, because the sea-level pressure was entered in hPa.

Sea-level pressure calculations. Converting pressure in the opposite direction – that is, from the *in situ* pressure at some station ASL, down to its equivalent sea-level pressure – is another matter.

Pressure is reported by thousands of weather stations around the world on an hourly basis. A basic problem arises because these weather stations are at different altitudes, and because elevation-based pressure changes are much larger than the *horizontal* pressure changes occurring with weather systems. If you go back and look at Figure 1.9, or any of the other figures in this book showing a Skew-T diagram, you'll see the close proximity of the 1,000 and 850 hPa pressure levels. Figure 9.4 shows typical thickness values for this layer in the range of about 1,300 meters. To be clear, that's 150 hPa across a distance of 1.3 km, or a *vertical pressure gradient* of about 115 hPa/km. (This incredible vertical pressure gradient force is balanced against gravity, resulting in a near-static condition, thus we can use the hydrostatic relation most of the time.) Compare this to a very windy day in the Midlatitudes, when the *horizontal pressure gradient* might be 1 hPa over a distance of 50 km, or 0.02 hPa/km. Combining these two, we see that vertical pressure gradients are usually at least 5,000 times greater than horizontal pressure gradients.

Because of this, given the differences in elevation from one weather station to the next, if the reported pressures weren't converted to sea level (or some other common reference elevation), the *vertical* pressure differences between the stations would completely mask the *horizontal* differences. If we simply contoured *in situ* pressures (also known as *station pressures*), our contours (isobars) would closely resemble the elevation contours of a topographic map, and it would be very difficult to identify systems such as cyclones and anticyclones. For this reason, all *in situ* pressures are converted ("reduced"[25]) to sea-level for inclusion in METAR

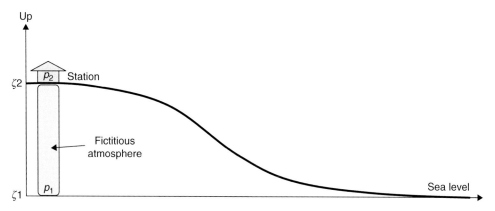

Figure 9.6. Sea-level pressure calculations involving a "fictitious atmosphere." The problem inherent in trying to compute sea-level pressure (p_1) from station pressure (p_2) is that there is no atmosphere stretching directly downward from the station's elevation to sea level, so there is no way to compute a scaling height (H) and use Equation (9.48).

observations. Station pressures are recorded, but are not included in the METAR observation.

If you didn't think about it, you might simply try to use (9.48) to compute the sea-level pressure from the station pressure. The problem with this is that the weather station is not suspended in the sky – it is sitting on solid ground, and there is no atmosphere directly beneath it stretching the distance between station elevation and sea level, so there's no way to compute a scaling height. Figure 9.6 illustrates the problem. To use (9.48), you have to construct a "fictitious atmosphere" beneath the station and use the mean virtual temperature of this fictitious atmosphere to compute the scaling height. From there, you could use the station elevation as ζ_2, sea level as ζ_1, and the station pressure as p_2. You would then invert (9.48) to solve for p_1, the pressure at sea level.

Let's assume we have a station with an elevation of 1,000 meters ASL. If the environmental lapse rate (γ_e) is equal to the Dry Adiabatic Lapse Rate (Γ_D), then the *in situ* temperature at the station is 10 °C cooler than an equivalent station at sea level. *This difference is of the same order as the diurnal variation in temperature*, so a suitable substitute for the vertical mean temperature through the fictitious atmosphere might be the average of the *in situ* (station) temperatures recorded over the last twelve hours.

Beginning with (9.48), which stated that:

$$p_2 = p_1 \exp\left[-\left(\frac{\zeta_2 - \zeta_1}{H}\right)\right]$$

we will take advantage of the fact that one geopotential meter is almost exactly the same as one linear meter. Using this we'll define ζ_2 as the station elevation (z_{STN}), and

ζ_1 as sea level (0 meters). In this case, p_1 is the sea-level pressure (*SLP*) and p_2 is the station pressure (p_{STN}). Using these, we can rewrite (9.48) as:

$$p_{STN} = SLP \exp\left[-\left(\frac{z_{STN}}{H} \right) \right] \tag{9.49}$$

Using the identity in (9.47), we can solve for *SLP* by:

$$SLP = p_{STN} \exp\left(\frac{z_{STN}}{H} \right) \tag{9.50}$$

Putting in the explicit definition (9.36) of the scaling height (and substituting the twelve-hour mean temperature for $\bar{T}_v$), we obtain:

$$\boxed{SLP = p_{STN} \exp\left(\frac{z_{STN} g_0}{R_d \bar{T}_{12hr}} \right)} \tag{9.51}$$

Example. Suppose we have a station at 154 meters ASL. Its current *in situ* temperature is 15 °C (288.16 K), and the temperature twelve hours ago was 5 °C (278.16 K). This means that the twelve-hour mean temperature is 10 °C (283.16 K). If the station pressure is 1000.0 hPa, then its equivalent sea-level pressure is given by:

$$SLP = (1000) \exp\left[\frac{(154)(9.8)}{(286.8)(283.16)} \right] = 1018.8$$

which has units of hPa, because the station pressure was entered in hPa.

Overall, this result is not very satisfying. For example, it isn't difficult to imagine circumstances in which the twelve-hour variation in temperature at a particular station is very different than the vertical mean temperature of the fictitious atmosphere. Nonetheless, this idea is the basis of sea-level pressure calculations that have been used worldwide for the better part of a century.

Sea-level pressure calculations using the Smithsonian Meteorological Tables. The method described by List (1949) is based on the concept of geopotential height (9.30), and is essentially an improved version of the relation shown in (9.51), converted for use with common logarithms ($\log_{10}$). It was developed long before the availability of electronic calculators, when most scientific calculations were accomplished using a slide rule and log tables. The steps for computing sea-level pressure are as follows:

Step 1: Compute an estimate of the mean virtual temperature through the fictitious column of air beneath the station by:

$$\boxed{\bar{t}_v \cong \bar{t} + L + c} \tag{9.52}$$

where $\bar{t}$ is the twelve-hour mean air temperature, L is a correction for station elevation, and c is a correction for the current dew point. All three terms on the RHS of (9.52) are in °C. $\bar{t}$ is computed by:

$$\bar{t} \equiv \frac{t_0 + t_{12}}{2} \tag{9.52a}$$

where t_0 is the current *in situ* air temperature, and t_{12} is the *in situ* air temperature recorded twelve hours ago [°C]. L attempts to determine the mean temperature half way up the fictitious air column, assuming a mean environmental lapse rate (γ_e) equal to *half* the Dry Adiabatic Lapse Rate (Γ_D), and is computed by:

$$L \equiv \left(\frac{z_{STN}}{2}\right)\left(\frac{\Gamma_D}{2}\right) = \frac{z_{STN}}{2}\frac{1}{200} = \frac{z_{STN}}{400} \tag{9.52b}$$

where z_{STN} is the station elevation [meters ASL], and the Dry Adiabatic Lapse Rate (Γ_D) $\cong .01$ °C/m, thus $\Gamma_D/2$ is 1/200. The value of the last term on the RHS of (9.52), c, is found in the Smithsonian Meteorological Table (SMT) No. 48a[26] and has typical values near sea level of a few °C.

Step 2: Compute the *adjusted* geopotential height difference between sea level and station elevation by:

$$\boxed{\Delta\tilde{\zeta} = \frac{T_0}{T_0 + \bar{t}_v}\Delta\zeta} \tag{9.53}$$

where $\Delta\tilde{\zeta}$ is the *adjusted* geopotential height difference [gpm], T_0 is the reference temperature (273.16 K), $\bar{t}_v$ is the mean virtual temperature computed in Step 1 [°C], and $\Delta\zeta$ is defined by:

$$\Delta\zeta = \zeta_{STN} - \zeta_{SL} \cong \zeta_{STN} \tag{9.53a}$$

All quantities shown in (9.53a) are in geopotential meters. The geopotential height of the station (ζ_{STN}) can be estimated from the station elevation [meters] and the station latitude [degrees] using SMT No. 50[27].

Step 3: Compute the *adjusted* geopotential height of the station above the 1,100 hPa (used because the atmospheric pressure is never this high on Earth) pressure level by:

$$\boxed{\tilde{\zeta}_{STN} = 67.442\, T_0\, log_{10}\left(\frac{1100}{p_{STN}}\right)} \tag{9.54}$$

which has units of geopotential meters, and p_{STN} is the *in situ* (station) pressure in hPa.

Step 4: Compute the adjusted geopotential height [gpm] of sea level above 1,100 hPa by:

$$\boxed{\tilde{\zeta}_{SL} = \tilde{\zeta}_{STN} - \Delta\tilde{\zeta}} \tag{9.55}$$

Step 5: In the final step, we begin with the same form as (9.54), with sea-level values substituted for station values:

$$\tilde{\zeta}_{SL} = 67.442\,T_0\, log_{10}\left(\frac{1100}{SLP}\right) \tag{9.56}$$

Dividing through by the constants on the RHS yields:

$$\frac{\tilde{\zeta}_{SL}}{67.442\,T_0} = log_{10}\left(\frac{1100}{SLP}\right) \tag{9.57}$$

The inverse function of $log_{10}(x)$ is 10^x, so:

$$10^{\left(\frac{\tilde{\zeta}_{SL}}{67.442\,T_0}\right)} = \frac{1100}{SLP} \tag{9.58}$$

or:

$$SLP \times 10^{\left(\frac{\tilde{\zeta}_{SL}}{67.442\,T_0}\right)} = 1100 \tag{9.59}$$

Finally, we solve for SLP by:

$$\boxed{SLP = (1100)\,10^{\left(\frac{-\tilde{\zeta}_{SL}}{67.442\,T_0}\right)}} \tag{9.60}$$

Example. Suppose we have a station at 1000 m ASL and 45 °N, with a recorded pressure of 900 hPa. The current *in situ* temperature is 15 °C, and the temperature twelve hours ago was 5 °C. The current dew point is 0 °C. What is the equivalent sea-level pressure?

According to SMT 48a, c is 0.8 °C, so, by (9.52),

$$\bar{t}_v \cong \underbrace{\left(\frac{15+5}{2}\right)}_{\bar{t}} + \underbrace{\left(\frac{1000}{400}\right)}_{L} + \underbrace{0.8}_{c} = 13.3\,°C$$

By (9.53), the adjusted geopotential height difference between sea level and station elevation ($\Delta\tilde{\zeta}$) is:

$$\Delta \tilde{\zeta} = \left(\frac{273.16}{273.16 + 13.3} \right) \underbrace{(1000)}_{\Delta \tilde{\zeta}} = 953.57 \, gpm$$

where the value of $\Delta \zeta$ came from SMT 50. By (9.54), the adjusted geopotential height of the station above 1100 hPa ($\tilde{\zeta}_{STN}$) is:

$$\tilde{\zeta}_{STN} = 67.442 \times (273.16) \times log_{10} \left(\frac{1100}{900} \right) = 1605.52 \, gpm$$

The adjusted geopotential height of sea level above 1100 hPa is given by (9.55):

$$\tilde{\zeta}_{STN} - \Delta \tilde{\zeta} = 1605.52 - 953.57 = 651.95 \, gpm$$

And finally, we compute the equivalent sea-level pressure using (9.60):

$$SLP = (1100)10^{\left(\frac{-\tilde{\zeta}_{SL}}{67.442 T_0} \right)} = (1100) \times 10^{\left(\frac{-651.95}{67.442 \times 273.16} \right)} = 1013.92$$

which has units of hPa, because the reference pressure (1100) is also in hPa.

9.6 Summary of Atmospheric Statics

In this chapter we examined gravity in detail, including the variations in gravity caused by latitude, altitude ASL, and centrifugal force. From there we defined geopotential (the potential energy of an object in a gravity field) and geopotential height (elevation corrected for local gravity). The Hypsometric Equation was derived, defining units of geopotential meters, as well as the quantity thickness (the vertical distance between two adjacent pressure levels). Thickness can be used to forecast precipitation type – that is, for estimating whether a given storm system will produce rain, snow, or other types of precipitation. The Hypsometric Equation was inverted to derive an expression for pressure at any altitude, which forms the basis of sea-level pressure calculations. The SMTs method of computing sea-level pressure was described.

$$g^* = \frac{Gm_e}{\left(R_p + \Delta R \cos(\varphi) + z \right)^2}$$

where,

g^* = Newtonian gravitational acceleration [m/s^2]

G = Universal Gravitational Constant (6.673×10^{-11} Nm^2/kg^2)

m_e = mass of Earth (5.988×10^{24} kg)

R_p = polar radius of Earth (6.35691×10^6 m)

ΔR = difference between polar and equatorial radius (2.148×10^4 m)

φ = latitude [°]

z = elevation above surface [m]

$$a_v = \Omega^2 \left(R_p + \Delta R cos(\varphi) + z \right) cos^2(\varphi)$$

where,

a_v = outward-pointing centrifugal acceleration [m/s²]

Ω = angular rotation of Earth (7.292 × 10⁻⁵ s⁻¹)

$$g = \frac{Gm_e}{\left(R_p + \Delta R cos(\varphi) + z \right)^2} - \Omega^2 \left(R_p + \Delta R cos(\varphi) + z \right) cos^2(\varphi)$$

where,

g = local effective gravitational acceleration [m/s²]

$$g = g_\varphi \left(1 - 3.14 \times 10^{-7} z \right)$$

where,

g_φ = local effective gravity at latitude φ at sea level [m/s²]

$$d\Phi \equiv g dz$$

where,

$d\Phi$ = geopotential [m²/s²]

$$\zeta \equiv \frac{1}{g_0} \int_0^z g dz$$

where,

ζ = geopotential height [gpm]

g_0 = reference gravity (9.8 m/s²)

$$\Delta \zeta \cong \frac{R_d \overline{T}_v}{g_0} ln\left(\frac{p_1}{p_2} \right)$$

where,

$\Delta \zeta$ = thickness of layer between p_1 and p_2 [gpm]

R_d = individual gas constant for dry air (286.8 J/(kg K))

$\overline{T}_v$ = mean virtual temperature in layer between p_1 and p_2 [K]

p_1, p_2 = boundaries of layer ($p_1 > p_2$) [any unit]

$$\Delta \zeta \cong Hln\left(\frac{p_1}{p_2} \right)$$

where,

H = scaling height [gpm]

The scaling height of the atmosphere is defined by:

$$H \equiv \frac{R_d \overline{T}_v}{g_0}$$

By inverting the Hypsometric Equation, we derived an equation for computing the pressure at any geopotential height, given the scaling height, two geopotential heights, and the pressure at one of them:

$$p_2 = p_1 \exp\left[-\left(\frac{\zeta_2 - \zeta_1}{H}\right)\right]$$

where,

  p_2 = pressure at geopotential height ζ_2 [hPa]
  p_1 = pressure at geopotential height ζ_1 [hPa]

This formed by the basis for routine computations of a station's equivalent sea-level pressure, that is, its observed station pressure reduced to the equivalent pressure the station would have if it were moved to sea level:

$$\mathrm{SLP} = p_{\mathrm{STN}} \exp\left(\frac{z_{\mathrm{STN}} g_0}{R_d \overline{T}_{12\mathrm{hr}}}\right)$$

where,

  SLP = equivalent sea-level pressure [hPa]
  p_{STN} = *in situ* (station) pressure [hPa]
  z_{STN} = station elevation [m ASL]
  $\overline{T}_{12\mathrm{hr}}$ = twelve-hour mean *in situ* temperature [K]

The method for computing SLP in the SMTs is very similar, but makes additional corrections for dew point, and the mean temperature in a fictitious vertical column of air between station elevation and sea level. Appendix 8 contains the tabular data needed to complete the calculations.

Practice Problems

1. Show that, by combining the improved form of the Ideal Gas Law involving virtual temperature (8.7) with the hydrostatic relation (5.77), the definition of geopotential height (9.30):

$$\zeta \equiv \frac{1}{g_0} \int_0^z g\,dz$$

can be rewritten by:

$$\Delta\zeta = -\frac{R_d}{g_0} \int_{p_1}^{p_2} T_v \frac{dp}{p}$$

2. Calculate the vertical pressure gradient in a static atmosphere, assuming altitude at sea level, latitude of 45 °N, and an air mass density of 1.2 kg/m³. Perform a unit analysis to show that your results have the correct units.

3. Use the expression for effective gravity to compute values for the latitudes and elevations in the following text. List your answers to at least three decimal places of precision. Useful quantities:

$m_e = 5.988 \times 10^{24}$ kg
$G = 6.673 \times 10^{-11}$ Nm²kg⁻²
$\Omega \approx 2\pi/86164$ s⁻¹

Lat [°N]	Elevation [m]	g [m/s²]
0	10	
10	10	
20	100	
30	100	
40	1000	
50	1000	
60	5000	
70	5000	
80	10^4	
90	10^4	

4. Write a function to compute the effective gravity for Jupiter for any latitude and altitude that includes both the Newtonian and centrifugal effects. The mass of Jupiter is 1.899×10^{27} kg. Jupiter rotates on its axis in ten hours. Its polar radius is 66,854 km, and its equatorial radius is 71,492 km.[28]

5. Use the function you derived for No. 4 to complete the following table for Jovian gravity:

Lat [°N]	Elevation [km]*	g [m/s²]
0	1000	
10	10,000	
20	10,000	
30	10,000	
40	1000	
50	1000	
60	5000	
70	5000	
80	10^4	
90	10^4	

* These are distances above the visible surface of the planet.

6. What is the definition of the scaling height?

7. Calculate the scaling height of an atmosphere with mean temperature of 10 °C, mean dew point of 5 °C, and pressure 900 hPa. Assume that the latitude is 45 °N.

8. Estimate the scaling height (H) for the near-surface Martian atmosphere given the following composition:

Gas	Fraction by Mass [%]	Mole weight [kg kmol^{-1}]
CO_2	95.32	44.01
N_2	2.70	28.02
Ar	1.60	39.94
Kr	0.15	83.80
O_2	0.13	32.00
CO	0.07	28.01
H_2O	0.03	18.02
Ne	0.00025	20.18

and:
Mean equatorial surface temperature: $1\,°C$
Mean surface gravitational acceleration: 3.71 m/s^2

9. Given the scaling height you just derived for Mars, and "sea-level pressure" of 7 hPa, estimate the altitude (in "geo"potential meters) of the 1 hPa pressure surface.

10. Back to Earth. Determine the thickness of a layer [gpm] with mean temperature of $2\,°C$, mean dew point of $-3\,°C$, at 45 °N, between pressures 1000 and 925 hPa.

11. Calculate the mean virtual temperature [$°C$] of a layer from 1000 and 850 hPa, if $\Delta\zeta$ is 1300 gpm.

12. Calculate the mean virtual temperature [$°C$] of a layer from 850 hPa to 700 hPa, if $\Delta\zeta = 1550$ gpm.

13. Calculate the pressure at an elevation of 1000 gpm ASL, if the sea-level pressure is 990 hPa, the mean layer temperature is $15\,°C$, and the mean layer dew point is $10\,°C$. Assume a latitude of 45 °N.

14. Use the SMTs to determine the sea-level pressure, given:
Station elevation = 100 m
Station pressure = 970 hPa
Current temperature = $30\,°C$
Twelve hour temperature = $18\,°C$
Current dew point = $2\,°C$
Latitude = 30 °N

15. Use the SMTs to determine the sea-level pressure, given:
Station elevation = 1000 m
Station pressure = 900 hPa
Current temperature = $10\,°C$
Twelve hour temperature = $0\,°C$
Current dew point = $2\,°C$
Latitude = 45 °N

10

Model and Standard Atmospheres

10.1. Introduction

On many Skew-T diagrams, especially the old paper versions, two types of information are included that we haven't mentioned yet: A standard height scale, parallel to the vertical pressure scale, *and* a standard temperature profile, usually plotted as a heavy line in the center of the diagram. In this chapter we will discuss the origin of the standard atmosphere used to generate both of these data types. We'll also discuss an application of the standard atmosphere used in aviation.

10.2. Model Atmospheres, Altimeter Setting, and Pressure Altitude

We'll begin with three possibilities for creating a simplified model of the vertical structure of Earth's atmosphere and a practical application of one of them.

A homogeneous atmosphere is one where the density is constant with height. In functional notation,

$$\boxed{\rho(z) = \rho_0} \tag{10.1}$$

where the LHS is functional notation for the density (ρ) at height z, and the RHS (ρ_0) indicates the density at $z = 0$ (sea level). In this case, the hydrostatic relation (5.77), namely:

$$\frac{dp}{dz} = -\rho g$$

becomes:

$$\frac{dp}{dz} = -\rho_0 g \tag{10.2}$$

The vertical extent of the homogeneous atmosphere can be determined by rearranging and integrating (10.2). First, we multiply through by dz:

228

$$dp = -\rho_0 g\, dz \qquad (10.3)$$

Next, we integrate the LHS from sea-level pressure (p_0) to 0, and the RHS from sea-level elevation ($z = 0$) to the top of the atmosphere (H), that is:

$$\int_{p_0}^{0} dp = -\rho_0 g \int_{0}^{H} dz \qquad (10.4)$$

which has been simplified somewhat by also assuming that the effective gravitational acceleration (g) is a constant. Performing both integrals results in:

$$0 - p_0 = -\rho_0 g (H - 0) \qquad (10.5)$$

or:

$$-p_0 = -\rho_0 g H \qquad (10.6)$$

Canceling the negative sign on both sides of (10.6) yields:

$$p_0 = \rho_0 g H \qquad (10.7)$$

and solving for H yields the scaling height for a homogeneous atmosphere:

$$\boxed{H \equiv \frac{p_0}{\rho_0 g}} \qquad (10.8)$$

The Ideal Gas Law (8.7) states that:

$$p = \rho R_d T_v$$

If the atmosphere has a constant density, this becomes:

$$p = \rho_0 R_d T_v \qquad (10.9)$$

and at sea level:

$$p_0 = \rho_0 R_d T_{v0} \qquad (10.10)$$

where p_0 indicates the sea-level pressure and T_{v0} indicates the virtual temperature at sea level. If we now substitute the RHS of (10.10) into the RHS of (10.8), we obtain:

$$H \equiv \frac{p_0}{\rho_0 g} = \frac{\rho_0 R_d T_{v0}}{\rho_0 g} = \frac{R_d T_{v0}}{g} \qquad (10.11)$$

which you should compare to (9.36).

If pressure decreases with height (and it must, because at higher elevations, there is less remaining atmosphere overhead), and density remains constant, then (10.9) states that temperature *must* also decrease. In other words:

$$\frac{dp}{dz} = \rho_0 R_d \frac{dT_v}{dz} \tag{10.12}$$

We can make this somewhat more manageable by substituting temperature (T) for the virtual temperature (T_v), that is:

$$\frac{dp}{dz} = \rho_0 R_d \frac{dT}{dz} \tag{10.13}$$

and by also substituting the RHS of (10.2) into the LHS of (10.13):

$$-\rho_0 g = \rho_0 R_d \frac{dT}{dz} \tag{10.14}$$

or:

$$-g = R_d \frac{dT}{dz} \tag{10.15}$$

and finally:

$$\boxed{\frac{dT}{dz} = \frac{-g}{R_d}} \tag{10.16}$$

If we substitute the mean value of g for the Midlatitudes obtained in Chapter 9 (9.24c), and numerical value for the dry air gas constant computed in Chapter 2 (2.65), we see that:

$$\frac{dT}{dz} = \frac{-9.81 \left[\frac{m}{s^2}\right]}{286.8 \left[\frac{J}{kg\,K}\right]} = -0.0342 \left[\frac{K}{m}\right] = -34.2 \left[\frac{K}{km}\right] \tag{10.17}$$

There are three important points to consider about (10.17). The first is that it *isn't real*, because it assumes that pressure falls with height but density remains constant, which is only possible with an *incompressible fluid* (which is approximately true of water, but certainly not air). The second is that the implied *lapse rate* (34.2 K/km) is *superadiabatic*; that is, it cools faster with height than the Dry Adiabatic Lapse Rate (Γ_D), which has a numerical value of about 10 K/km (5.88). This means that a homogeneous atmosphere is *absolutely unstable*.[1] (The lowest part of the troposphere, known as the Planetary Boundary Layer, sometimes becomes absolutely unstable on hot summer days, but never to this degree.) The final point is the question of whether the units on the RHS of (10.16) really come out to K/m, which is the physically correct unit shown at the second intermediate point in (10.17). A unit analysis settles this question by:

$$\frac{\dfrac{m}{s^2}}{\dfrac{J}{kg\,K}} = \frac{\dfrac{m}{s^2}}{\left(kg\dfrac{m^2}{s^2}\right)} = \frac{\dfrac{m}{s^2}}{\left(\dfrac{kg}{kg}\right)\left(\dfrac{m^2}{s^2}\right)\left(\dfrac{1}{K}\right)} = \left(\dfrac{m}{s^2}\right)\left(\dfrac{s^2}{m^2}\right)\left(\dfrac{K}{1}\right) = \dfrac{K}{m} \tag{10.18}$$

where the definition for Joules (J) was taken from Chapter 1, Table 1.4.

An isothermal atmosphere is one in which the temperature is constant with height. This is a reasonable approximation for lower part of Earth's stratosphere. In other words:

$$\boxed{\gamma_e = -\frac{dT}{dz} = 0} \tag{10.19}$$

where γ_e indicates the environmental (*in situ*) temperature lapse rate. The hydrostatic relation (5.77) states that:

$$\frac{dp}{dz} = -\rho g$$

and the Ideal Gas Law (8.7) states that:

$$p = \rho R_d \bar{T}_v$$

where we have substituted $\bar{T}_v$ for T_v, because the atmosphere is *isothermal*. Solving the latter for density (ρ), and further simplifying by substituting $\bar{T}$ for $\bar{T}_v$, yields:

$$\rho = \frac{p}{R_d \bar{T}} \tag{10.20}$$

which we then substitute into (5.77) to obtain:

$$\frac{dp}{dz} = -\frac{p}{R_d \bar{T}} g \tag{10.21}$$

Moving all pressure terms to the LHS and multiplying through by dz results in:

$$\frac{dp}{p} = -\frac{g}{R_d \bar{T}} dz \tag{10.22}$$

which we then integrate between p_0 (a reference pressure, such as sea level) and p on the LHS, and 0 (the bottom of the layer, corresponding to p_0) and z on the RHS. Because the atmosphere is isothermal, everything except dz can be pulled out of the integral on the RHS, that is:

$$\int_{p_0}^{p} \frac{dp}{p} = -\frac{g}{R_d \overline{T}} \int_{0}^{z} dz \tag{10.23}$$

Performing both integrals results in:

$$ln(p) - ln(p_0) = -\frac{g}{R_d \overline{T}}(z - 0) \tag{10.24}$$

which simplifies to:

$$ln\left(\frac{p}{p_0}\right) = -\frac{g}{R_d \overline{T}} z \tag{10.25}$$

Next, we exponentiate both sides of (10.25):

$$\exp\left[ln\left(\frac{p}{p_0}\right)\right] = \exp\left(-\frac{g}{R_d \overline{T}} z\right) \tag{10.26}$$

or:

$$\frac{p}{p_0} = \exp\left(-\frac{g}{R_d \overline{T}} z\right) \tag{10.27}$$

Finally, multiplying through by p_0 (and using functional notation on the LHS) results in:

$$\boxed{p(z) = p_0 \exp\left(-\frac{g}{R_d \overline{T}} z\right)} \tag{10.28}$$

which is an expression describing the variation in pressure p with height z in an isothermal atmosphere. This can be simplified somewhat further by using the definition of the scaling height given by (9.36), namely:

$$H \equiv \frac{R_d \overline{T}_v}{g_0}$$

and substituting $\bar{T}$ for $\bar{T}_v$:

$$H \cong \frac{R_d \bar{T}}{g_0} \tag{10.29}$$

Substituting this into (10.28) results in:

$$\boxed{p(z) = p_0 \exp\left(-\frac{z}{H}\right)} \tag{10.30}$$

which you should compare to (9.49) and (9.50). Like the real atmosphere, the pressure (and therefore density, using the Ideal Gas Law) of the isothermal model atmosphere approaches zero asymptotically.

A constant lapse-rate atmosphere is one in which the temperature decreases (or increases) linearly with height. This model provides reasonable analogs for the lowest three kilometers of the troposphere (i.e., below about 700 hPa), as well as the upper part of the stratosphere. In this case, the temperature (T) at any height z is computed by:

$$\boxed{T(z) = T_0 - \gamma_e z} \tag{10.31}$$

where T_0 is the temperature at the bottom of the layer (such as at sea level) [K], γ_e is the environmental lapse rate [K/m], and z is the elevation above the bottom of the layer [m]. Once again, we're also interested in how the *pressure* in the constant lapse-rate layer varies with height. Beginning with the combined hydrostatic relation and Ideal Gas Law shown in (10.22):

$$\frac{dp}{p} = -\frac{g}{R_d \bar{T}} dz$$

and replacing $\bar{T}$ with $T(z)$, we obtain:

$$\frac{dp}{p} = -\frac{g}{R_d T(z)} dz \tag{10.32}$$

But by definition, $T(z)$ varies linearly with height according to the relation shown in (10.31). Substituting this identity into (10.32), we obtain:

$$\frac{dp}{p} = -\frac{g}{R_d (T_0 - \gamma_e z)} dz \tag{10.33}$$

which must be integrated. On the LHS, we integrate from p_0 to p, and on the RHS, we integrate from 0 to z; that is:

$$\int_{p_0}^{p} \frac{dp}{p} = -\frac{g}{R_d} \int_{0}^{z} \frac{dz}{(T_0 - \gamma_e z)} \tag{10.34}$$

where we once again assume g is constant. This is not so simple to integrate, but one of your calculus courses[2] taught you about something called *u-du substitution*. Now you'll get a chance to use it.

To integrate the RHS of (10.34), let's assume that:

$$u = T_0 - \gamma_e z \tag{10.35a}$$

Because T_0 and γ_e are constants,

$$du = d(T_0 - \gamma_e z) = dT_0 - d(\gamma_e z) = -\gamma_e dz \tag{10.35b}$$

and therefore,

$$-\frac{1}{\gamma_e} du = dz \tag{10.35c}$$

This means that:

$$\frac{du}{u} = \frac{-\gamma_e dz}{T_0 - \gamma_e z} \tag{10.35d}$$

Going back to the RHS of (10.34), we have:

$$-\frac{g}{R_d} \int_{0}^{z} \frac{dz}{(T_0 - \gamma_e z)} \tag{10.36a}$$

Rewriting this in terms of *du/u* results in:

$$-\frac{g}{R_d} \left(-\frac{1}{\gamma_e} \right) \int_{0}^{z} \overbrace{\frac{-\gamma_e dz}{\underbrace{(T_0 - \gamma_e z)}_{u}}}^{du} \tag{10.36b}$$

where the extra factor of $(-1/\gamma_e)$ *outside* the integral is necessary to cancel the new factor of $-\gamma_e$ added *inside* the integral. (Combined, the two new terms are equal to unity.) Now the terms inside the integral amount to *du/u*, which is the derivative of *ln(u)*. Combining this result with (10.34), we have:

$$\int_{p_0}^{p} \frac{dp}{p} = -\frac{g}{R_d} \left(-\frac{1}{\gamma_e} \right) \int_{0}^{z} \frac{-\gamma_e dz}{(T_0 - \gamma_e z)} \tag{10.37}$$

or:

$$ln(p) - ln(p_0) = \frac{g}{R_d \gamma_e} \left[ln(T_0 - \gamma_e z) - ln(T_0) \right] \tag{10.38}$$

This can be rewritten as:

$$ln\left(\frac{p}{p_0}\right) = \frac{g}{R_d \gamma_e} ln\left(\frac{T_0 - \gamma_e z}{T_0}\right) \tag{10.39}$$

Using another property of logarithms, we can rewrite the RHS by:

$$ln\left(\frac{p}{p_0}\right) = ln\left[\left(\frac{T_0 - \gamma_e z}{T_0}\right)^{\frac{g}{R_d \gamma_e}} \right] \tag{10.40}$$

Exponentiating both sides we have:

$$\frac{p}{p_0} = \left(\frac{T_0 - \gamma_e z}{T_0}\right)^{\frac{g}{R_d \gamma_e}} \tag{10.41}$$

or finally:

$$\boxed{p(z) = p_0 \left(\frac{T_0 - \gamma_e z}{T_0}\right)^{\frac{g}{R_d \gamma_e}}} \tag{10.42}$$

which you should compare to (10.28) and (10.30) (they're quite different). Like the isothermal atmosphere, this relation also shows the pressure asymptotically approaching zero, but it's a *different* asymptote. Further, the function shown in (10.42) only produces physically realistic numerical answers for heights up to about 25 km.

Altimeter setting and pressure altitude are two parameters commonly used in aviation. Both are based on a constant lapse-rate atmosphere. Altimeter setting (ALSTG) is similar to sea-level pressure, but it lacks many of the small-order corrections of the latter quantity.[3] For stations above sea level, reducing station pressure to ALSTG is simply a matter of adding a fixed pressure correction based on the station's elevation.[4] ALSTG is usually reported in hPa or inches of mercury (in.Hg).[5]

Pressure altitude (PA) is a parameter used to tell pilots the *apparent* elevation of an airfield in a hypothetical fixed atmosphere. We imagine that the atmosphere never changes, but instead the station pressure changes because the *airfield* moves up and down in the atmosphere. PA is usually reported in feet or meters.

- If the ALSTG is equal to 1,013.25 hPa (29.92 in.Hg), which is standard atmospheric pressure at sea level, the PA is equal to the station elevation.

- Under *low* pressure conditions (ALSTG less than 1,013.25 hPa), the PA is *higher* than the station elevation, because the airfield has apparently moved *upward* in the atmosphere.
- Under *high* pressure conditions (ALSTG greater than 1,013.25 hPa), the PA is *lower* than the station elevation, because the airfield has apparently moved *downward* in the atmosphere.

A related quantity is *density altitude*, which once again relies on a standard and unchanging atmosphere and assumes that variations in dry-air *density* result from changes in the elevation of the airfield. Density is an important quantity in aviation because it is a primary control on the ability of aircraft wings to generate lift, and of engines to generate thrust.

PA is computed by solving (10.42) for z. Beginning with (10.41), we can invert the equation and solve for z. After a few steps, we find that:

$$ z_{PA} = \frac{T_0}{\gamma_e}\left[1 - \left(\frac{p}{p_0}\right)^{\frac{R_d \gamma_e}{g}}\right] \tag{10.43} $$

where z_{PA} is the PA, that is, the *apparent* elevation of the station in the U.S. Standard Atmosphere[6] (USSA) [m], T_0 is the temperature at sea level for the USSA (288.15 K), γ_e is the environmental lapse rate for the USSA (0.0065 K/m), p is the *in situ* station pressure [hPa], p_0 is the standard sea-level pressure for the USSA (1013.25 hPa), R_d is the individual gas constant for dry air for the USSA (287.1 J/(kg K)), and g is the effective gravity for the USSA (9.80665 m/s^2).

10.3. The U.S. Standard Atmosphere and the International Standard Atmosphere

Originally published in 1958 (then updated in 1962, 1966, and 1976), the USSA is a combination of climatological means and mathematical models of the atmosphere. Below 32 km ASL, the USSA and the International Standard Atmosphere (ISA) (published in 1976) are identical, and include descriptions of mean molecular weight, gravity, density, virtual temperature, temperature lapse rate, and stability, among other variables. The 1976 USSA/ISA describes the atmosphere in eight vertical layers (numbered zero through seven). The lowest layer is based at sea level, and the highest layer is based at the bottom of the heterosphere (near 85 km ASL). Some of the basic assumptions about Layers 0, 1, 2, and 3, describing the troposphere and stratosphere, are listed in the following text. Vertical temperature profiles for different seasons and latitudes, and the Tropics, Midlatitudes, and world as a whole, are shown in Figures 10.1 through 10.3.[7]

10.3.1. Basic Assumptions

Some of the basic USSA/ISA assumptions about the global atmosphere are:

Standard sea-level conditions:

- Temperature: T_0 is 15 °C (288.15 K)[8]
- Pressure: p_0 is 1013.25 hPa
- Density: ρ_0 is 1.225 kg/m³

Gravity. The variation in sea-level effective gravity with latitude (φ) is given by:

$$g_{\varphi,0} = 9.780356\left(1+0.0052885 sin^2\left(\varphi\right)-0.0000059 sin^2\left(2\varphi\right)\right)\left[\frac{m}{s^2}\right] \qquad (10.44)$$

and at sea level, at 45 °N, $g_0 = 9.80665$ m/s².

Composition and molar mass:

- The atmosphere is composed of pure, dry air, with a mean molecular mass ($\bar{M}$) of 28.964 kg/kmol, from the surface through the turbopause.[9] Compare this result to (2.63) and (2.64), which used only the first four gases listed in Table 2.1. The two results differ by about 0.02 percent.
- The Universal Gas Constant (R^*) is equal to 8314.32 J/(kmol K), which differs from the value computed in (2.43) by about 0.05 percent.
- The gas constant for the dry-air ensemble is:

$$R_d = \frac{R^*}{\bar{M}} = \frac{8314.32\left[\dfrac{J}{kmol\ K}\right]}{28.964\left[\dfrac{kg}{kmol}\right]} = 287.1\left[\frac{J}{kg\ K}\right] \qquad (10.45)$$

which differs from the value we have been using up until now (286.8 J/(kg K)) by about 0.10 percent. Because this difference is so small,[10] the calculations shown in Tables 10.1 and 10.2 are based on a value of 286.8 J/(kg K).

Virtual temperature is approximated by:

$$T_v \cong \frac{T}{1-0.379\dfrac{e}{p}} \qquad (10.46)$$

where T is the *in situ* temperature [K], e is the vapor pressure [hPa], and p is the total atmospheric pressure [hPa].

The atmosphere is an ideal gas, so that the Ideal Gas Law (8.7) applies, that is:

$$p = \rho R_d T_v$$

The atmosphere is in hydrostatic equilibrium, so that the hydrostatic relation applies (5.75), that is:

$$p = -\rho g z$$

where the elevation (z) is in *linear* meters.

10.3.2. Vertical Temperature, Pressure, and Density Variation
in the Troposphere (Layer 0)

Some of the USSA/ISA basic assumptions about the troposphere are:

Vertical temperature lapse rate. The troposphere is considered a constant lapse-rate layer, as described by (10.31):

$$T(z) = T_0 - \gamma_e z$$

In this case, we rewrite γ_e as β_0, a constant equal to 0.0065 K/m, that is:

$$\boxed{T(z) = T_0 - \beta_0 z} \tag{10.47}$$

where $T(z)$ is functional notation for temperature T as a function of elevation z, and T_0 is the standard temperature at sea level (288.15 K).

Vertical pressure gradient. The variation in pressure with height is also described in the 1976 USSA/ISA. The pressure variation is derived by equating the linear relation shown in (10.47) to a variation on the adiabatic temperature relation shown in (5.17), namely:

$$T_f = T_i \left(\frac{p_f}{p_i} \right)^{\frac{R_i}{c_p}}$$

We begin with (5.82), which defined the vertical temperature variation of an unsaturated parcel as:

$$\frac{dT}{dz} = -\frac{g}{c_p}$$

Both sides have units of a vertical lapse rate, so, working with the RHS, we can generalize this somewhat by writing the unit equivalence (indicated by square brackets) as:

$$\left[\frac{g}{c_p} \right] = [\gamma] \tag{10.48}$$

where γ on the RHS indicates any vertical lapse rate. We can take this a step further and state that, if the LHS of (10.48) has units of *any* lapse rate, it must also have the same units as the *environmental* lapse rate, that is:

$$\left[\frac{g}{c_p}\right]=\left[\gamma_e\right] \tag{10.49}$$

In (10.47), we stated that γ_e in the lower 11 km of the atmosphere is equal to β_0, so we can write:

$$\left[\frac{g}{c_p}\right]=\left[\beta_0\right] \tag{10.50}$$

Near sea level, we can write this somewhat more specifically by substituting standard sea-level gravity (10.44):

$$\left[\frac{g_0}{c_p}\right]=\left[\beta_0\right] \tag{10.51}$$

If this is true, then:

$$\left[\frac{R_d g_0}{c_p}\right]=\left[R_d \beta_0\right] \tag{10.52}$$

must also be true, and therefore:

$$\left[\frac{R_d}{c_p}\right]=\left[\frac{R_d \beta_0}{g_0}\right] \tag{10.53}$$

must *also* be true. In fact, if you complete the analysis be reducing both sides to their base SI units, you'll find that they are both are unitless. Note that the LHS of (10.53) is the exponent on the RHS of (5.17).

  This is not to say that the *numerical values* of both sides of (10.52) are the same. *They are not.* By plugging in the known values of the base variables shown, we obtain:

$$\frac{R_d}{c_p}=\frac{286.8}{1003.8}=0.2857 \tag{10.54a}$$

where the value of R_d comes from (2.65), and the value of c_p comes from (4.62). On the RHS of (10.53):

$$\frac{R_d \beta_0}{g_0}=\frac{(286.8)(0.0065)}{9.80665}=0.1901 \tag{10.54b}$$

where the value of β_0 is defined in (10.47), and g_0 is defined by (10.44).

Now we set our two expressions for vertical temperature variation in the tropo-sphere equal to each other:

$$T_0 - \beta_0 z = T_0 \left(\frac{p}{p_0} \right)^{\frac{R_d \beta_0}{g_0}} \tag{10.55}$$

where the LHS came from (10.47), and the RHS came from (5.17; shown in the preceding text), with the following substitutions:

- T_0 (standard sea-level temperature) was substituted for T_i.
- p (pressure as a function of z) was substituted for p_f.
- p_0 (standard sea-level pressure) was substituted for p_i.
- $R_d \beta_0 / g_0$ was substituted for R/c_p, based on the analysis shown. (The environmental lapse rate is usually not the same as Γ_D.)

First, we divide through by T_0, resulting in:

$$\frac{T_0 - \beta_0 z}{T_0} = \left(\frac{p}{p_0} \right)^{\frac{R_d \beta_0}{g_0}} \tag{10.56}$$

or:

$$1 - \frac{\beta_0 z}{T_0} = \left(\frac{p}{p_0} \right)^{\frac{R_d \beta_0}{g_0}} \tag{10.57}$$

Next, we raise both sides to the inverse of the exponent on the RHS:

$$\left(1 - \frac{\beta_0 z}{T_0} \right)^{\frac{g_0}{R_d \beta_0}} = \left[\left(\frac{p}{p_0} \right)^{\frac{R_d \beta_0}{g_0}} \right]^{\frac{g_0}{R_d \beta_0}} \tag{10.58}$$

or:

$$\left(1 - \frac{\beta_0 z}{T_0} \right)^{\frac{g_0}{R_d \beta_0}} = \frac{p}{p_0} \tag{10.59}$$

And finally, we multiply through by p_0:

$$p(z) = p_0 \left(1 - \frac{\beta_0 z}{T_0} \right)^{\frac{g_0}{R_d \beta_0}} \tag{10.60}$$

Table 10.1. *Pressure, temperature and density as a function of elevation for USSA and ISA.*

Pressure values in column 2 were computed using the relation shown in (10.60). Temperatures were computed using (10.47). Density was computed using untruncated values from columns 2 and 3, a dry air gas constant of 286.8 $Jkg^{-1}K^{-1}$, and the Ideal Gas Law. All pressures and temperatures shown have been rounded to the nearest whole number value. Density is reported to the nearest 0.01 kgm^{-3}. Range of elevations was chosen to coincide with valid ranges of (10.47) and (10.60).

Elevation [m]	Pressure [hPa]	Temperature [K]	Density [kg m^{-3}]
0	1,013	288	1.23
500	955	285	1.17
1000	899	282	1.11
1500	845	278	1.06
2000	795	275	1.01
2500	745	272	0.96
3000	701	269	0.91
3500	657	265	0.86
4000	616	262	0.82
4500	577	259	0.78
5000	540	256	0.74
5500	505	252	0.70
6000	472	249	0.66
6500	440	246	0.62
7000	410	243	0.59
7500	382	239	0.56
8000	356	236	0.53
8500	331	233	0.50
9000	307	230	0.47
9500	285	226	0.44
10,000	264	223	0.41
10,500	244	220	0.39
11,000	226	217	0.36

where we have changed p to the functional notation $p(z)$. Table 10.1, showing the standard heights of pressures at regular *elevation* intervals, was generated using the relation shown in (10.60).

We can also turn this around to compute the height of a certain pressure surface. Beginning with (10.55), which states:

$$T_0 - \beta_0 z = T_0 \left(\frac{p}{p_0} \right)^{\frac{R_d \beta_0}{g_0}}$$

we subtract T_0 from both sides, and multiply through by -1:

$$\beta_0 z = T_0 - T_0 \left(\frac{p}{p_0} \right)^{\frac{R_d \beta_0}{g_0}} \tag{10.61}$$

Next, we use the reverse Distributive Law on the RHS to obtain:

$$\beta_0 z = T_0 \left[1 - \left(\frac{p}{p_0} \right)^{\frac{R_d \beta_0}{g_0}} \right] \tag{10.62}$$

and divide through by β_0, yielding:

$$\boxed{z(p) = \frac{T_0}{\beta_0} \left[1 - \left(\frac{p}{p_0} \right)^{\frac{R_d \beta_0}{g_0}} \right]} \tag{10.63}$$

which you should compare to (10.43), the relationship defining PA. $z(p)$ is functional notation to indicate "the elevation z of pressure p." Table 10.2, showing the standard elevations of pressures at regular *pressure* intervals, was generated using the relation shown in (10.63).

The average vertical pressure gradient can be estimated using data from Table 10.2. For example, between 1000 and 700 hPa:

$$\frac{dp}{dz} \cong \frac{\Delta p}{\Delta z} = \frac{p_{700} - p_{1000}}{z_{700} - z_{1000}} = \frac{700 - 1000}{3010 - 111} \left[\frac{hPa}{m} \right] = -\frac{300}{2899} \left[\frac{hPa}{m} \right] = -0.103 \left[\frac{hPa}{m} \right] \tag{10.64}$$

that is, there is a decrease of 1.03 hPa for every ten vertical meters.

10.3.3. Vertical Temperature Variation in the Stratosphere (Layers 1, 2, and 3)

The 1976 USSA/ISA models of the stratosphere are:

11 km to 20 km above sea level (Layer 1; the lower part of the stratosphere) – the atmosphere is *isothermal*, where:

$$T(z) = 216.65\,K = -56.5°C \tag{10.65}$$

Compare this to the last row of Table 10.1.

Table 10.2. *Altitudes, temperatures, and densities of pressure surfaces for USSA and ISA.*

Altitude values in column 2 were computed using the relation shown in (10.63). Temperatures were computed using (10.47). Density was computed using untruncated values from column 3, a dry air gas constant of 286.8 $Jkg^{-1}K^{-1}$, and the Ideal Gas Law. All elevations and temperatures shown were rounded to the nearest whole number value. Density is reported to the nearest 0.01 kgm^{-3}. Range of pressures chosen to match to usual pressure range shown on a Skew-T diagram. Elevations above 11 km are beyond the range of (10.63), and are highlighted in gray. Temperatures and densities above 11 km are not computed.

Pressure [hPa]	Elevation [m]	Temperature [K]	Density [kg m^{-3}]
1013.25	0	288	1.23
1000	111	287	1.21
950	540	285	1.16
900	988	988	1.11
850	1456	279	1.06
800	1947	276	1.01
750	2464	272	0.96
700	3010	269	0.91
650	3588	265	0.86
600	4203	261	0.80
550	4861	257	0.75
500	5570	252	0.69
450	6339	247	0.64
400	7180	241	0.58
350	8111	235	0.52
300	9157	229	0.46
250	10,355	221	0.39
200	11,766		
150	13,500		
100	15,787		

20 km to 32 km (Layer 2; the middle of the stratosphere) – the atmosphere is once again at a constant lapse rate, described by:

$$T(z) = T_{20km} - \beta_2(z - z_2) \tag{10.66}$$

where T_{20km} is the temperature at the top of Layer 1 [K], β_2 is −0.001 K/m (meaning that temperature *increases* with increasing height), z is elevation [m], and z_2 is the elevation of the bottom of this layer (20,000 m).

32 km to 47 km (Layer 3; the top of the stratosphere) – the atmosphere is still at a constant lapse rate, but now warms at a greater rate, described by:

$$T(z) = T_{32km} - \beta_3 (z - z_3) \tag{10.67}$$

where T_{32km} is the temperature at the top of Layer 2 [K], β_3 is −0.0028 K/m (temperature increases with height), z is elevation [m], and z_3 is the elevation of the bottom of this layer (32,000 m). *The USSA combines Layers 2 and 3 into a single layer with an environmental lapse rate equal to β_2. The ISA retains separate Layers 2 and 3.*

10.3.4. Vertical Temperature Profiles

In 1966, NASA published the *U.S. Standard Atmosphere Supplement*, which contained vertical temperature profiles based on radiosonde data for the lowest 30 km and observations from rocketsondes for the region above that. Some interpolated and modeled data were also included to fill in data voids.

Separate Northern Hemisphere Midlatitude temperature profiles were generated for January and July (between 30 and 75 °N), and a single annual temperature profile was created for the Tropics (at 15 °N). Figure 10.1 shows the January profiles for the Midlatitudes. The figure also shows the "Spring/Fall" profile originally published in NASA (1962), which represented mean conditions for all of the Midlatitudes. The January profiles were created by combining radiosonde data for the three-month period of December, January, and February. Figure 10.2 shows both the July Midlatitude profiles – based on data for the three-month period of June, July, and August – and the annual profile for the Tropics.

The four well-known, temperature-defined regions of the atmosphere are clearly visible in Figures 10.1 and 10.2. The troposphere (USSA/ISA Layer 0) is the lowest layer, extending from the surface through 11 km. Variations in the height of the tropopause are also evident, with the lowest visible (near 8 km ASL) in the January, 75 °N profile, and the highest visible (near 17 km ASL) in the July, 30 °N, and annual (tropical), 15 °N profiles. The stratosphere (USSA/ISA Layers 1–3) begins as an isothermal region, directly above the tropopause, and transitions into a negative-lapse rate (positive vertical temperature gradient) region between about 20 km and about 47 km. The rate of warming with height increases toward the upper third of the stratosphere, reflecting the transition from Layer 2 to Layer 3. The tropical profile does not reflect the isothermal region (Layer 2), which is directly above the tropopause indicated in the Midlatitude profiles. The stratopause (base of Layer 4) can be seen near 50 km. The mesosphere (Layer 5) is also visible between about 50 and about 80 km, and the mesopause (Layer 6) can be seen near 85 km. The thermosphere (Layer 7) is visible above about 90 km.

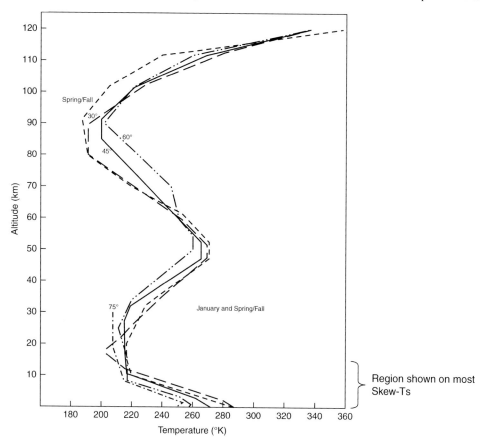

Figure 10.1. USSA, vertical temperature profile for winter (January), Spring/Fall. Temperatures are indicated on the Absolute (Kelvin) scale; altitudes are in kilometers. Mean vertical profiles are shown for January at latitudes 30, 45, 60, and 75 °N. Spring/Fall profile is the same as the National Aeronautics and Space Administration (1962) Midlatitude Profile. The vertical region discussed in text is highlighted in yellow (National Aeronautics and Space Administration, 1966).

Figure 10.3 shows the standard overall (global) annual temperature profile from the National Aeronautics and Space Administration (1976). Temperature in the figure is listed as "Molecular Temperature," which is defined by:

$$T_M = T\left(\frac{M_0}{M}\right) \tag{10.68}$$

where T_M is the molecular temperature, T is the kinetic (thermodynamic) temperature, defined by the relation shown in (3.35), M_0 is the mean molar mass of the dry-air gas ensemble at sea level, and M is the mean molar mass of the dry-air gas ensemble at the altitude in question. Both temperatures are on the Absolute (Kelvin) scale; both molar masses are in units of kg/kmol. Below the turbopause, the primary gases constituting

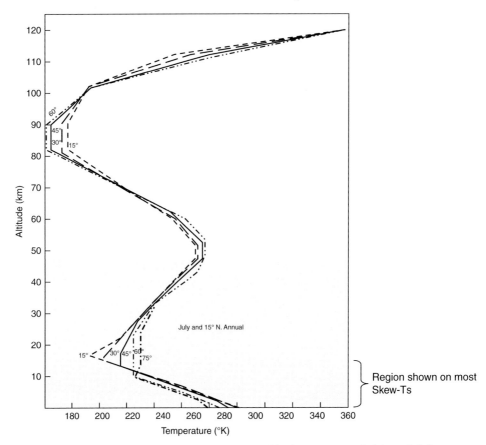

Figure 10.2. USSA, vertical temperature profile for summer (July) and full year (annual) at 15 °N. Temperatures are indicated on the Absolute (Kelvin) scale; altitudes are in kilometers. July vertical profiles are shown for latitudes 30, 45, 60, and 75 °N. The annual profile is shown for 15 °N. The vertical region discussed in the text is highlighted in yellow (National Aeronautics and Space Administration, 1966).

the dry-air ensemble are well mixed, so that $M_0 \cong M$, and therefore, $T_M \cong T$. The four temperature-defined regions of the atmosphere are again evident in Figure 10.3. In The USSA, the stratosphere's two upper layers (Layer 2 with $\beta_2 = -0.001$ K/m, and Layer 3 with $\beta_3 = -0.0028$ K/m) are simplified into a single layer with an overall lapse rate of -0.001 K/m. The ISA version of the standard atmosphere retains separate and distinct Layers 2 and 3. Figure 10.4 shows a Skew-T with the USSA temperature profile through 100 hPa, as well as standard heights for selected pressure surfaces.

10.4. Summary of Model and Standard Atmospheres

In this chapter, we described several models for approximating the vertical variation of temperature in the real atmosphere. These included *homogeneous* atmospheres

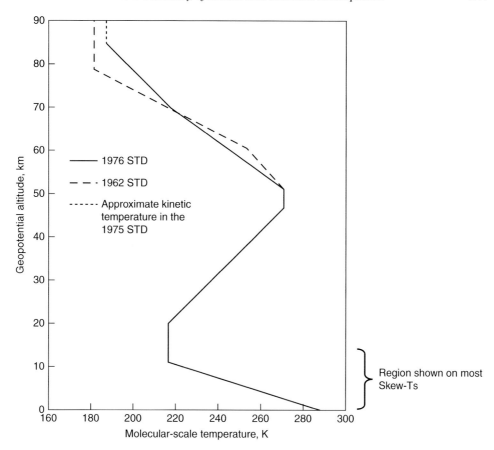

Figure 10.3. USSA, vertical temperature profile. Temperatures are indicated on the Absolute (Kelvin) scale; altitudes are in geopotential kilometers. Below the turbopause ($\cong 85$ km), the *molecular temperature* does not differ appreciably from the *kinetic (thermodynamic) temperature*, with the latter defined by (3.35). The vertical region discussed in the text is highlighted in yellow. (National Aeronautics and Space Administration, 1976.)

(those with a constant density, independent of elevation), *isothermal* atmospheres (those with a constant temperature, independent of elevation), and *constant lapse-rate* atmospheres (those with a temperature changing at a fixed, linear rate with respect to height). We concluded that the homogeneous atmosphere is superadiabatic; that is, its environmental lapse rate (γ_e) exceeds Γ_D. We also concluded that it is not a realistic description of an atmosphere composed of compressible gases. The remaining two models were serviceable and appeared to be reasonable approximations of specific regions of the real atmosphere.

For the *homogeneous* atmosphere:

$$\rho(z) = \rho_0$$

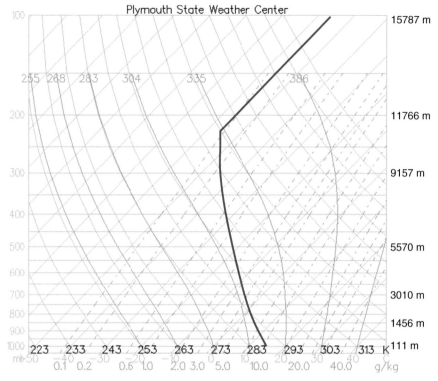

Figure 10.4. USSA on a Skew-T. The temperature is drawn for the region between 1000 and 100 hPa. The base of the stratosphere is clearly visible at 11 km ASL, where the temperature profile becomes isothermal. Standard elevations for selected pressure surfaces are shown in black on the vertical axis to the right of the figure. Both data types are based on Table 10.2. Background image is courtesy of Plymouth State University meteorology program (2012).

where,

$\rho(z)$ = density as a function of height (z) [kg/m³]

ρ_0 = sea-level density [kg/m³]

$$H \equiv \frac{p_0}{\rho_0 g}$$

where,

H = scaling height of homogeneous atmosphere [gpm]

p_0 = sea-level pressure [Pa]

g = gravitational acceleration (9.81 m/s²)

$$\frac{dT}{dz} = \frac{-g}{R_d} = \frac{-9.81 \left[\dfrac{m}{s^2} \right]}{286.8 \left[\dfrac{J}{kg\,K} \right]} = -0.0342 \left[\frac{K}{m} \right] = -34.2 \left[\frac{K}{km} \right]$$

where,

   dT/dz = vertical temperature gradient [K/m]
   R_d = individual gas constant for dry air (286.8 J/(kg K))

For the *isothermal* atmosphere:

$$\gamma_e = -\frac{dT}{dz} = 0$$

where,

   γ_e = environmental lapse rate [K/m]

$$p(z) = p_0 \exp\left(-\frac{g}{R_d \overline{T}} z\right)$$

where,

   $p(z)$ = pressure at elevation z [hPa]
   p_0 = sea-level pressure [hPa]
   $\overline{T}$ = temperature of isothermal atmosphere [K]
   z = elevation [m]

$$p(z) = p_0 \exp\left(-\frac{z}{H}\right)$$

where,

   H = scaling height of isothermal atmosphere [gpm]

For the *constant lapse-rate* atmosphere:

$$T(z) = T_0 - \gamma_e z$$

where,

   $T(z)$ = temperature at elevation z [K]
   T_0 = sea-level temperature [K]
   γ_e = environmental lapse rate [K/m]
   z = elevation [m]

$$p(z) = p_0 \left(\frac{T_0 - \gamma_e z}{T_0}\right)^{\frac{g}{R_d \gamma_e}}$$

where,

   $p(z)$ = pressure at elevation z [hPa]
   p_0 = sea-level pressure [hPa]
   T_0 = sea-level temperature [K]

ALSTG and PA are quantities used by aviators and are based on a standardized constant lapse-rate troposphere. ALSTG is an approximation of the sea-level pressure, and PA is the equivalent altitude of an airfield in the standard atmosphere, modeling variations in station pressure variations with changes in elevation.

$$z_{PA} = \frac{T_0}{\gamma_e} \left[1 - \left(\frac{p}{p_0} \right)^{\frac{R_d \gamma_e}{g}} \right]$$

where,

z_{PA} = pressure altitude [m ASL]
T_0 = standard sea-level temperature (288.15 K)
γ_e = standard temperature lapse rate (0.0065 K/m)
p = *in situ* station pressure [hPa]
p_0 = standard sea-level pressure (1013.25 hPa)
g = standard gravity (9.80665 m/s^2)

We also reviewed the USSA and the ISA, including basic assumptions about sea-level temperature, pressure, and density; chemical composition of the atmosphere, its mean molar mass, and corrections to the individual gas constant for dry air; vertical lapse rates in the lowest four layers (including the troposphere and three regions of the stratosphere); and vertical variations in pressure and density. For the troposphere, we have:

$$T(z) = T_0 - \beta_0 z$$

where,

$T(z)$ = temperature at elevation z [K]
T_0 = standard sea-level temperature (288.15 K)
β_0 = temperature lapse rate (0.0065 K/m)
z = elevation [m]

$$p(z) = p_0 \left(1 - \frac{\beta_0 z}{T_0} \right)^{\frac{g_0}{R_d \beta_0}}$$

where,

$p(z)$ = pressure at elevation z [hPa]
p_0 = standard sea-level pressure (1013.25 hPa)
g_0 = standard gravity (9.80665 m/s^2)
R_d = individual gas constant for dry air (287.1 J/(kg K))

$$z(p) = \frac{T_0}{\beta_0} \left[1 - \left(\frac{p}{p_0} \right)^{\frac{R_d \beta_0}{g_0}} \right]$$

where,

$z(p)$ = elevation of pressure p [m]

Practice Problems

1. List five important characteristics of the USSA.
2. Explain why the USSA is conditionally unstable in the lowest 11 km.
3. Assuming USSA, calculate $p(z)$ for the following values of z, if $p_0 = 1013.25$ hPa, $T_0 = 15\,°C$. Assume $g = 9.80665\ ms^{-2}$.
   a. $z = 1300$ m; $p(z) =$ ——————————— [hPa]
   b. $z = 3000$ m; $p(z) =$ ——————————— [hPa]
   c. $z = 5500$ m; $p(z) =$ ——————————— [hPa]
   d. $z = 9300$ m; $p(z) =$ ——————————— [hPa]
4. Begin with (10.42), and show that the PA (z_{PA}) for a station in a standard atmosphere is given by:

$$z_{PA} = \frac{T_0}{\gamma_e} \left[1 - \left(\frac{p}{p_0} \right)^{\frac{R_d \gamma_e}{g}} \right]$$

5. Assuming USSA, calculate $z(p)$ for the following values of p, if $p_0 = 1013.25$ hPa, $T_0 = 15\,°C$. Assume $g = 9.80665\ ms^{-2}$.
   a. $p = 925$ hPa; $z(p) =$ ——————————— [m]
   b. $p = 975$ hPa; $z(p) =$ ——————————— [m]
   c. $p = 790$ hPa; $z(p) =$ ——————————— [m]
   d. $p = 320$ hPa; $z(p) =$ ——————————— [m]
6. Use the SMTs to determine the sea-level pressure, given:
   Station elevation = 100 m
   Station pressure = 970 hPa
   Current temperature = 30 °C
   Twelve hour temperature = 18 °C
   Current dew point = 2 °C
   Latitude = 40 °N
7. Calculate the PA for the same station. Assume $\gamma = \beta_0$.
8. The elevation of the Mount Washington Observatory (KMWN), in New Hampshire, USA, is 1910 m ASL. What is its mean station pressure in the U.S. Standard Atmosphere?
9. Compute the vertical temperature lapse rate for a homogeneous atmosphere. Discuss the implications for stability indicated by your results.
10. Given an isothermal atmosphere, compute the pressure at an elevation of 2500 meters, with mean temperature of 15 °C, $p_0 = 1013.25$ hPa, and $g = 9.80665\ ms^{-2}$.
11. Given a constant lapse rate atmosphere equal to the USSA's Layer 0, and a surface temperature of 25 °C, compute the temperature at an elevation of 5000 meters above sea level.

12. Combining the relationships in the Ideal Gas Law (2.37), the description of a constant lapse-rate atmosphere (10.47), and the vertical pressure variation (10.60), derive an expression for density (ρ) as a function of height (z).

   a. Perform a unit analysis to show that your derived function has units of density [kg/m³].

   b. Using your derived function and USSA values for the constants, complete the following table (enter density values to four decimal places of precision):

Elevation [m]	Density [kg/m³]
0	
1000	
2000	
3000	
4000	
5000	
6000	
7000	
8000	
9000	
10,000	

   c. Compare your results to the values shown in Table 10.1 and comment on the differences.

11

Stability

11.1. Introduction

The term *stability* has many definitions. In this text we'll limit our scope to the most basic form of stability used in practical meteorology, sometimes called *static stability*, which is a measure of the buoyancy of a parcel in the surrounding atmosphere. If a parcel becomes positively buoyant, no external lifting force is needed for it to rise to great heights in the troposphere. If there's sufficient moisture in the lower troposphere, and an "exhaust mechanism" (divergence) in the upper troposphere, then strong upward vertical motion can result in severe weather.

Over the last sixty years, we've learned to recognize the potential for severe weather in the atmosphere hours before it actually occurs, by looking at radiosonde data, usually plotted on a thermodynamic diagram. In this chapter, we will review some of these methods. In Chapter 12 (the final chapter of this book), we'll add a few more mathematical tools and put them into action predicting thunderstorms, severe thunderstorms, and tornadoes.

11.2. Static Stability

At its base, stability is a statement about the vertical distribution of density in a column of fluid, such as air. If the fluid is distributed in such a way that the densest material is at the bottom of the column, and the lightest fluid is at the top, then the column is *statically stable* – gravity alone will not change the vertical distribution of the fluid. Furthermore, a statically stable atmosphere implies the near total absence of vertical motions, unless those motions are imposed by some external force.[1] A *denser* fluid sample, if lifted, will be surrounded by lighter fluid, and *sink* back to the point where gravity and buoyancy are in equilibrium. A *lighter* fluid sample, if it is caused to descend, will be surrounded by denser fluid, and *rise* back to the point where gravity and buoyancy are in equilibrium. By contrast, if the fluid is vertically homogenous – that is, with a fixed density,[2] independent of elevation – then the column is *statically unstable*, and very large vertical motions can take place.

If the fluid in the column is initially a mixture of two fluids with different densities, then after a sufficient amount of time passes, the two different fluids will separate, creating a *stratified* condition where the denser substance is at the bottom of the column and the thinner substance is at the top. (Think of oil and vinegar in Italian salad dressing, or the ocean underlying the atmosphere.) In this case we say that the two fluids have become *decoupled*; that is, motion in the overlying lighter fluid does not influence the underlying denser fluid very much. Stratification and decoupling are extreme forms of static stability that commonly occur in the atmosphere: temperature inversions of all kinds routinely create decoupled shallow layers of air near the surface, separating the overlying free troposphere from important interactions with the surface. This will be discussed in greater detail in the following text.

Assessment of static stability can be done using density. So that the comparison of air samples higher in the column to those selected from lower in the column is physically consistent, the densities of all samples should be converted to some standard basis, such as a "potential density" (ρ_0) all the samples of the air would have if they were brought to a fixed reference level. Using this approach, one could assume a fixed atmospheric pressure, such as 1000 hPa, and then convert the *in situ* temperatures for all air samples in the column to their potential temperatures (5.27). From here, the equivalent 1000-hPa densities for the air samples can be computed using their potential temperatures (in place of *in situ* temperatures) in the Ideal Gas Law (2.37). In other words:

$$\boxed{p_0 = \rho_0 R_d \theta} \tag{11.1}$$

where p_0 is equal to 10^5 [Pa], ρ_0 is the potential density of the air from a selected level [kg/m³], R_d is the individual gas constant for the dry-air ensemble (286.8 J/(kg K)),[3] and θ is the potential temperature of the air sample [K]. If the potential density of the air column decreases with height, then it is statically stable.

Another approach is to simply use potential temperature. If pressure is held constant, then the Ideal Gas Law (11.1) indicates that density and potential temperature are inversely proportional. With this in mind, the statements about the vertical potential temperature (isentropic) gradient shown in Table 11.1 can be used to describe static stability conditions.[4]

In the real atmosphere, water vapor changes the density of air. This is the reason for using the *virtual temperature* (8.5–8.6c) instead of simple temperature in many of the calculations, such as thickness (9.35 and 9.37). Similarly, using the equivalent potential temperature (θ_e) (8.26) in Table 11.1, rather than the simple potential temperature (θ), gives a slightly better result.

This formulation is good conceptually, but as written, it doesn't permit us to answer three important questions: (1) If the atmosphere becomes unstable, how much vertical

Table 11.1. *Basic definitions of static stability in terms of the vertical isentropic gradient for an unsaturated atmosphere.*

ρ_0 is the density a parcel would have if brought adiabatically to 1000 hPa

Stability Condition	Isentropic Gradient	Vertical Variation of θ	Vertical Variation of ρ_0
Statically unstable	$\dfrac{d\theta}{dz} < 0$	θ decreases with z	ρ_0 increases with z
Statically neutral	$\dfrac{d\theta}{dz} = 0$	θ independent of z	ρ_0 independent of z
Statically stable	$\dfrac{d\theta}{dz} > 0$	θ increases with z	ρ_0 decreases with z

acceleration will a parcel experience, (2) what is the quantitative *degree* of instability, and (3) what kind of sensible weather will result?

11.3. The Parcel Method and Vertical Acceleration

In our simplest model of vertical motion, we assume that the environment remains in hydrostatic balance, but isolated[5] parcels rising and falling within the environment are out of balance, making them either positively or negatively buoyant. This is called the *parcel method* of assessing stability, and it makes the following assumptions:

- Parcels are *isolated* and retain their identity during the lifting process. They do not exchange mass or energy with the surrounding environmental air.
- The vertical movement of parcels does not disturb the surrounding environmental air.
- All processes affecting a parcel's temperature are either dry or moist adiabatic.
- At every instant during the lifting process, the vertical pressure gradient force acting on a parcel is equal to the vertical pressure gradient force on the surrounding environment.

The first two assumptions are problematic. The concepts of closed and isolated parcels (separated from the surrounding environment by a membrane that only allows the passage of energy in the former, and nothing in the latter) seem to be rather fantastic constructs to begin with, but we have been using them all the way through, and they seem to produce results that are serviceable. The second assumption is just as difficult to believe. It's hard to imagine a rising bubble of air moving upward through the environment, with no ambient air rushing in to fill the empty void left behind by the departed parcel. In more advanced models of convection, some effort is made to compensate for these problems.[6] The third assumption is also flawed but to a much lesser extent. There were very few unwarranted assumptions made in the derivation of the Dry Adiabatic Lapse Rate, other than those inherent in the application of the Ideal

Gas Law. There were several simplifying assumptions made in the derivation of the Moist Adiabatic Lapse Rate but most of them seemed fairly reasonable. The fourth assumption is essentially correct.

We'll begin by describing the environment, which we assume is in hydrostatic balance. In Chapter 5, we derived the hydrostatic relation (5.77), which states that:

$$\frac{dp}{dz} = -\rho g$$

where dp/dz is the vertical pressure gradient [Pa/m], ρ is the density of air ($\cong 1.23 \text{ kg/m}^3$ near sea level),[7] and g is effective gravitational acceleration ($\cong 9.81$ m/s^2 in the Midlatitudes).[8] A unit analysis of this equation will show that both sides have units of force per unit volume [N/m^3]. Thus, if volume is held constant, it is equivalent to a balance of forces equation.

The hydrostatic relation represents a balance between the vertical pressure gradient force (pushing the mass of the atmosphere away from Earth's surface) and effective gravity (pulling the mass of the atmosphere toward Earth's surface). If the forces exactly balance, then there is no net force acting on the environmental air. This relationship can be rewritten in a more useful form by adding ρg to both sides, resulting in:

$$\frac{dp}{dz} + \rho g = 0 \tag{11.2}$$

If a parcel is *not* in hydrostatic balance, then there *is* a net, unbalanced force acting on it. This unbalanced force may be the result of the difference between the density of the environmental air and the density of the air in the isolated parcel. In this case, the equivalent equation for the parcel is:

$$\frac{dp}{dz} + \acute{\rho} g = -\acute{\rho}\frac{d^2 z}{dt^2} \tag{11.3}$$

where the accent symbol above density indicates that it refers to the *parcel*, and the new term on the RHS consists of the parcel density ($\acute{\rho}$) and the second derivative of vertical position with respect to time, that is, the vertical acceleration. The LHS of (11.3) represents the vertical pressure gradient force pushing the parcel upward (dp/dz), and the force of gravity pulling the parcel downward ($\acute{\rho} g$). Note that we use p, not $\acute{p}$, and are therefore assuming the pressure gradient force acting on the parcel and the pressure gradient force acting on the environment are identical. The RHS of (11.3) represents the remaining net force from the imbalance between the upward vertical pressure gradient imposed by the environment, and the gravitational force acting on the mass of the parcel.

Recall the basic kinematic definition that states:

$$a \equiv \frac{dw}{dt} = \frac{d^2z}{dt^2} \qquad (11.4)$$

where a is acceleration [m/s²], w is the linear vertical velocity [m/s], and z is vertical position [m]. We can rewrite (11.3) in terms of changes to the vertical velocity by substituting the center definition shown in (11.4), yielding:

$$\frac{dp}{dz} + \acute{\rho}g = -\acute{\rho}\frac{dw}{dt} \qquad (11.5)$$

Multiplying through by −1 results in:

$$-\left(\frac{dp}{dz} + \acute{\rho}g\right) = \acute{\rho}\frac{dw}{dt} \qquad (11.6)$$

or:

$$-\frac{dp}{dz} - \acute{\rho}g = \acute{\rho}\frac{dw}{dt} \qquad (11.7)$$

The objective is to compute the vertical acceleration by comparing the densities of the environmental air and the parcel air, so we'll add (11.7) to (11.2) in order to eliminate the common terms:

$$0 = +\frac{dp}{dz} + \rho g \qquad \left.\right\}(11.2)$$

$$+\left(\acute{\rho}\frac{dw}{dt} = -\frac{dp}{dz} - \acute{\rho}g\right) \left.\right\}(11.7)$$

$$\overline{\qquad\qquad\qquad\qquad} \qquad (11.8)$$

$$\acute{\rho}\frac{dw}{dt} = \rho g - \acute{\rho}g$$

or:

$$\frac{dw}{dt} = \frac{\rho g - \acute{\rho}g}{\acute{\rho}} \qquad (11.9)$$

or finally:

$$\boxed{\frac{dw}{dt} = g\left(\frac{\rho - \acute{\rho}}{\acute{\rho}}\right)} \qquad (11.10)$$

where *dw/dt* is the vertical acceleration acting on the isolated parcel [m/s²], *g* is the effective gravitational acceleration [m/s²], ρ is the density of a sample of environmental air outside the parcel [kg/m³], and $\acute{\rho}$ is the density of the air inside the parcel [kg/m³]. Note that the units inside the parenthesis on the RHS cancel. According to (11.10):

- If the density of the environmental air is greater than the density of the parcel air, there is a net upward (positive) acceleration.
- If the density of the environmental air is lower than the density of the parcel air, there is a net downward (negative) acceleration.

This doesn't necessarily imply that *motion* (velocity) will be upward or downward. The relation only describes the force acting to *change* the velocity. Positive (upward) vertical motion may still occur, even in the presence of a negative (downward) acceleration, if some other force (such as a front or a topographic obstruction) is responsible for the initial positive velocity.

Vertical acceleration as a function of virtual temperature. Densities are not usually reported in base radiosonde data, nor are they plotted on thermodynamic diagrams, so it would be useful to have an expression for the vertical acceleration of a parcel in terms of more common quantities. The obvious choice is virtual temperature, because it's related to density through the Ideal Gas Law.

According to (8.7):

$$p = \rho R_d T_v \tag{11.11a}$$

for the environment, and:

$$p = \acute{\rho} R_d \acute{T}_v \tag{11.11b}$$

for the parcel. Inverting these to solve for density, we have:

$$\rho = \frac{p}{R_d T_v} \tag{11.12a}$$

and:

$$\acute{\rho} = \frac{p}{R_d \acute{T}_v} \tag{11.12b}$$

Substituting (11.12a) and (11.12b) into (11.10), we obtain:

$$\frac{dw}{dt} = g \left[\frac{\left(\dfrac{p}{R_d T_v}\right) - \left(\dfrac{p}{R_d \acute{T}_v}\right)}{\left(\dfrac{p}{R_d \acute{T}_v}\right)} \right] \tag{11.13}$$

After canceling p and R_d, we have:

$$\frac{dw}{dt} = g\left[\frac{\left[\left(\dfrac{1}{T_v}\right) - \left(\dfrac{1}{\acute{T}_v}\right)\right]}{\left(\dfrac{1}{\acute{T}_v}\right)}\right] \tag{11.14}$$

which is a bit awkward. We can clean this up by creating a common denominator for all the terms on the RHS:

$$\frac{dw}{dt} = g\left[\frac{\left[\left(\dfrac{\acute{T}_v}{\acute{T}_v}\right)\left(\dfrac{1}{T_v}\right) - \left(\dfrac{T_v}{T_v}\right)\left(\dfrac{1}{\acute{T}_v}\right)\right]}{\left(\dfrac{T_v}{T_v}\right)\left(\dfrac{1}{\acute{T}_v}\right)}\right] \tag{11.15}$$

Pulling the common denominator out of the square brackets yields:

$$\frac{dw}{dt} = g\frac{\left(\dfrac{1}{T_v \acute{T}_v}\right)}{\left(\dfrac{1}{T_v \acute{T}_v}\right)}\left[\frac{\acute{T}_v - T_v}{T_v}\right] \tag{11.16}$$

Simplifying the compound fraction results in:

$$\frac{dw}{dt} = g\frac{\left(T_v \acute{T}_v\right)}{\left(T_v \acute{T}_v\right)}\left[\frac{\acute{T}_v - T_v}{T_v}\right] \tag{11.17}$$

which finally results in:

$$\boxed{\frac{dw}{dt} = g\left(\frac{\acute{T}_v - T_v}{T_v}\right)} \tag{11.18}$$

Once again we can ask the question of whether it is really necessary to use the virtual temperature, because computing it is complex. Would using the *in situ* temperature make a big difference? Let's take the example of environmental air with *in situ* temperature of $-15\,°C$ (258.16 K) and dew point of $-20\,°C$ (253.16 K). The corresponding parcel has a temperature of $-10\,°C$ (263.16 K) and is saturated. Ambient pressure is 500 hPa. By (7.46), the environmental vapor pressure is:

$$e = e_0 \exp\left[\frac{l_v}{R_v}\left(\frac{1}{T_0} - \frac{1}{T_d}\right)\right]$$

$$= (611.12)\exp\left[\left(\frac{2.5366 \times 10^6}{461.2}\right)\left(\frac{1}{273.16} - \frac{1}{253.16}\right)\right] = 124.54 Pa$$

where the value of the latent heat of vaporization (l_v) came from Table 4.2, and the value of R_v came from (2.68). By (7.67), the environmental water vapor mixing ratio is:

$$w = \frac{e\varepsilon}{p-e} = \frac{(124.54 \times 0.622)}{(5 \times 10^4 - 124.54)} = 0.00155 \frac{kg_{vapor}}{kg_{dry\,air}}$$

The environmental virtual temperature is then computed by (8.5):

$$T_v = T\left(\frac{1+\frac{w}{\varepsilon}}{1+w}\right) = (258.16)\left(\frac{1+\frac{0.00155}{0.622}}{1+0.00155}\right) = 258.40 K$$

For the *parcel*:

$$\acute{e} = e_0 \exp\left[\frac{l_v}{R_v}\left(\frac{1}{T_0} - \frac{1}{T_d}\right)\right]$$

$$= (611.12)\exp\left[\left(\frac{2.5247 \times 10^6}{461.2}\right)\left(\frac{1}{273.16} - \frac{1}{263.16}\right)\right] = 285.36\ Pa$$

therefore:

$$\acute{w} = \frac{e\varepsilon}{p-e} = \frac{(285.36 \times 0.622)}{(5 \times 10^4 - 285.36)} = 0.00357 \frac{kg_{vapor}}{kg_{dry\,air}}$$

and:

$$\acute{T}_v = \acute{T}\left(\frac{1+\frac{\acute{w}}{\varepsilon}}{1+\acute{w}}\right) = (263.16)\left(\frac{1+\frac{0.00357}{0.622}}{1+0.00357}\right) = 263.73\ K$$

We then plug these into (11.18) to obtain:

$$\frac{dw}{dt} = g\left(\frac{\acute{T}_v - T_v}{T_v}\right) = 9.81\left[\frac{m}{s^2}\right]\left(\frac{263.73 - 258.40}{258.40}\right) = 0.202\left[\frac{m}{s^2}\right]$$

If we simply substitute the *in situ* parcel and environment temperatures for the corresponding virtual temperatures, we find that:

$$\frac{dw}{dt} \cong g\left(\frac{\acute{T}-T}{T}\right) = 9.81\left[\frac{m}{s^2}\right]\left(\frac{263.16-258.16}{258.16}\right) = 0.190\left[\frac{m}{s^2}\right]$$

The difference between the two results is about 5 percent. The acceleration computed using the virtual temperature is a little larger, because it includes the effects of water vapor on both the parcel and the environmental air, but an error of only 5 percent is probably acceptable in most applications. In settings where numerical models are run hundreds of hours into the future, this error could become problematic.

Vertical acceleration as a function of specific volume. Vertical acceleration can also be expressed in terms of the difference between the specific volumes of the ambient environmental air and of the air inside the parcel. Beginning with the definition of specific volume from Table 1.4, we can rewrite the density of the ambient air and the parcel air by:

$$\alpha \equiv \frac{1}{\rho} \therefore \rho = \frac{1}{\alpha} \tag{11.19a}$$

and:

$$\acute{\alpha} = \frac{1}{\acute{\rho}} \therefore \acute{\rho} = \frac{1}{\acute{\alpha}} \tag{11.19b}$$

where the primed quantities correspond to the isolated parcel. Substituting these into (11.10) we obtain:

$$\frac{dw}{dt} = g\left(\frac{\frac{1}{\alpha}-\frac{1}{\acute{\alpha}}}{\frac{1}{\acute{\alpha}}}\right) \tag{11.20}$$

Creating a common denominator results in:

$$\frac{dw}{dt} = g\left[\frac{\left(\frac{\acute{\alpha}}{\acute{\alpha}}\right)\frac{1}{\alpha}-\left(\frac{\alpha}{\alpha}\right)\frac{1}{\acute{\alpha}}}{\left(\frac{\alpha}{\alpha}\right)\frac{1}{\acute{\alpha}}}\right] \tag{11.21}$$

or:

$$\frac{dw}{dt} = g \frac{\left(\dfrac{1}{\acute{\alpha}\,\alpha}\right)}{\left(\dfrac{1}{\acute{\alpha}\,\alpha}\right)} \left[\frac{\acute{\alpha} - \alpha}{\alpha}\right] \tag{11.22}$$

and therefore:

$$\frac{dw}{dt} = g \left(\frac{\acute{\alpha}\,\alpha}{\acute{\alpha}\,\alpha}\right) \left[\frac{\acute{\alpha} - \alpha}{\alpha}\right] \tag{11.23}$$

and finally:

$$\boxed{\frac{dw}{dt} = g \left(\frac{\acute{\alpha} - \alpha}{\alpha}\right)} \tag{11.24}$$

which states that the vertical acceleration of a parcel is positive if the parcel's specific volume is greater than an equivalent volume of ambient air. This relationship forms the basis of our estimates of potential energy stored in a vertical temperature profile, or *Convectively Available Potential Energy*, which we will return to in Chapter 12.

11.4. Qualitative Estimates of Stability Using a Thermodynamic Diagram

Equation (11.18) shows that the vertical acceleration of a parcel at a given vertical point in the atmosphere is proportional to the difference between the virtual temperature of the environmental air around the parcel, and the virtual temperature inside the parcel. If the virtual temperature of the parcel is higher, the parcel experiences a positive (upward) vertical acceleration. If the vertical acceleration is positive, we classify the atmosphere as unstable. If it is negative, we classify it as stable.

The analysis that followed the derivation of (11.18) showed that, with only a small loss in accuracy, temperature can be substituted for virtual temperature. This idea forms the basis of computing stability using a *stability index*. The purpose of a stability index is to quantitatively compare lifted parcels to the surrounding environment. In practice, this means working with observed radiosonde data, usually visualized on a thermodynamic diagram (or analyzed automatically by a computer), and selecting parcels to move vertically. (Sometimes, an entire layer of the lower atmosphere is lifted). A parcel near the surface is usually chosen. Once the parcel is moved to one or more reference pressures, it is compared to the environmental temperature. The temperature difference, or some quantity derived from it, is then compared to an empirically derived table of weather outcomes. This will be discussed in greater detail in the following text.

Lifting parcels. Recall that, in our simplified model of the atmosphere, an isolated parcel cools adiabatically by expanding as it rises. Because heat is conserved in a purely adiabatic process, the relation shown in (5.2) applies:

$$-c_v \, dT = p \, d\alpha$$

where c_v is the specific heat of dry air at constant volume (717.0 J/(kg K)),[9] dT is a small change in the parcel's internal temperature [K], p is the pressure [Pa], and $d\alpha$ is a small change in the parcel's specific volume [m³/kg]. *This is a special case of the First Law of Thermodynamics,* and it implies:

- If the parcel expands ($d\alpha > 0$), work is performed on the environment ($dw > 0$), and the parcel temperature will decrease ($dT < 0$).
- If the parcel is compressed ($d\alpha < 0$), the environment performs work on the parcel ($dw < 0$), and the parcel temperature will increase ($dT > 0$).

After a certain amount of algebra and calculus, we arrived at an expression for the Dry Adiabatic Lapse Rate (Γ_D), which describes the rate at which an unsaturated parcel cools as it rises. According to (5.83) and (5.84):

$$\Gamma_D = -\frac{dT}{dz} = \frac{g}{c_p} = \frac{9.81\left[\frac{m}{s^2}\right]}{1003.8\left[\frac{J}{kg\,k}\right]} = 0.00977\left[\frac{K}{m}\right]$$

where g is effective gravitational acceleration ($\cong 9.81$ m/s² in the Midlatitudes[10]), and c_p is the specific heat of dry air at constant pressure (1003.8 J/(kg K)).[11] In round numbers, this means that an unsaturated parcel of air cools at a rate of about 10 K/km as it rises. The numerical value of Γ_D is *approximately* constant with height, and the small variation that does occur is due to elevation-dependent variations in g and temperature-dependent variations in c_p.[12]

If a parcel is not saturated before it starts to rise, and it cools adiabatically until its internal temperature drops down to its internal dew point (i.e., it reaches the *Lifted Condensation Level,* or LCL), the condensation of its water vapor load begins to release latent heat. In our simplified model of this extremely complex scenario, we assumed:

- An isolated compound parcel with water vapor coexisting in equilibrium with liquid water;
- As the vapor condenses, the resulting liquid drops remain inside the parcel, so that when they evaporate again, they can take up exactly the same amount of latent heat that they released when they condensed. This makes the process reversible and allows the heat to be conserved;

- The combined mass of the parcel's vapor and dry air is equal to one kilogram;
- All latent heat released during condensation is taken up by the parcel's *dry air*, and none of it goes into the parcel's remaining load of water vapor;
- The total atmospheric pressure (p) is much greater than the saturation vapor pressure (e_s), so that $p - e_s \cong p$;
- The *approximation* for the saturation mixing ratio (w_s) is an *equality* given by $e_s \varepsilon / p$, and equivalently, $e_s = p w_s / \varepsilon$. Epsilon ($\varepsilon$) is the ratio of the molar mass of water vapor to the molar mass of the dry air ensemble, equal to 0.622; and,
- Temperature varies only as a function of elevation.

After more algebra and calculus (involving the first *and* second laws of thermodynamics; see Chapter 8 and Appendix 7), we arrived at an expression for the Moist Adiabatic Lapse Rate (Γ_M), which describes the rate at which the *saturated* parcel cools as it rises. According to (8.28):

$$\Gamma_M = \Gamma_D \frac{\left(\dfrac{l_v w_s}{R_d T} + 1 \right)}{\left(\dfrac{l_v^2 \varepsilon w_s}{c_p R_d T^2} + 1 \right)}$$

Unlike the dry rate, the Moist Adiabatic Lapse Rate is not a constant, but in fact varies dramatically, primarily because of temperature variations. Variation in Γ_M with temperature was examined following the derivation of the expression, and it was shown that, in warm temperature regions (10 °C in the first example), Γ_M is about half Γ_D. As the parcel's temperature decreases, it becomes progressively less capable of carrying water vapor (as $T \downarrow$, $e_s \downarrow$), and the amount of latent heat released by condensation steadily decreases. This causes Γ_M to increase; that is, the cooling rate of a saturated parcel under very cold conditions increases as the temperature decreases (as $T \downarrow$, $\Gamma_M \uparrow$), because the offsetting of cooling by expansion from latent heat release is reduced. By the time the parcel temperature reaches $\cong -40$ °C, the amount of water vapor it can carry converges toward zero, and Γ_M converges to Γ_D.[13]

In our simplified model of the atmosphere, a rising parcel of air can only cool by one of these two rates. Below the LCL, it cools dry adiabatically. Above the LCL, if lift continues, it cools moist adiabatically. The method for lifting a parcel to the LCL was discussed in Chapter 8, Figure 8.4.

Stability classes and the environmental lapse rate. The *environmental lapse rate* (γ_e) is a result of many factors, among them the horizontal mixing of air masses, advection of air from different directions at different altitudes, diabatic heating and cooling, and evaporation and condensation. Above any given radiosonde station in the Midlatitudes, γ_e in the troposphere varies by time of day, time of year, and vertical location. On a hot summer afternoon, the lowest kilometer of the atmosphere may cool with height at a rate exceeding Γ_D, which then usually slows significantly in the middle troposphere. On a cold winter morning, the lowest few hundred meters often experience a negative environmental lapse rate, meaning temperature increases with height, then reverses itself and begins cooling with height in the middle troposphere.

In all cases, we can classify the stability implied by the environmental lapse rate by comparing the resulting temperature at a given altitude to the temperature of a parcel (usually, but not always from the surface) lifted to the same altitude.

- If the lifted parcel is *warmer* than the environmental air, then by (11.18), the parcel experiences a *positive* (upward) vertical acceleration. Further, using the Ideal Gas Law, this implies that the density of the parcel is lower than the density of the environmental air, meaning it is positively buoyant, and by (11.10) it is positively accelerated. *In this case, we classify the atmosphere as unstable.*
- If the lifted parcel is *colder* than the environmental air, then by (11.18), the parcel experiences a *negative* (downward) vertical acceleration. Further, using the Ideal Gas Law, this implies that the density of the parcel is higher than the density of the environmental air, meaning it is negatively buoyant, and by (11.10) it is negatively accelerated. *In this case, we classify the atmosphere as stable.*
- If the lifted parcel is *the same temperature* as the environmental air, then by (11.18), the parcel experiences a *net zero* vertical acceleration. Further, using the Ideal Gas Law, this implies that the density of the parcel is the same as the density of the environmental air, meaning it is neutrally buoyant, and by (11.10) it receives a net zero acceleration. *In this case, we classify the atmosphere as neutral. Remember that this does not necessarily imply zero vertical motion – just zero vertical acceleration.*

Figure 11.1 shows a Skew-T with the dry (Γ_D) and moist (Γ_M) rates applied to a surface parcel at the USSA/ISA standard 1000 hPa temperature of about 13.5 °C.[14] The two parcel lapse rates represent the extremes of the parcel's thermodynamic path through the ambient atmosphere. If completely saturated at the surface, it will follow the moist rate on the right. If completely dry, it will follow the dry rate on the left. Most parcels will follow an intermediate path between the two, because they will rise dry adiabatically (cooling rapidly) until reaching saturation at the LCL, and then rise moist adiabatically (cooling more slowly).

Figure 11.1 also shows five hypothetical *environmental* lapse rates, labeled γ_{au}, γ_{dn}, γ_c, γ_{mn}, and γ_{as} (dashed red lines). As a rising parcel moves upward in these different atmospheres, there are five very different results. The following discussion is summarized in Table 11.2.

Absolutely unstable atmosphere. To the left of the dry rate, the environmental temperature falls off more rapidly than the temperature of either an unsaturated parcel (following Γ_D) or a saturated parcel (following Γ_M). That means, in all cases, the relation in (11.18) becomes:

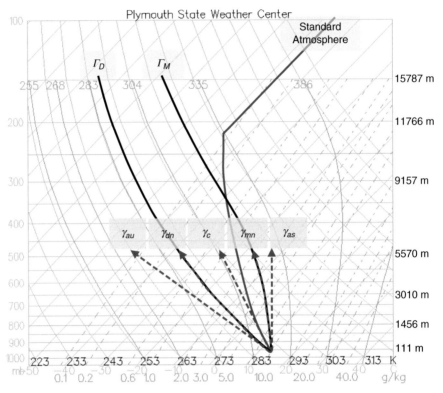

Figure 11.1. Static stability classes. Dry (Γ_D) and moist (Γ_M) adiabatic rates (black lines) are shown applied to a parcel with USSA/ISA standard 1000 hPa temperature of $\cong 13.5\,°C$, representing extremes of a parcel's thermodynamic pathway, as it moves upward through environmental air. Five sample environmental lapse rates (dashed red lines) are shown. γ_{au} indicates an absolutely unstable atmosphere; γ_{dn}, a dry neutral atmosphere; γ_c a conditionally stable atmosphere; γ_{mn}, a moist neutral atmosphere; γ_{as}, an absolutely stable atmosphere. The vertical temperature profile of the USSA/ISA is also shown (solid red), as are standard heights for selected pressure levels (black numbers on right). Background image courtesy of Plymouth State University meteorology program (2012).

$$\frac{dw}{dt} = g\left(\frac{T'_v - T_v}{T_v}\right) \cong g\left(\frac{T' - T}{T}\right) > 0 \qquad (11.25)$$

That is, the parcel receives a thermodynamically driven positive (upward) acceleration, independent of any outside lifting force. Because the saturation condition of the parcel is unimportant (11.25 is true for both unsaturated and saturated parcels), we refer to this condition as *absolutely unstable*. The environmental lapse rate associated with an absolutely unstable condition can easily be recognized as one that exceeds the dry rate, that is, is *superadiabatic*, or

$$\boxed{\Gamma_M < \Gamma_D < \gamma_{au}} \qquad (11.26)$$

where the subtext *au* indicates *absolutely unstable*. If the environmental lapse rate exceeds 34.2 K/km,[15] the density of the environmental air actually *increases* with height, triggering *autoconvection* – a condition where vertical motion (initially downward, followed by upward return flow) begins spontaneously, without the necessity of an external lifting force.[16]

Recall that dry adiabats are also isentropes (lines of equal potential temperature),[17] so an environmental lapse rate to the left of Γ_D means the potential temperature is decreasing with height, or $d\theta/dz < 0$, which is the first condition listed in Table 11.1. Absolutely unstable conditions occur in the Midlatitudes on hot summer days, when diabatic heating raises the surface temperature to a much greater degree than the overlying atmosphere. They are usually visible in late afternoon soundings, affecting the lowest kilometer of the troposphere.

Figure 11.2 shows a temperature profile recorded at Albuquerque, New Mexico (KABQ), a station in the high desert of the southwestern United States. The RAOB was taken in mid-April, and shows a superadiabatic lapse rate in the lowest 250 meters of the profile. Above that, the environmental lapse rate slows somewhat, becoming approximately equal to the Dry Adiabatic Lapse Rate for the next one and a half vertical kilometers. Figure 11.3 shows the meteogram for KABQ for the twenty-four-hour period prior to the valid time of the RAOB plotted in Figure 11.2. The surface weather observations (METAR) summarized by the meteogram indicate mostly clear conditions for the majority of the sunlight hours during the day prior to the RAOB. From this we can conclude that the diabatic heating of Earth's surface by the Sun was probably the cause of the very unstable conditions in the lowest two kilometers of the troposphere above KABQ.

Dry neutral atmosphere. A closely related case occurs when the environmental lapse rate is exactly equal to the Dry Adiabatic Lapse Rate (Γ_D). In this case, a lifted, *unsaturated* parcel (cooling at Γ_D) will have the same temperature as the surrounding environmental air, and by (11.18):

$$\left(\frac{dw}{dt}\right)_D = g\left(\frac{\acute{T}_v - T_v}{T_v}\right) \cong g\left(\frac{\acute{T} - T}{T}\right) = 0 \tag{11.27a}$$

will *not* experience a thermodynamically driven vertical acceleration. (The subscript *D* means that this applies to an unsaturated or "dry" parcel.) However, a lifted, *saturated* parcel (cooling at Γ_M) will be warmer than the surrounding air aloft, so that by:

$$\left(\frac{dw}{dt}\right)_M = g\left(\frac{\acute{T}_v - T_v}{T_v}\right) \cong g\left(\frac{\acute{T} - T}{T}\right) > 0 \tag{11.27b}$$

the saturated parcel *will* experience a positive (upward) vertical acceleration (subscript *M* implies a saturated or "moist" parcel). The relation shown in (11.27b) will

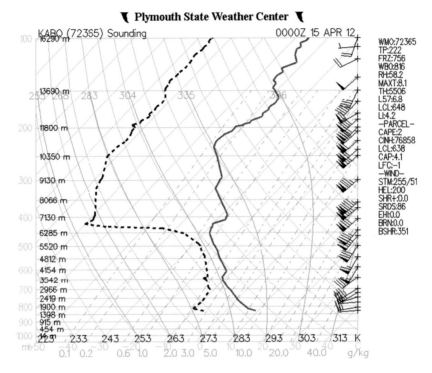

Figure 11.2. Absolutely unstable atmosphere. Plotted radiosonde observation for Albuquerque, New Mexico (KABQ/72365). (KABQ is the International Civil Aviation Organization identifier; 72365 is the World Meteorological Organization identifier.) The time zone is UTC – seven hours, so this RAOB was recorded at 1700 LST (late afternoon). The KABQ station elevation is about 1600 meters ASL. The superadiabatic region is visible in the lowest 25 hPa ($\cong 250$ m) above the station. *In situ* heights for several pressure surfaces are plotted (in black) next to the corresponding pressure on the left, vertical axis. Characters on the right represent the output from automated analysis routines. (Plot courtesy of Plymouth State University meteorology program, 2012.)

also be true for an initially unsaturated (dry) parcel, that, when lifted, reaches its saturation point (the LCL) and continues ascent through the Moist Adiabatic Lapse Rate.

In the dry-neutral case, the relationship between the parcel and environmental lapse rates is described by:

$$\boxed{\Gamma_M < \gamma_{dn} = \Gamma_D} \tag{11.28}$$

where the subtext *dn* indicates *dry neutral*. Because the dry adiabats represent constant potential temperature, an environmental lapse rate equal to Γ_D implies that $d\theta/dz = 0$, which is the second condition listed in Table 11.1.

Figure 11.4 shows a temperature profile recorded at Maniwaki, Quebec (CWMW), a station in southeastern Canada. The RAOB was taken in late May, and shows a dry-neutral environmental lapse rate in the lowest two kilometers of the profile.

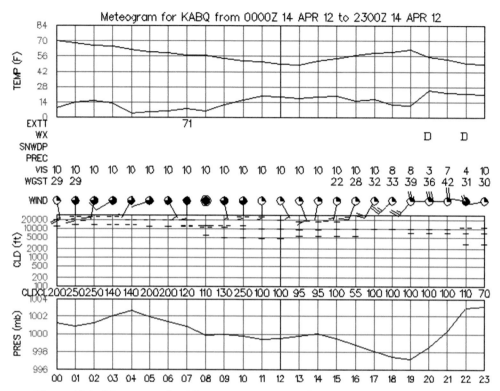

Figure 11.3. Meteogram for Albuquerque, New Mexico (KABQ). Total cloud cover is encoded in the station circles with the plotted wind barbs. (Meteogram courtesy of Plymouth State University meteorology program, 2012.)

Above that, near 780 hPa, the vertical temperature profile indicates a weak subsidence inversion. Figure 11.5 shows the synoptic surface analysis valid twelve hours prior to the time of the RAOB plotted in Figure 11.4. The surface analysis indicates the presence of an anticyclone in the southeastern part of Hudson's Bay, with a ridge extending southward into the Midwest part of the United States. Anticyclones are usually associated with cloudless skies; so once again, we can conclude that the low-level instability shown in Figure 11.4 was caused by the diabatic heating of Earth's surface.

Conditional atmosphere. *In between the dry and moist rates*, the environmental temperature falls off more slowly than the temperature of an unsaturated parcel (following Γ_D), but more rapidly than the temperature of saturated parcel (following Γ_M). That means, for *unsaturated* parcels, (11.18) becomes:

$$\left(\frac{dw}{dt}\right)_D = g\left(\frac{\acute{T}_v - T_v}{T_v}\right) \cong g\left(\frac{\acute{T} - T}{T}\right) < 0 \qquad (11.29a)$$

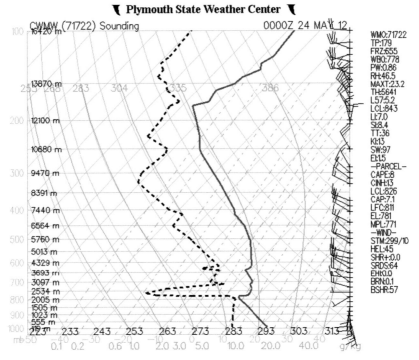

Figure 11.4. Dry neutral atmosphere. Plotted radiosonde observation for Maniwaki, Quebec (CWMW/71722). The time zone is UTC – five hours, making this RAOB representative of late May 2012, at 1900 LST. The dry neutral region is visible from the surface to about 780 hPa. (Plot courtesy of Plymouth State University meteorology program, 2012.)

and for *saturated* parcels, (11.18) becomes:

$$\left(\frac{dw}{dt}\right)_M = g\left(\frac{\acute{T}_v - T_v}{T_v}\right) \cong g\left(\frac{\acute{T} - T}{T}\right) > 0 \tag{11.29b}$$

As with the dry-neutral environmental lapse rate, the thermodynamically driven acceleration the parcel receives experiences is dependent on its saturation condition, but in this case:

- If the parcel is *saturated*, it becomes warmer than the environment as it rises, and it experiences a *positive* (upward) acceleration.
- If the parcel is *unsaturated*, it becomes colder than the environment as it rises, and it experiences a *negative* (downward) acceleration.

Because the sign of the acceleration is dependent on the saturation of the parcel, we refer to this condition as *conditionally stable*, or sometimes simply *conditional*. The environmental lapse rate associated with a conditional stability state is related to the parcel lapse rates by:

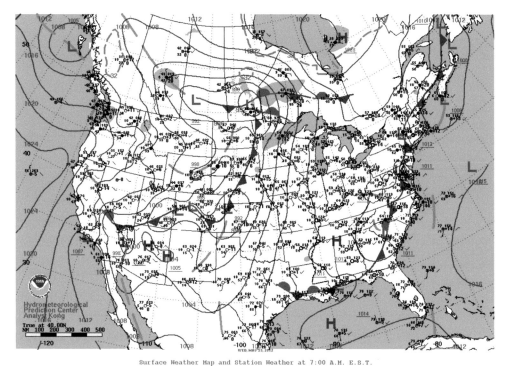

Surface Weather Map and Station Weather at 7:00 A.M. E.S.T.

Figure 11.5. Surface analysis valid, May 23, 2012, 1200 UTC (0700 EST), twelve hours prior to the RAOB plotted in Figure 11.4. An anticyclone is centered over northern Quebec, with a ridge extending southward into the Midwest United States. (Graphic courtesy of Hydrometeorological Prediction Center, 2012.)

$$\boxed{\Gamma_M < \gamma_c < \Gamma_D} \tag{11.30}$$

where the subtext *c* indicates *conditional*. The environmental lapse rate is to the right of Γ_D, so the potential temperature is increasing with height or $d\theta/dz > 0$, which is the third condition listed in Table 11.1. In this case, however, the role of moisture and the latent heat it releases when it condenses becomes determinative. With an environmental lapse rate of $\cong 6.5$ K/km, the USSA/ISA standard troposphere (shown as solid red line in Figure 11.1) is conditionally stable.

Moist neutral atmosphere. The next case occurs when the environmental lapse rate is exactly equal to the Moist Adiabatic Lapse Rate (Γ_M). In this case, a lifted, *unsaturated* parcel (cooling at Γ_D) will have a lower temperature than the surrounding environmental air, and by:

$$\left(\frac{dw}{dt}\right)_D = g\left(\frac{\acute{T}_v - T_v}{T_v}\right) \cong g\left(\frac{\acute{T} - T}{T}\right) < 0 \tag{11.31a}$$

will experience a thermodynamically driven negative (downward) vertical acceleration. However, a lifted, *saturated* parcel (cooling at Γ_M) will have the same temperature as the surrounding air aloft, so that by (11.18):

$$\left(\frac{dw}{dt}\right)_M = g\left(\frac{\acute{T}_v - T_v}{T_v}\right) \cong g\left(\frac{\acute{T} - T}{T}\right) = 0 \qquad (11.31b)$$

the saturated *will not* experience a vertical acceleration. In this case, the relationship between the parcel and environmental lapse rates is described by:

$$\boxed{\Gamma_M = \gamma_{mn} < \Gamma_D} \qquad (11.32)$$

where the subtext *mn* indicates *moist neutral*. Because the dry adiabats represent constant potential temperature, an environmental lapse rate greater than Γ_D implies that $d\theta / dz > 0$, which is the third condition listed in Table 11.1.

Figure 11.6 shows a temperature profile recorded at Gray, Maine (KGYX), a station in the northeastern United States, about 25 km north of Portland. The RAOB was taken in late May and shows a moist-neutral environmental lapse rate in the six and a half km vertical region between 700 and 300 hPa. Above about 700 hPa, the vertical temperature profile stabilizes further. Figure 11.5 shows the synoptic surface analysis valid at the same time RAOB was recorded, and it indicates that Gray was in the warm sector of a small wave cyclone propagating northward along a quasistationary front.

Absolutely stable atmosphere. To the right of the moist rate, the environmental temperature falls off more slowly than both that of an unsaturated parcel (following Γ_D) and a saturated parcel (following Γ_M). That means, in all cases, the relation in (11.18) becomes:

$$\frac{dw}{dt} = g\left(\frac{\acute{T}_v - T_v}{T_v}\right) \cong g\left(\frac{\acute{T} - T}{T}\right) < 0 \qquad (11.33)$$

In this case, the parcel receives a thermodynamically driven negative (downward) acceleration, independent of outside lifting forces. Because the saturation condition of the parcel is unimportant (11.33 is true for both unsaturated and saturated parcels), we refer to this condition as *absolutely stable*. The environmental lapse rate associated with an absolutely stable condition is related to the parcel lapse rates by:

$$\boxed{\gamma_{as} < \Gamma_M < \Gamma_D} \qquad (11.34)$$

where the subtext *as* indicates *absolutely stable*. Because the environmental lapse rate is also to the right of Γ_D, indicating that the potential temperature is increasing with height, or $d\theta/dz > 0$, this also corresponds to the third condition listed in Table 11.1.

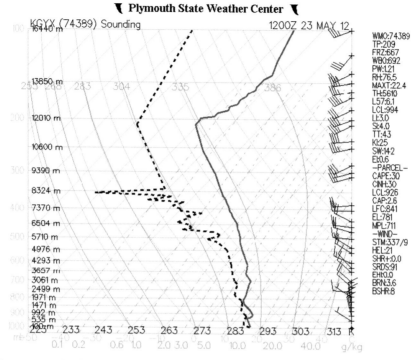

Figure 11.6. Moist neutral atmosphere. Plotted radiosonde observation for Gray, Maine (KGYX/74389). The time zone is UTC – five hours, making this RAOB representative of late May 2012, at 0700 LST. The moist neutral region is visible from 700 hPa to about 350 hPa. (Plot courtesy of Plymouth State University meteorology program, 2012.)

Temperature inversions are always absolutely stable regions, but an inversion isn't necessary to produce absolute stability. With an environmental lapse rate either equal to zero, or actually negative (indicating that the temperature *increases* with height), the entire USSA/ISA stratosphere is absolutely stable (Figure 11.1). In Figure 11.6, the region above KGYX from about 260 to about 210 hPa is absolutely stable, that is, the environmental lapse rate is less than the Moist Adiabatic Lapse Rate, but it is not a temperature inversion. Because absolutely stable temperature profiles impose a negative acceleration on air parcels, ventilation of the air near Earth's surface to greater heights is suppressed, and poor air quality is often the result in urban areas.

Figures 11.7, 11.9, and 11.10 show regions of absolute stability resulting from inversions. The first is a radiation inversion, which is the most common type. The RAOB was recorded above Wilmington, Ohio (KILN), on May 23, 2012, at 1200 UTC (0700 LST). Figure 11.5, which shows the synoptic surface analysis valid at the same time as the RAOB, indicates that KILN is under the influence of an anticyclone

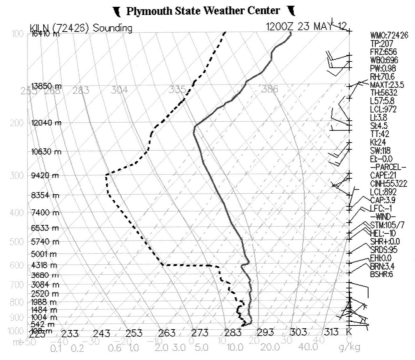

Figure 11.7. Absolutely stable atmosphere: Radiation inversion. Plotted radiosonde observation for Wilimington, Ohio (KILN/72426). The time zone is UTC – five hours, making this RAOB representative of late May 2012, at 0700 LST. The radiation inversion is visible in the lowest 25 hPa of the sounding. (Plot courtesy of Plymouth State University meteorology program, 2012.)

centered in southeastern Canada. Anticyclones are usually associated with partly cloudy or clear skies, and the RAOB plotted on the Skew-T indicates an unsaturated atmosphere above the surface. Overnight radiational cooling reduced air temperatures near Earth's surface to a greater degree than the overlying atmosphere, causing the low-level temperatures to fall, and leaving the region above about 950 hPa unaffected. Note that the surface (2-m) temperature has dropped down to the dew point, inducing a saturated condition. The corresponding meteogram (Figure 11.8) shows light winds and clear skies during the early morning hours, resulting in near saturation conditions (note the temperature and dew point plots) and haze.

Figure 11.9 shows a frontal inversion. The RAOB was recorded above Pittsburgh, Pennsylvania (KPIT), on May 23, 2012, at 1200 UTC (0700 LST). Figure 11.5 shows the corresponding synoptic surface analysis and indicates that the quasistationary front creating the inversion is in the eastern half of the state. Frontal inversions represent the transition zone between two air masses, with the colder air mass below, and the warmer one above.[18] Unlike the radiation inversion discussed in the preceding

❧ Plymouth State Weather Center ❧

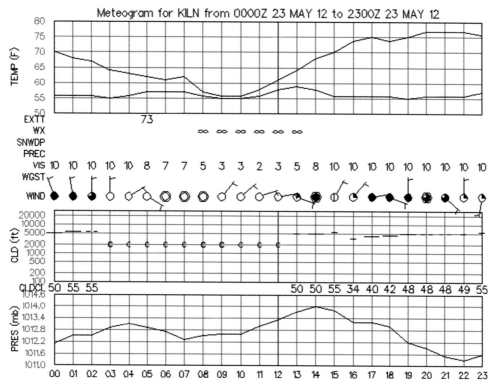

Figure 11.8. Meteogram for Wilmington, Ohio (KILN). Total cloud cover is encoded in the station circles with the plotted wind barbs. Present weather symbols are indicated on the line labeled "wx." (Meteogram courtesy of Plymouth State University meteorology program, 2012.)

text, the frontal inversion is *not* surface based. The base of the inversion is visible approximately 25 hPa above the surface, and the top of the inversion is near 925 hPa, or, about 50 hPa above the surface. Another difference between the radiation inversion and the frontal inversion is the vertical saturation profile. In the radiation inversion, the highest saturation point is on the surface, where temperatures were coldest. In the frontal inversion, the highest saturation point is between the base and the top of the inversion, that is, in the transition zone between the two air masses, where the two dissimilar air masses are mixing.

Anticyclones produce subsidence inversions, because convergence between the middle troposphere and the tropopause results in excess mass, which sinks toward the surface, inducing downward vertical motion. All downward motion occurs along a dry adiabat, unlike upward motion, which may follow either a dry or a moist adiabat. Sinking results in compression, which results in warming, so, even if the

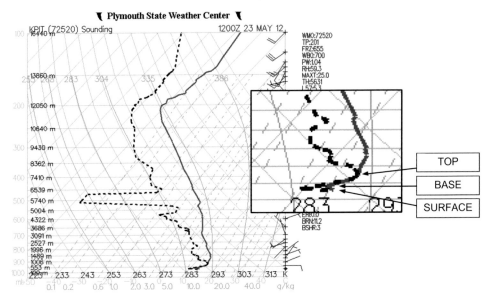

Figure 11.9. Absolutely stable atmosphere: Frontal inversion. Plotted radiosonde observation for Pittsburgh, Pennsylvania (KPIT/72520). The time zone is UTC – five hours, making this RAOB representative of late May 2012, at 0700 LST. The frontal inversion is visible between about 975 and about 925 hPa. The inset panel (on right) shows this region in detail. "BASE" means base of the inversion; "TOP" means top of the inversion. (Plot courtesy of Plymouth State University meteorology program, 2012.)

high-altitude parcel is initially saturated, downward motion opens up a difference between the parcel's temperature and dew point. A subsidence inversion is visible in the Maniwaki, Quebec profile (Figure 11.4), near 780 hPa. It is associated with a migratory high-pressure system centered east of Hudson Bay, in northern Quebec province, Canada (Figure 11.5). Another is visible in the Wilmington, Ohio profile, near 600 hPa (Figure 11.7). These RAOBs clearly illustrate an identifying feature of subsidence inversions: they are generally very dry at the top, moister near the base, and rarely surface based.

The stronger the downward motion, the greater the difference between the dew point and temperature. It follows that *stronger* anticyclones will produce *greater* subsidence, more warming, and a correspondingly greater amount of drying. Figure 11.10 shows a subsidence inversion resulting from the subsidence beneath the North Pacific High, which is a component of the semipermanent subtropical ridge centered near 30 °N. The subtropical ridge is strongest and farthest north during the meteorological summer. The RAOB shown in Figure 11.10 was recorded above Oakland, California (KOAK), which is located near 35 °N, on July 5, 2011, at 0000 UTC (1600 LST). Figures 11.11 and 11.12 show the corresponding surface and 500 hPa synoptic analyses (valid at 1200 UTC, or 0400 LST, the same day).

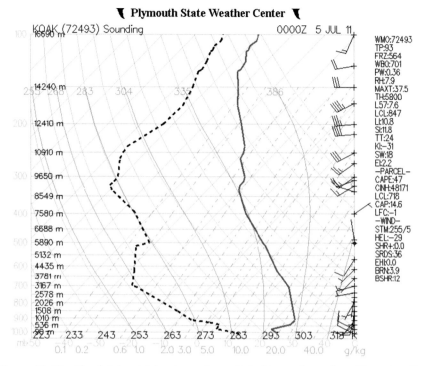

Figure 11.10. Absolutely stable atmosphere: Subsidence inversion. Plotted radio-sonde observation for Oakland, California (KOAK/72493). The time zone is UTC – eight hours, making this RAOB representative of early July 2011, at 0400 LST. The subsidence inversion is visible between about 975 and 900 hPa. (Plot courtesy of Plymouth State University meteorology program, 2012.)

11.5. Quantitative Estimates of Stability Using a Thermodynamic Diagram

Since the middle of the twentieth century, *stability indexes* have been used to esti-mate stability and make provisional forecasts of the sensible weather that will result. Very detailed publications are available in the public domain that demonstrate a large variety of practical stability and severe weather forecasting techniques that can be performed with Skew-T diagrams. One of them, called *Notes on Analysis and Severe-Storm Forecasting Procedures of the Air Force Global Weather Central*, dis-cusses the motivation for developing these methods in its introduction. The author, Robert Miller (writing as Chief Scientist at the Air Force Global Weather Central's Analysis and Forecasting Section) explains:

In March of 1948 at Tinker Air Force Base, the author, working under Lt. Col. E.J. Fawbush, experienced a tornado that struck the base without warning and caused severe damage. The Commanding General of the Oklahoma City Air Materiel Area asked us to try to find a way to forecast such occurrences. We analyzed many past cases, searched the existing tornado lit-erature and found a report by Showalter and Fulks of the US Weather Bureau most helpful. A week later when synoptic conditions similar to the previous storm appeared we forecast a tornado to hit the base, and fortunately for us, the forecast verified (Miller, 1975).

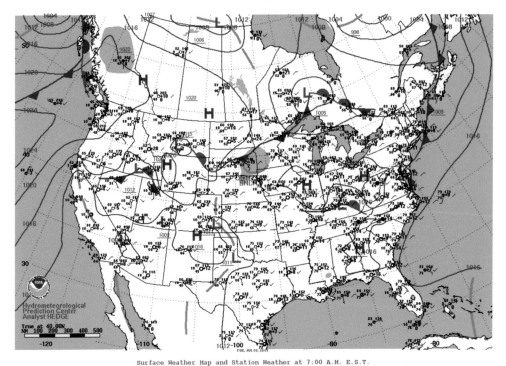

Surface Weather Map and Station Weather at 7:00 A.M. E.S.T.

Figure 11.11. Surface analysis valid, July 5, 2011, 1200 UTC (0400 LST), twelve hours after the RAOB plotted in Figure 11.10. A weak thermal trough of low pressure extends northwestward through the inland valley of California from the warm-core surface low centered south of Las Vegas, Nevada. The surface center North Pacific High is centered west of the left edge of the graphic. (Graphic courtesy of Hydrometeorological Prediction Center, 2012.)

Another U.S. Air Force publication, *The Use of the Skew-T Diagram in Analysis and Forecasting*,[19] lists many practical applications of the thermodynamic principles we have derived in this book. The U.S. National Weather Service has similar publications. Some of these methods are described in the following text. The examples discussed are not exhaustive, but are meant to illustrate the range in *complexity* (from very simple to rather complex) and *application* (from general to specific).

These methods are illustrated using a real-world event that occurred on April 14, 2012. Figure 11.13 shows the synoptic surface analysis for 1200 UTC on the fourteenth. There was a substantial severe weather outbreak across the states of Oklahoma, Kansas, Nebraska, and Iowa later in the day, triggered (in part) by a warm front bisecting the interior United States from eastern Colorado through western Kentucky. Warm, moist, unstable maritime tropical air was flooding into the South, out of the Gulf of Mexico, while cool polar air dominated the region north of the warm front. Figure 11.14 shows the 500 hPa analysis for the same time. An upper-level low was centered over southern Nevada, with divergent southwesterly flow extending from

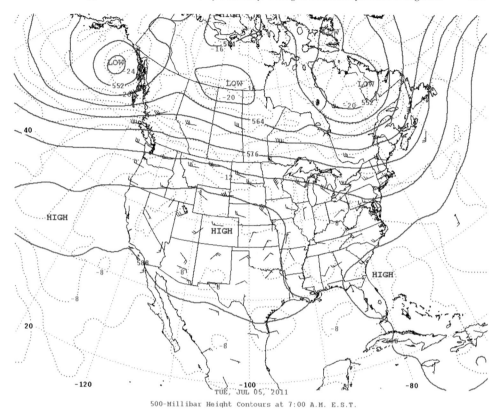

500-Millibar Height Contours at 7:00 A.M. E.S.T.

Figure 11.12. 500 hPa analysis valid, July 5, 2011, 1200 UTC (0400 LST), twelve hours after the RAOB plotted in Figure 11.10. The subtropical ridge can be seen extending from west to east across the central United States. (Graphic courtesy of Hydrometeorological Prediction Center, 2012.)

Baha, California, through the upper Midwest. The severe weather reports from the Storm Prediction Center for April 14 are shown in Figure 11.15. Much of the stability analysis that follows is based on radiosonde observations from Dodge City, Kansas (KDDC), located in the southwestern part of the state. The location of the KDDC is shown in Figure 11.16.

Showalter Stability Index.[20] The Showalter Stability Index (SI) is a general purpose index that is very simple to compute. This stability index uses the conditions at 850 hPa to represent a low-level, moisture-laden layer of air that is forced aloft by an external lifting force. (The choice of 850 hPa is somewhat arbitrary, and users of this method should consider using a point higher in the atmosphere for high altitude stations.) A typical lifting scenario would be the approach of a cold front or dry line, bringing it into collision with a warm, moist air mass. The SI assumes that the 850 hPa parcel is lifted to 500 hPa, and a temperature comparison is made to determine if the atmosphere will provide an additional vertical acceleration because of the

Table 11.2. *Expanded definitions of static stability. Five stability conditions by environmental lapse rate.*

Isentropic gradient refers to environmental potential temperature variation with height. Vertical acceleration refers to the net effect of the environment on a lifted parcel, and was computed using (11.18). Γ_D and Γ_M refer to the dry and moist parcel lapse rates, respectively, and γ refers to the environmental lapse rates. The USSA/ISA standard lapse rate (shown as solid red line in Figure 11.1) indicates a conditional stability condition in the troposphere (highlighted), and an absolutely stable condition in the stratosphere.

Stability Condition	Isentropic Gradient	Vertical Acceleration of Parcel		Lapse Rate Relationship
		Unsaturated (Dry)	Saturated (Moist)	
Absolutely Unstable	$\dfrac{d\theta}{dz} < 0$	$\left(\dfrac{dw}{dt}\right)_D > 0$	$\left(\dfrac{dw}{dt}\right)_M > 0$	$\Gamma_M < \Gamma_D < \gamma_{au}$
Dry Neutral	$\dfrac{d\theta}{dz} = 0$	$\left(\dfrac{dw}{dt}\right)_D = 0$	$\left(\dfrac{dw}{dt}\right)_M > 0$	$\Gamma_M < \gamma_{dn} = \Gamma_D$
Conditional	$\dfrac{d\theta}{dz} > 0$	$\left(\dfrac{dw}{dt}\right)_D < 0$	$\left(\dfrac{dw}{dt}\right)_M > 0$	$\Gamma_M < \gamma_c < \Gamma_D$
Moist Neutral	$\dfrac{d\theta}{dz} > 0$	$\left(\dfrac{dw}{dt}\right)_D < 0$	$\left(\dfrac{dw}{dt}\right)_M = 0$	$\Gamma_M = \gamma_{mn} < \Gamma_D$
Absolutely Stable	$\dfrac{d\theta}{dz} > 0$	$\left(\dfrac{dw}{dt}\right)_D < 0$	$\left(\dfrac{dw}{dt}\right)_M < 0$	$\gamma_{as} < \Gamma_M < \Gamma_D$

resulting density difference between the lifted parcel and the surrounding environment. If the parcel is warmer (lighter) than the environment, then the atmosphere is unstable at 500 hPa, and the parcel experiences a positive (upward) acceleration, as shown in (11.18). The temperature difference is then compared to a table of climatologically observed sensible weather outcomes.

To compute the SI, use a Skew-T to lift the observed 850 hPa temperature along a dry adiabat, and the 850 hPa dew point along a mixing ratio line, until the lines intersect. The intersection point is the 850 hPa LCL (LCL_{850}). Continue to lift the saturated parcel (along a moist adiabat) until reaching 500 hPa. Note both the environmental temperature (T_{500}) and the parcel temperature ($\acute{T}_{500}$) in degrees Celsius. The SI is then computed by:

$$SI = T_{500} - \acute{T}_{500} \qquad\qquad (11.35)$$

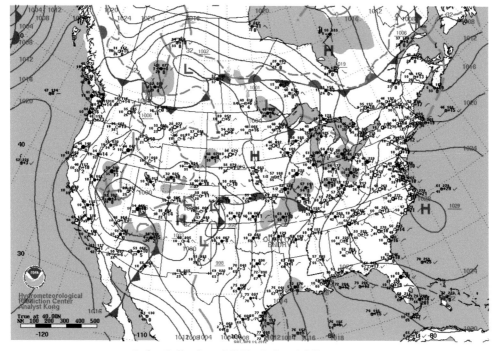

Surface Weather Map and Station Weather at 7:00 A.M. E.S.T.

Figure 11.13. Surface analysis valid, April 14, 2012, 1200 UTC. Warm, moist maritime tropical air from the Gulf of Mexico dominates the Southeast, as well as Texas, Oklahoma, and Kansas. The Polar Front extends from a low centered in eastern Colorado through western Kentucky. Cool, saturated polar air is north of the front. (Graphic courtesy of Hydrometeorological Prediction Center, 2012.)

Table 11.3 shows a *generalized* expected outcome for a given Showalter's value, but it should be recalibrated with locally observed data before using it as a forecast method. Even with local recalibration, the forecast should be considered provisional – *just one tool among many*. The synoptic and mesoscale conditions in the forecast area must also be carefully evaluated.

Table 11.3 lists "severe" thunderstorms as one possible outcome. According to the U.S. National Weather Service,[21] a thunderstorm is defined as "severe" when any of the following conditions occur:

- Any tornadic activity (including a *tornado*, which touches down on land, a *water spout*, which touches down on water, or a *funnel cloud*, which does not touch the surface); or
- Hailstones 1 inch ($\cong$ 2.5 cm) or greater in diameter; or
- Wind speeds or gusts[22] of 50 kts ($\cong$ 58 statute miles per hour, or 93 km/h).

Dodge City, Kansas (KDDC) is approximately 800 meters ASL (it is a high-altitude station). Its location is shown in Figure 11.16. The KDDC radiosonde observations for April 14, 2012, 1200 UTC (0600 LST) and April 15, 2012, 0000 UTC (1800 LST)

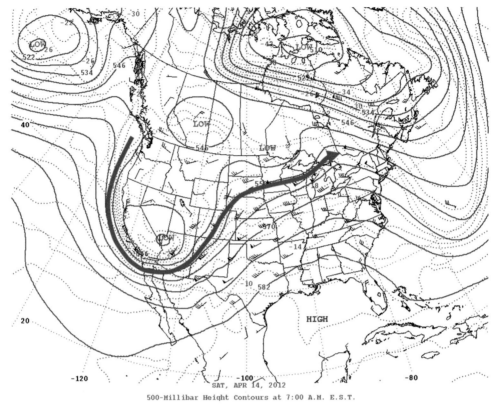

Figure 11.14. 500 hPa analysis valid, April 14, 2012, 1200 UTC. An upper-level low was centered over southern Nevada, with divergent southwesterly flow extending from Baha, California, through the upper Midwest. The axis of the relevant segment of the Polar Jet Stream is highlighted with a heavy red arrow. (Graphic courtesy of Hydrometeorological Prediction Center, 2012.)

are shown in Figures 11.17 and 11.18, along with the computation of the Showalter Stability Index. The low-level parcel was lifted from 850 hPa for simplicity. The 1200 UTC sounding indicated an SI value of +0.2 (Figure 11.16), indicating a preliminary forecast of "thunderstorms probable." The first tornado report came in at about 1730 UTC (1130 LST),[23] and by the end of the day, there were more than 120 tornado reports over the Great Plains, east of Dodge City (Figure 11.15). In addition, there were more than eighty reports of severe convective winds and 140 reports of severe hail. The 0000 UTC sounding (six hours after the first tornado was reported) indicated an SI value of −7.0 (Figure 11.18), for a preliminary forecast of "tornado possible."

Figure 11.19 shows contoured values of the SI based on observed soundings at many locations, all valid at 0000 UTC. The region highlighted in red corresponds to SI values of −6 or less, indicating provisional forecast guidance of "tornadoes possible." There is a good general correspondence between the SI forecast tornado area shown in Figure 11.19 and the locations of observed tornadoes shown in Figure 11.15.

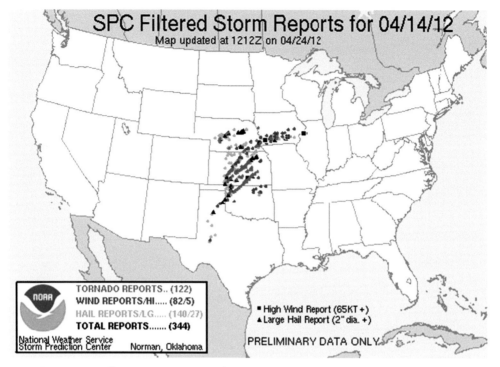

Figure 11.15. Severe storm reports for April 14, 2014. There were more than 120 reports of tornadoes in the region east of KDDC in one day. (Plot courtesy of Storm Prediction Center, 2012.)

In this case, the morning's Showalter Stability Index indicated the likelihood of convective activity during the day, but it certainly didn't suggest the *severity* of the coming day. The cool temperatures of early morning made the profile appear too stable. It wasn't until the evening sounding became available, shortly after the peak heating of the day, that the possibility of tornadoes was indicated by the SI. This is clearly too late to be of practical use to forecasters. Further, as Figure 11.19 indicates, the areal extent of the potential tornado region is underestimated. There are at least two approaches that can be taken to improve the accuracy of provisional forecasts made from the morning sounding: (1) Base the stability calculation on a predicted high temperature, which implies using *in situ* data from nearer the surface, or (2) use numerical models to generate full forecast profiles of temperature and dew point, on which stability computations can be performed. The latter approach is taken in applications such as *BUFKIT*, which ingests model data from the National Centers for Environmental Prediction (NCEP) and gives the user many options for visualizing and analyzing the forecast profiles. The former approach is discussed next.

Lifted Index.[24] This stability index uses the conditions at the surface (i.e., 2 m), or the average conditions in a 1000-m deep layer near the surface, rather than the arbitrarily

Table 11.3. *Provisional forecasts based on computed Showalter Stability Index.*

SI Value	Outcome
+3 < SI	Convective activity unlikely
+1 < SI ≤ +3	Showers probable Thunderstorms possible
−3 < SI ≤ +1	Thunderstorm probable
−6 < SI ≤ −3	Severe thunderstorm possible
SI ≤ −6	Tornado possible

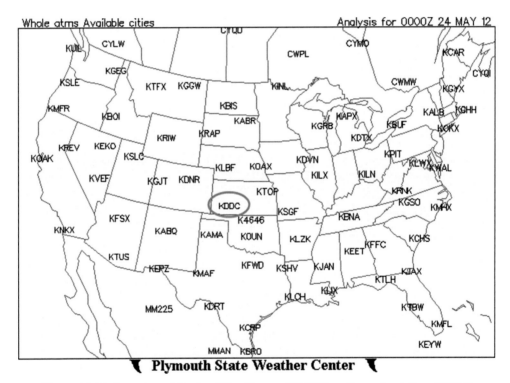

Figure 11.16. Location of Dodge City, Kansas (KDDC) radiosonde station. KDDC is located in southwestern Kansas, at 37.76 °N, −99.97 °E, approximately 800 meters ASL. Its time zone is UTC − six hours. (Plot courtesy of Plymouth State University meteorology program, 2012.)

chosen 850 hPa parcel used in the SI. The Lifted Index (LI) assumes that the parcel near the surface is lifted to 500 hPa, where the temperature comparison is made. If the parcel is warmer (lighter) than the environment, then the atmosphere is unstable at 500 hPa and the parcel experiences a positive (upward) acceleration, as shown in

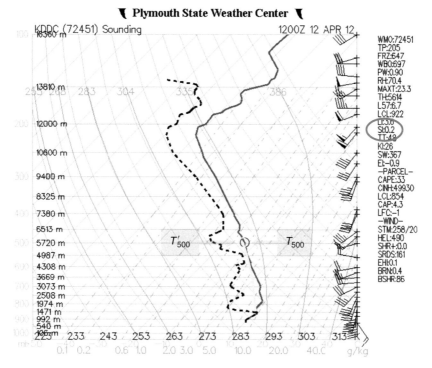

Figure 11.17. KDDC RAOB for April 14, 2012, 1200 UTC (0600 LST). Path of 850 hPa parcel shown in yellow, with 500 hPa parcel temperature (T_{500}) circled (The 850 hPa parcel was already saturated, so it wasn't necessary to lift it dry adiabatically to the LCL_{850}). The 500 hPa environmental temperature (T_{500}) is circled in red. The parcel temperature is slightly cooler than the environmental temperature at 500 hPa. The automated analysis (shown on right) indicates the SI value is +0.2 (circled in blue). (Plot courtesy of Plymouth State University meteorology program, 2012.)

(11.18). The temperature difference is then compared to a table of climatologically observed sensible weather outcomes. LI values occurring with given types of convective weather are usually slightly lower than the corresponding SI values.

To compute the LI using the surface parcel method (LI_p), use a Skew-T to lift the surface temperature along a dry adiabat, and the surface dew point along a mixing ratio line, until the lines intersect. The intersection point is the surface parcel's LCL (LCL_{SFC}). Continue to lift the saturated parcel (along a moist adiabat) until reaching 500 hPa. Note, both the environmental temperature (T_{500}) and the parcel temperature ($\acute{T}_{500}$) are in degrees Celsius. The LI is then computed by:

$$\boxed{LI = T_{500} - \acute{T}_{500}}$$
(11.36)

which is identical to (11.35). Table 11.4 shows a *generalized* list of expected outcomes for a given LI value, but, like the SI, it should be recalibrated with locally observed

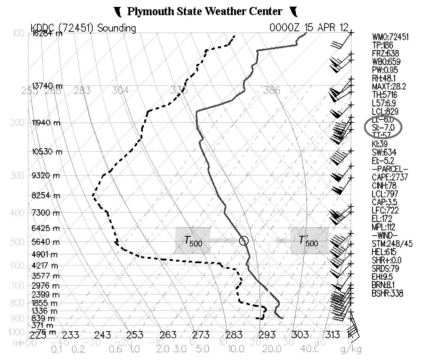

Figure 11.18. KDDC RAOB for April 15, 2012, 0000 UTC (1800 LST). Path of 850 hPa parcel shown in yellow, with 500 hPa parcel temperature (T_{500}) circled. The 850 hPa parcel was lifted dry adiabatically to the LCL_{850}, then moist adiabatically to 500 hPa. The 500 hPa environmental temperature (T_{500}) is circled in red. The parcel temperature is warmer than the environmental temperature at 500 hPa. The automated analysis (shown on right) indicates the SI value is −7.0 (circled in blue). (Plot courtesy of Plymouth State University meteorology program, 2012.)

data before using it as a forecast method, and – like the other stability indexes – it should considered a *provisional* forecast.

The LI_p can be improved somewhat – first by forecasting the high temperature for the day (T_{max}) and then by lifting the resulting forecast parcel to 500 hPa, rather than using the observed surface parcel. This corrects for the inaccuracy noted in the case study used to demonstrate the SI, which occurred because the RAOB was taken near sunrise, the coolest part of the day, causing the stability calculation to underestimate the buoyancy of the low-level parcel. One way to forecast the 2-m maximum temperature from the morning sounding utilizes the *in situ* temperature near 850 hPa and the estimated total cloud cover for the day. (Table 11.5 shows the interpretation of standard METAR total cloud-cover codes.) The method assumes that the temperature at 850 hPa will remain relatively unaffected by daytime heating and that vertical motions in the atmosphere will create a mixed layer between the surface and 850 hPa. To estimate T_{max} by this method:[25]

Table 11.4. *Provisional forecasts based on computed Lifted Index.*

LI Value	Outcome
$+2 < \mathrm{LI}$	Convective activity unlikely
$0 < \mathrm{LI} \leq +2$	Showers probable
	Thunderstorms possible
$-4 < \mathrm{LI} \leq 0$	Thunderstorm probable
$-7 < \mathrm{LI} \leq -4$	Severe thunderstorm possible
$\mathrm{LI} \leq -7$	Tornado possible

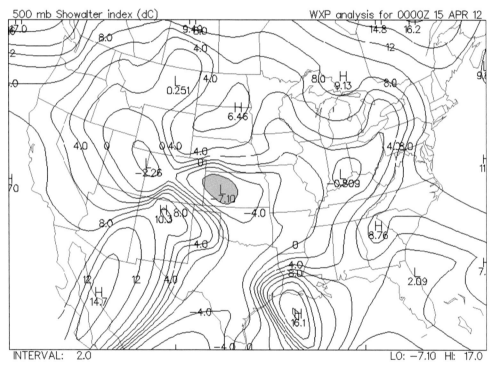

Figure 11.19. Contoured values of Showalter Stability Index for 15 April 2012, 0000 UTC. Region highlighted in red corresponds to $\mathrm{SI} \leq -6$, indicating provisional forecast guidance "tornadoes possible." Compare to tornado reports shown in Figure 11.15. (Plot courtesy of Plymouth State University meteorology program, 2012.)

- If there *is no* inversion between 1200 and 1800 m above ground level (the usual range of elevation of 850 hPa for stations near sea level):
  - ➤ *and* if the sky condition is forecasted to be predominantly clear to scattered, take the 850 hPa temperature to the surface using a *dry* adiabat.
  - ➤ *and* if the sky condition is forecast to be predominantly broken to overcast, take the 850 hPa temperature to the surface using a *moist* adiabat.

Table 11.5. *Interpretations of total cloud cover codes used in METAR reports.*

Code	Total Cloud Cover (8[ths])
CLR/SKC[26]	0
FEW	1–2
SCT	3–4
BKN	5–7
OVC	8

- If there *is* an inversion between 1,200 and 1,800 m AGL:
  - ➢ identify the temperature at the warmest point in the inversion, and take it to the surface using a *dry* adiabat.

The LI can be further improved by using an *average* temperature and dew point in the 1000-m deep layer near the surface, rather than the values at a discrete elevation. In this case, the forecaster estimates the mean temperature in the lowest 100 hPa[27] of the profile by using an isotherm and "equal areas." The mean dew point is estimated in the same region, using a mixing ratio line and equal areas. (This is demonstrated in Figure 11.20.) The mean temperature ($\bar{T}$) and dew point ($\bar{T}_d$) are then moved vertically from 50 hPa above the surface to the LCL (using a dry adiabat and mixing ratio line, respectively). If the LCL is below 500 hPa, the remainder of the vertical movement is completed along a moist adiabat. From here, the moist-layer LI (LI_{ml}) is computed using (11.36), and a provisional forecast is made using Table 11.4.

Figure 11.21 shows the LI_{ml} computation for KDDC on April 15, 2012 0000 UTC (1800 LST). The computation begins with the mean low-level parcel computed in Figure 11.20, which has $\bar{T} \cong 19.5\,°C$ and $\bar{T}_d \cong 15.0\,°C$. The mean parcel is lifted from 50 hPa ($\cong 500$ m) above the surface to the LCL near 805 hPa. From here, vertical ascent continues along a moist adiabat to 500 hPa, where the parcel cools to $\acute{T}_{500} \cong -3.5\,°C$. The environmental temperature (T_{500}) is $\cong -11.5\,°C$. Using (11.36), the LI is −8, which, by comparison to Table 11.5, indicates a strong possibility of tornadoes. Figure 11.21 also shows the simple parcel-method LI (computed by automation) with a value of only −6.0, corresponding to severe thunderstorms. Clearly, by going to the added effort of estimating the mean temperature and dew point in the lowest 100 hPa of the atmosphere, we've gotten a stronger indicator of the tornadoes that occurred in the area during the day (Figure 11.15).

K Index.[28] This stability index (K Index, or KI) uses the conditions at 850, 700, and 500 hPa to estimate the probability of *air mass thunderstorms*. According to the *American Meteorological Society Glossary of Meteorology*, an air mass thunderstorm is one produced by local convection in a statically unstable atmosphere. It is initiated

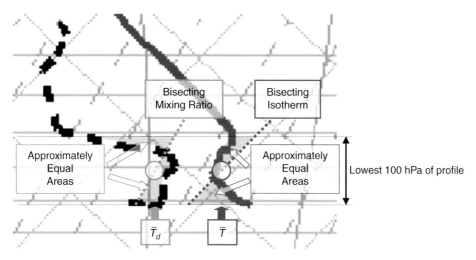

Figure 11.20. Enlarged view of lower portion of KDDC RAOB for April 15, 2012, 0000 UTC (1800 LST). The observed temperature profile is a thick solid red line. The observed dew point profile is thick dashed black line. The temperature in the lowest 100 hPa of profile is bisected with an isotherm, dividing approximately equal areas to the left and right of the line. The dew point is similarly divided using a mixing ratio line. The estimated mean temperature ($\bar{T}$) of $\cong 19.5\,°C$ and mean dew point ($\bar{T}_d$) of $\cong 16\,°C$ are circled at 50 hPa ($\cong 500$ m) above the surface for use in computing LI_{ml}. (Plot courtesy of Plymouth State University meteorology program, 2012.)

by solar heating of Earth's surface, not by an external lifting mechanism, such as a front or a dry line. The temperature difference between the environment at 850 hPa and 500 hPa is used to estimate the environmental lapse rate, and the 850 hPa dew point is used to estimate the amount of water vapor in the layer near the surface. The dew point depression ($= T - T_d$) at 700 hPa is used to estimate the vertical depth of the moist air. A Skew-T isn't needed to compute this index. One simply uses the equation:

$$KI = \left(T_{850} + T_{d_{850}}\right) - \left(T_{700} - T_{d_{700}}\right) - T_{500} \qquad (11.37)$$

and compares the result to Table 11.6 for a probability forecast. (All temperatures are in Celsius.) The results shown in Table 11.6 correspond to best fits between computed KI and resulting sensible weather in the Midwest and Great Plains regions of the United States. Before using this index anywhere outside this region, a careful study should be completed that recalibrates the index values with the corresponding convective weather.

Table 11.7 shows partial numerical sounding data for Dodge City, Kansas (KDDC), valid April 15, 2012 at 0000 UTC (1800 LST)[29] from the surface (at 905 hPa) through 500 hPa. Plugging the data from the table into (11.37) results in:

Table 11.6. *Provisional forecasts based on computed K Index.*

KI Value	Probability of Air Mass Thunderstorms [%]
KI < 15	0
15 ≤ KI ≤ 20	20
20 < KI ≤ 25	20–40
25 < KI ≤ 30	40–60
30 < KI ≤ 35	60–80
35 < KI ≤ 40	80–90
40 < KI	≅ 100

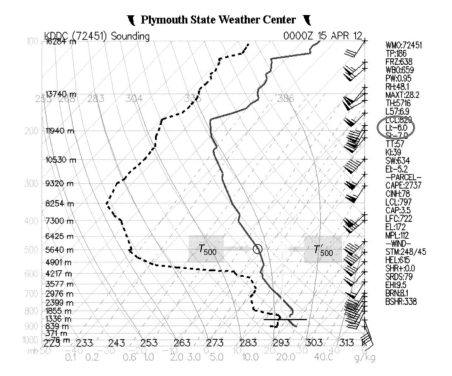

Figure 11.21. KDDC RAOB for April 15, 2012, 0000 UTC (1800 LST). The pathway of the mean parcel in lowest 100 hPa of the profile (computed in Figure 11.19) is shown in yellow. The thin black horizontal line near the base of the profile corresponds to the point of origin of the mean parcel, about 50 hPa (≅ 500 m) above ground level. The LI, *which was* computed using the parcel method by automated routine, is circled in blue on right. (Plot courtesy of Plymouth State University meteorology program, 2012.)

Table 11.7. *Partial numerical sounding data for KDDC, April 15, 2012, 0000 UTC.*[i]

Data are shown from the surface (905 hPa) through 500 hPa. T is the *in situ* temperature; T_d the dew point; and T_w, the wet-bulb temperature. Data needed to compute the KI are highlighted in gray. Table cells containing data needed to compute the TT index are highlighted and have a heavy outline. Table cells containing data needed to compute the SW index only have a heavy outline but no highlighting.

p [hPa]	z [m ASL]	T [°C]	T_d [°C]	T_w [°C]	Wind Direction [°]	Wind Speed [kt]
905	790	23.4	17.4	19.2	155	12
903	809	23.0	15.0	17.6	155	13
898	858	22.0	17.0	18.5	156	16
862	1215	18.8	15.5	16.5	160	40
850	1336	18.6	15.8	16.7	175	39
828	1561	17.8	14.2	15.4	188	38
820	1644	18.2	11.2	13.7	193	37
798	1877	17.2	5.2	10.1	206	37
717	2778	9.0	0.0	4.2	215	46
700	2976	7.0	0.0	3.3	215	44
677	3249	4.2	−0.5	1.8	215	47
621	3944	−1.9	−5.8	−3.7	219	56
606	4138	−2.9	−10.9	−6.1	223	58
593	4310	−4.3	−12.3	−7.4	225	60
592	4323	−4.3	−12.3	−7.4	225	60
586	4405	−4.5	−19.5	−9.2	225	62
574	4566	−4.9	−30.9	−11.1	225	67
561	4746	−5.3	−39.3	−11.9	225	72
560	4760	−5.5	−39.5	−12.1	225	72
531	5175	−8.3	−46.3	−14.3	225	80
501	5625	−11.3	−51.3	−16.6	221	85
500	5640	−11.5	−50.5	−16.8	220	85

[i] PSU Weather Center (2012a).

$$KI = (18.6 + 15.8) - (7.0 - 0.0) - (-11.5) = 38.9$$

which corresponds to an 80 to 90 percent chance of air mass thunderstorms (Table 11.6). Figure 11.22 shows contoured values of the KI based on radiosonde observations at many locations, all valid at April 15, 2012, 0000 UTC. The region of KI values greater than thirty, corresponding to provisional forecast guidance of at 60 percent chance of air mass thunderstorms or greater, is highlighted in red. In

❧ Plymouth State Weather Center ❧

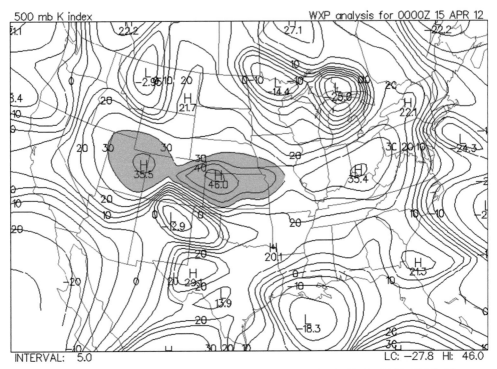

Figure 11.22. Contoured values of K Index for April 15, 2012, 0000 UTC. The region highlighted in red corresponds to KI ≥ 30, corresponding to provisional forecast guidance of a 40 percent or greater chance of air mass thunderstorms. (Plot courtesy of Plymouth State University meteorology program, 2012.)

spite of the fact that this index is meant for use in predicting ordinary air mass thunderstorms, the region highlighted in red roughly corresponds to the region of tornado reports shown in Figure 11.15.

Total Totals index.[30] The Total Totals (TT) index is the sum of two older indexes called the Vertical Totals (VT) index and the Cross Totals (CT) index, and is used to estimate the qualitative probability of severe thunderstorms (as defined in the preceding text). The former is defined by:

$$\boxed{VT = T_{850} - T_{500}} \tag{11.38}$$

where T_{850} and T_{500} are the environmental temperatures at 850 hPa and 500 hPa, respectively [°C]. The VT index parameterizes the vertical temperature lapse rate between the two levels. The CT index parameterizes the influence of low-level moisture and cold upper-level temperatures by:

$$\boxed{CT = T_{d_{850}} - T_{500}} \tag{11.39}$$

Table 11.8. *Provisional forecast guidance based on the Total Totals index.*

TT Value	Probability of Severe Thunderstorms
$TT < 44$	None
$44 \leq TT < 50$	Weak
$50 \leq TT \leq 55$	Moderate
$55 < TT$	Strong

where $T_{d_{850}}$ is the environmental dew point at 850 hPa [°C]. Combining these two to compute the TT:

$$TT = VT + CT \tag{11.40}$$

or:

$$TT = \left(T_{850} - T_{500}\right) + \left(T_{d_{850}} - T_{500}\right) \tag{11.41}$$

and therefore:

$$\boxed{TT = \left(T_{850} + T_{d_{850}}\right) - 2T_{500}} \tag{11.42}$$

The result of the computation is then compared to the provisional forecast guidance in Table 11.8, and comes with the usual cautions about recalibrations for local use and careful consideration of the synoptic conditions.

Using the sample data for KDDC shown in Table 11.7, the TT index at Dodge City, Kansas, on April 14, 2012, at 0000 UTC was:

$$TT = \left(18.6 + 15.8\right) - 2\left(-11.5\right) = 57.4$$

and therefore indicated a "strong" probability of severe thunderstorms. Figure 11.23 shows contoured values of the TT index based on radiosonde observations at many locations, all valid at April 15, 2012, 0000 UTC. The two regions of TT values greater than fifty-five, corresponding to provisional forecast guidance of a "strong" chance of severe thunderstorms, are highlighted in red. The western region (centered over Utah) did not experience any severe weather, as shown in Figure 11.15. The eastern region (Colorado, Kansas, and Nebraska) *did*, although farther east than indicated in the contoured region.

Severe Weather Threat Index.[31] Up until now, the stability indexes discussed in this section have been based exclusively on the thermodynamic buoyancy arguments

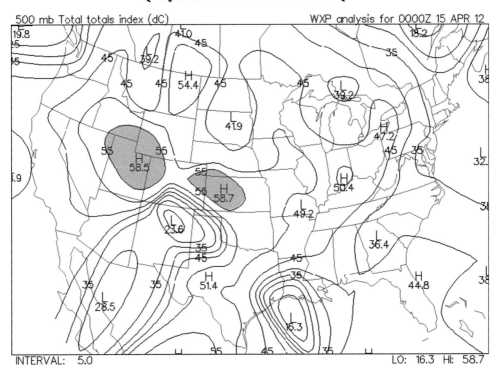

Figure 11.23. Contoured values of TT index for April 15, 2012, 0000 UTC. Region highlighted in red corresponds to TT ≥ 55, corresponding to provisional forecast guidance of a "strong" probability of severe thunderstorms. Compare to severe storm reports shown in Figure 11.15. (Plot courtesy of Plymouth State University meteorology program, 2012.)

presented in the previous section. The influence of *kinematic* effects, such as twisting and shearing in the atmosphere, has been ignored.[32] But it is known that *wind shear* is a necessary ingredient in the atmosphere for thunderstorms to be long-lived, and grow large enough to become severe *supercells*. Without wind shear, the thunderstorm downdraft develops in the same place as the updraft and chokes off the convection in a relatively short period of time. When wind shear is present, the updraft and downdraft are separated.

The Severe Weather Threat (SW or SWEAT) index is a successful attempt to combine thermodynamically driven instability and dynamically driven vertical motion into a single parameter. Shear between 850 and 500 hPa is explicitly considered. The purpose of the SW index is to evaluate the potential for supercell thunderstorms and *tornado-producing* supercell thunderstorms in a given vertical profile. It is specifically designed to predict severe thunderstorms, and should not be used to infer the potential for ordinary (air mass) thunderstorms. The SW index consists of five terms that are linearly added:

Table 11.9. *Provisional forecast guidance based on the Severe Weather Threat index.*

SW Value	Severe Weather Potential
SW < 300	None
300 ≤ SW < 400	Severe Thunderstorms
400 ≤ SW	Tornadoes

$$SW = 12D + 20(TT - 49) + 2W_{850} + W_{500} + 125(S - 0.20)$$

(11.43)

where:

- D is the 850 hPa dew point [°C], if the latter is a positive number. If the 850 hPa dew point is less than zero, D is set to zero. If the surface elevation is above the 850 hPa pressure level, replace the 850 hPa data with the temperature and dew point at the point 100 hPa above the surface, in this term and in the next. *This term parameterizes the importance of moisture in the lower troposphere.*
- TT is the Total Totals index discussed above, if TT is 49 or greater. If TT is less than or equal to 49, the entire term 20(TT − 49) is set to zero. *This term capitalizes on the success of the TT.*
- W_{850} is the 850 hPa wind speed [kts]. *This term parameterizes the importance of low-level jets in severe storm development.*
- W_{500} is the 500 hPa wind speed [kts]. *This term parameterizes the importance of upper-tropospheric jets in severe storm development.*
- The last term is the *Shear Term*, where S is the sine of the angle computed by subtracting the 850 hPa wind direction from the 500 hPa wind direction. *This term parameterizes the importance of wind shear in severe storm development.* The entire *Shear Term* is set to zero if any of the following conditions are *not* true:
  - ➢ 850 hPa wind direction is from 130 to 250 degrees.
  - ➢ 500 hPa wind direction is from 210 to 310 degrees.
  - ➢ 500 hPa wind direction − 850 hPa wind direction is positive, indicating that the wind is *veering*[33] with height.
  - ➢ Both the 850 hPa and 500 hPa wind speeds are at least 15 knots.

Once the SW index is computed, the result is compared to the provisional guidance shown in Table 11.9, with the usual caveats.

Using the sample data for KDDC shown in Table 11.7, we can compute the SW index for Dodge City, Kansas, on April 14, 2012, at 0000 UTC. The 850 hPa dew point is more than zero, so D is equal to the 850 hPa dew point. The TT value from the same profile (computed above) is 57.4, which is more than 49, so the second term is 20(*TT* − 49). Last, we check the requirements for the Shear Term. The 850 hPa wind direction is 175°, which is in the correct range of wind directions. The 500 hPa

❧ Plymouth State Weather Center ❧

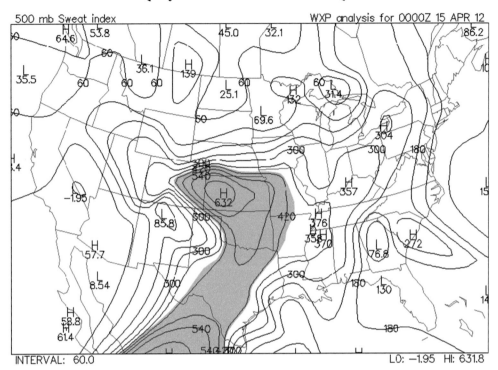

Figure 11.24. Contoured values of SW index for April 15, 2012, 0000 UTC. The region highlighted in red corresponds to SW ≥ 400, indicating a provisional forecast guidance of "tornadoes possible." Compare to severe storm reports shown in Figure 11.15. (Plot courtesy of Plymouth State University meteorology program, 2012.)

wind direction is 220°, so it is also in the correct directional range. The wind rotates clockwise (veers) from 850 hPa to 500 hPa, and both wind speeds are more than fifteen knots. Satisfying these conditions means we do not set the Shear Term to zero. Plugging the numbers for KDDC into (11.43) results in:

$$SW = 12(15.8) + 20(57.4 - 49) + 2(39) + (85) + 125\left[\sin(220 - 175) - 0.20\right] \cong 584$$

which, by Table 11.9, indicates the potential for tornadoes at KDDC. Figure 11.24 shows contoured values of the SW index based on radiosonde observations at many locations, all valid at April 15, 2012, 0000 UTC. The region of SW values greater than 400, corresponding to provisional forecast guidance of a "tornadoes possible," is highlighted in red. There appear to be two main centers: One to the south, centered over Mexico, and one farther north, centered over Kansas. The influence of the TT index can be seen in an extension of the northern center, westward into Utah. The northern center also exhibits a troughlike extension toward the east and south of the maximum over western Kansas. A qualitative comparison to the storm reports shown in Figure 11.15 indicates a good correlation between SW values higher than 400 and observed tornadoes, at least in the interior United States.

11.6. Summary of Stability

In this chapter, we defined stability as a measure of an isolated parcel's buoyancy in ambient air, driven by density differences between the two. The density difference results when a parcel is lifted and cools either dry or moist adiabatically, depending on its saturation condition. The vertical acceleration of the lifted parcel is given by three relations:

$$\frac{dw}{dt} = g\left(\frac{\rho - \acute{\rho}}{\acute{\rho}}\right)$$

where,

dw/dt = Vertical acceleration [m/s^2]
g = Gravitational acceleration ($\cong$ 9.81 m/s^2)
ρ = Density of environmental air [kg/m^3]
$\acute{\rho}$ = Density of lifted parcel [kg/m^3]

$$\frac{dw}{dt} = g\left(\frac{\acute{T}_v - T_v}{T_v}\right) \cong g\left(\frac{\acute{T}_v - T}{T}\right)$$

where,

$\acute{T}_v$ = Virtual temperature of lifted parcel [K]
T_v = Virtual temperature of environmental air [K]
$\acute{T}$ = Temperature of lifted parcel [K]
T = Temperature of environmental air [K]

$$\frac{dw}{dt} = g\left(\frac{\acute{\alpha} - \alpha}{\alpha}\right)$$

where,

$\acute{\alpha}$ = Specific volume of lifted parcel [m^3/kg]
α = Specific volume of environmental air [m^3/kg]

The second relation (involving T) was examined closely and analyzed for environmental lapse rates in five different regimes. These regions define absolutely unstable conditions (which include superadiabatic environmental lapse rates, resulting in autoconvection in the extreme case), dry neutral conditions, the conditional state (that depends on the saturation condition of the parcel, and is the mean state of the troposphere), moist neutral conditions, and, finally, absolutely stable conditions (which, in the extreme, are associated with temperature inversions).

Several stability indexes were then discussed. These were:

$$\text{SI} = T_{500} - \acute{T}_{500}$$

where,

    SI = Showalter Stability Index [°C], indicating general instability.

    T_{500} = 500 hPa temperature of ambient air [°C]

    $\acute{T}_{500}$ = 500 hPa temperature of lifted parcel [°C]

$$LI = T_{500} - \acute{T}_{500}$$

where,

    LI = Lifted Index [°C], computed by one of several means and indicating general instability.

$$KI = \left(T_{850} + T_{d_{850}}\right) - \left(T_{700} - T_{d_{700}}\right) - T_{500}$$

where,

    KI = K Index [°C], indicating the probability of air mass thunderstorms.

    T_{850} = 850 hPa temperature of ambient air [°C]

    $T_{d_{850}}$ = 850 hPa dew point of ambient air [°C]

    T_{700} = 700 hPa temperature of ambient air [°C]

    $T_{d_{700}}$ = 700 hPa dew point of ambient air [°C]

    T_{500} = 500 hPa temperature of ambient air [°C]

$$TT = \left(T_{850} + T_{d_{850}}\right) - 2T_{500}$$

where,

    TT = Total Totals index [°C], indicating the probability of severe thunderstorms.

$$SW = 12D + 20\left(TT - 49\right) + 2W_{850} + W_{500} + 125\left(S - 0.20\right)$$

where,

    SW = Severe Weather Threat Index [unitless], indicating the potential for severe thunderstorms and tornadoes

    D = Parameter dependent on 850 hPa T_d [°C]

    TT = Total Totals index

    W_{850} = 850 hPa wind speed [kt]

    W_{500} = 500 hPa wind speed [kt]

    S = Sine of the difference between the 500 hPa and 850 hPa wind directions

Practice Problems

1. Calculate the vertical acceleration of a parcel, if ρ_{env} = 1.20 kg m^{-3}, and ρ_{parcel} = 1.10 kg m^{-3}.
2. Calculate the vertical acceleration of a parcel, if $T_{v\ env}$ = 0 °C, and $T_{v\ parcel}$ = 5 °C.
3. Plot the temperature and dew point profiles from the following sounding on a Skew-T:

LEV	PRES	HGHT	TEMP	DEWP	RH	DD	WETB	DIR	SPD	THETA	THE-V	THE-W	THE-E	W
	mb	m	C	C	%	C	C	deg	knt	K	K	K	K	g/kg
SFC	1012	10	21.6	20.6	94	1.0	20.9			293.7	296.5	293.6	337.3	15.26
1	1000	106	21.2	20.1	93	1.1	20.5	175	21	294.4	297.0	293.6	337.1	14.97
2	971	362	21.2	18.7	86	2.5	19.5	188	24	296.8	299.4	293.7	337.6	14.11
3	925	782	18.6	16.4	87	2.2	17.1	210	30	298.3	300.7	293.2	335.5	12.78
5	914	885	18.6	13.7	73	4.9	15.4	211	30	299.4	301.3	292.0	331.1	10.84
6	901	1008	17.4	13.8	79	3.6	15.1	212	29	299.3	301.4	292.2	331.7	11.07
7	882	1190	16.2	11.4	73	4.8	13.2	213	29	299.9	301.7	291.2	328.2	9.64
8	873	1277	15.4	13.9	91	1.5	14.4	213	29	300.0	302.1	292.7	333.7	11.51
9	850	1503	14.0	12.4	90	1.6	13.0	215	28	300.8	302.8	292.3	332.3	10.70
10	835	1653	13.0	11.7	92	1.3	12.2	215	27	301.3	303.2	292.3	332.0	10.40
11	792	2097	11.0	7.1	77	3.9	8.7	217	24	303.7	305.2	291.1	327.8	8.01
12	765	2387	9.4	8.4	93	1.0	8.8	218	21	305.0	306.7	292.3	332.3	9.08
13	718	2911	6.6	5.1	90	1.5	5.7	219	18	307.6	309.0	292.0	331.0	7.69
15	700	3120	5.6	1.9	77	3.7	3.6	220	16	308.7	309.9	291.2	328.0	6.28
16	686	3285	4.4	0.0	73	4.4	2.1	222	17	309.1	310.2	290.7	326.4	5.58
17	676	3405	3.8	-3.2	60	7.0	0.4	224	18	309.8	310.6	289.9	323.7	4.47
18	652	3697	1.0	-5.0	64	6.0	-1.8	227	20	309.8	310.6	289.5	322.5	4.04
19	643	3809	0.4	-8.6	51	9.0	-3.4	229	21	310.4	311.0	288.8	320.2	3.10
21	625	4036	-1.1	-10.1	50	9.0	-4.8	232	23	311.2	311.7	288.8	320.3	2.84
23	606	4283	-2.3	-15.3	36	13.0	-7.0	235	25	312.6	312.9	288.3	318.9	1.92
25	570	4766	-5.7	-17.7	38	12.0	-9.6	241	29	314.1	314.4	288.6	319.6	1.67

(*continued*)

LEV	PRES	HGHT	TEMP	DEWP	RH	DD	WETB	DIR	SPD	THETA	THE-V	THE-W	THE-E	W
	mb	m	C	C	%	C	C	deg	knt	K	K	K	K	g/kg
27	551	5032	−7.9	−19.9	37	12.0	−11.6	245	31	314.5	314.8	288.5	319.3	1.43
29	527	5377	−10.7	−25.7	28	15.0	−14.5	250	34	315.2	315.4	288.2	318.3	0.89
30	517	5524	−11.7	−35.7	12	24.0	−16.2	252	35	315.7	315.8	287.8	317.0	0.35
36	500	5780	−12.7	−19.7	56	7.0	−14.8	255	37	317.6	317.9	289.6	323.0	1.60
37	493	5887	−13.5	−18.4	67	4.9	−15.0	256	38	317.9	318.2	289.9	324.0	1.82
39	480	6089	−14.7	−20.7	60	6.0	−16.4	258	41	318.8	319.1	290.0	324.0	1.53
40	473	6199	−15.5	−22.5	55	7.0	−17.4	259	43	319.2	319.4	289.9	323.7	1.33
43	450	6572	−17.1	−55.1	2	38.0	−21.1	262	48	321.7	321.8	289.3	321.9	0.05
44	436	6807	−18.7	−62.7	1	44.0	−22.5	264	51	322.6	322.6	289.6	322.7	0.02
45	400	7440	−23.7	−60.7	2	37.0	−26.5	270	60	324.2	324.2	290.0	324.3	0.03
48	370	8006	−28.1	−48.1	13	20.0	−30.0	265	62	325.6	325.7	290.6	326.2	0.13
49	345	8505	−32.7	−40.7	45	8.0	−33.6	260	64	326.0	326.1	290.9	327.2	0.32
50	300	9480	−39.1	−46.1	47	7.0	−39.6	250	67	330.3	330.3	292.0	331.0	0.20
52	250	10 700	−49.3	−57.3	39	8.0	−49.6	245	74	332.8	332.8	292.5	333.0	0.07
54	200	12 120	−60.7	−67.7	39	7.0	−60.8	245	96	336.6	336.6	293.5	336.7	0.02
55	185	12 603	−64.3	−71.3	38	7.0	−64.4	245	113	338.4	338.4	293.9	338.5	0.01
57	180	12 770	−65.1	−72.1	37	7.0	−65.1	245	114	339.8	339.8	294.3	339.8	0.01
58	175	12 943	−63.9	−70.9	38	7.0	−64.0	247	108	344.5	344.5	295.4	344.5	0.02
59	169	13 157	−65.3	−72.3	37	7.0	−65.4	248	101	345.6	345.6	295.7	345.7	0.01
60	150	13 890	−62.3	−69.3	39	7.0	−62.4	255	75	362.8	362.8	299.2	362.9	0.02

Go to the following webpage for the Skew-T tutorial: http://vortex.plymouth.edu/~stmiller/stmiller_content/Tutorials/skew_t.html

4. Forecast T_{MAX} and T_{MIN} using the clear-scattered (CLR-SCT) method. (Scattered means 50 percent or less cloud cover.)

$T_{MAX} = $ —————— °F = —————— °C

$T_{MIN} = $ —————— °F = —————— °C

5. The LCL is the height at which visible cloud *bases* appear when a parcel is lifted mechanically. The level of free convection (LFC) is the height to which the parcel must be lifted to begin auto-ascent (and may be on the surface in an absolutely unstable condition). The equilibrium level (EL) is the height at which the temperature of the ascending parcel becomes colder than the environmental temperature, and can be interpreted as the convective cloud *top*. Use T_{MAX} and the plotted profiles to calculate:

LCL = —————— hPa (—————— ft AGL) (parcel method)

LFC = —————— hPa (—————— ft AGL)

EL = —————— hPa (—————— ft AGL)

6. If the LFC is on the surface, convective clouds may develop from surface heating alone. If this occurs, clouds bases will be at the *convective* condensation level (CCL). Calculate the CCL using the parcel method, and your forecast for T_{MAX}:

CCL = —————— hPa (—————— ft AGL)

7. Use T_{MAX} and the plotted profiles to calculate:

Surface parcel LI = ——————; Thunderstorm forecast ——————

850 hPa parcel SSI = ——————; Thunderstorm forecast ——————

8. Compute the K Index, TT index, and SWEAT index:

K Index = ——————; Air mass thunderstorm forecast ——————

Total Totals Index = ——————; Severe thunderstorm forecast ——————

SWEAT index = ——————; Severe weather potential ——————

9. Calculate and plot the wet-bulb temperature profile. You don't need to calculate every point – just the points centered near wet-bulb zero (WBZ).

Height of wet-bulb zero = —————— hPa (—————— ft AGL)

10. Use the T_2 method to determine the maximum wind gust from a thunderstorm.

Max gust = —————— kts

12

Severe Weather Applications

12.1. Introduction

We have discussed qualitative estimations of atmospheric stability based on the environmental lapse rate (γ_e), as well as quantitative assessments of stability using several indexes. The former were useful as a conceptual method for stressing the importance of the vertical density profile, and the latter were useful for obtaining provisional forecasts of convective activity, such as thunderstorms. One of these indexes[1] also included a term for shear, based on the wind directions at 850 and 500 hPa. But while stability indexes are fast and easy to compute,[2] they are based on *partial* information, that is, they do not explicitly take advantage of all the information in the profile. In this final chapter, we'll extend the analysis to include more of the thermodynamic data reported in radiosonde profiles. We'll also introduce a kinematic parameter called *helicity*, which further develops the inclusion of wind shear information in provisional forecasts of thunderstorms. Finally, we'll put the two together and apply them to the prediction of severe convective weather.

12.2. Energy Stored in a Vertical Profile

In Chapter 11, we introduced the Showalter Stability Index (SI), and demonstrated how to compute it using a sounding recorded above Dodge City, Kansas, on April 15, 2012, 0000 UTC. Figure 12.1 shows this again. The SI begins with a parcel at 850 hPa, and lifts it to 500 hPa (first using a dry adiabat and then using a moist adiabat after saturation is reached at the LCL). The temperature of the lifted parcel is then subtracted from the environmental temperature at 500 hPa, and a table is consulted to make a provisional forecast. *The objective of this method is to obtain an estimate of the lifted parcel's buoyancy at 500 hPa.* If it is warmer than the surrounding air, it is also less dense,[3] and therefore positively buoyant.

The SI and the LI (which is similar) do not take advantage of most of the information in the sounding, and may produce misleading provisional forecasts. For example,

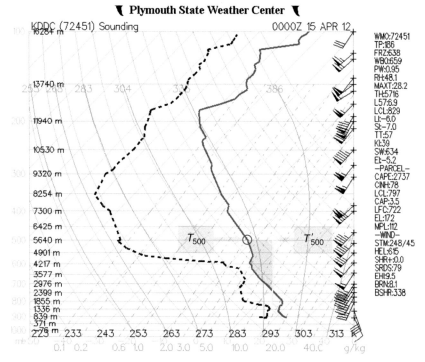

Figure 12.1. Demonstration of method for computing the Showalter Stability Index (SI), using KDDC RAOB for April 15, 2012, 0000 UTC (1800 LST). T_{500} refers to the environmental temperature at 500 hPa; T'_{500} refers to the temperature of the lifted parcel. The area highlighted in red corresponds to the region where the lifted parcel is positively buoyant; the blue area indicates negative buoyancy. (Plot courtesy of Plymouth State University meteorology program, 2012.)

the lifted parcel may have the same temperature as the environment at 500 hPa, but be much warmer than the environment at 700 hPa. The SI and LI would miss this. The K Index, the Total Totals index, and the Severe Weather Threat index make efforts to utilize data from additional levels, but they are still incomplete.

Computing potential energy. A related approach takes advantage of one of the basic characteristics of a good thermodynamic diagram (area is proportional to energy[4]) and evaluates the buoyant acceleration of the parcel through the entire vertical profile. The energy of interest is the potential energy that results from the difference between the density of a rising, isolated parcel[5] and the density of the environmental air. In some regions of the profile, the rising parcel is *warmer* (less dense) than the environmental air, and is therefore, *positively* buoyant (highlighted in red in Figure 12.1). In other regions, usually near Earth's surface and in the stratosphere, the rising parcel is *colder* (denser) than the environmental air, and is therefore, *negatively* buoyant (shown in blue in Figure 12.1).

The potential energy in the profile is computed by beginning with the basic definition of energy. According to Table 1.4:

$$E \equiv mv^2 \tag{12.1}[6]$$

where E is energy [J], m is mass [kg], and v is velocity (or speed, if considering scalar quantities) [m/s]. Substituting these units into (12.1) results in:

$$J \equiv (kg)\left(\frac{m^2}{s^2}\right) \tag{12.2}$$

Dividing through by mass results in:

$$\boxed{\frac{J}{kg} = \frac{m^2}{s^2}} \tag{12.3}$$

where the LHS is "specific energy" (i.e., energy per unit mass). The RHS represents the kinematic units equivalent to the thermodynamic units on the LHS.

One dimension of the Skew-T is pressure (p), which, according to Table 1.4, is defined by:

$$p \equiv \frac{F}{A} \tag{12.4}$$

where pressure is in Pascals [Pa], F is force [N], and A is the area over which the force is acting [m^2]. Some additional manipulation shows that:

$$\left[\frac{p}{\rho}\right] = [\alpha p] = \frac{m^2}{s^2} = \frac{J}{kg} \tag{12.5}$$

where the identity on the far RHS comes from (12.3). (The details of this unit analysis are left to the student to solve in one of the practice problems at the end of this chapter.) The total potential energy per unit mass through a vertical region of a profile is obtained by vertically integrating (12.5), and substituting the *difference* between the specific volume of the lifted parcel and the environmental air:

$$\boxed{E \equiv -\int_{p_i}^{p_f} (\acute{\alpha} - \alpha) \, dp} \tag{12.6}$$

where E is energy per unit mass [J/kg], $\acute{\alpha}$ is the specific volume of the lifted parcel [m^3/kg], α is the specific volume of the environmental air [m^3/kg], and

p_i and p_f are the pressures at the base and top of the region of the integration, respectively [Pa].

Because specific volume isn't a standard parameter shown on a thermodynamic diagram, we need to rewrite (12.6) in more conventionally available terms. The relation shown in (8.7) states that:

$$p = \rho R_d T_v$$

where p is the total atmospheric pressure [Pa], ρ is density [kg/m^3], R_d is the individual gas constant for dry air (286.8 J/(kg K)), and T_v is the virtual temperature [K]. Rewriting this in terms of the specific volume yields:

$$p\alpha = R_d T_v \tag{12.7}$$

or:

$$\alpha = \frac{R_d T_v}{p} \tag{12.8a}$$

for the environmental air, and:

$$\acute{\alpha} = \frac{R_d \acute{T}_v}{p} \tag{12.8b}$$

for the lifted parcel. Substituting these identities into (12.6) results in:

$$E = -\int_{p_i}^{p_f} \left(\frac{R_d \acute{T}_v}{p} - \frac{R_d T_v}{p} \right) dp \tag{12.9}$$

Pulling out the constant R_d, and combining the common denominator of pressure yields:

$$E = -R_d \int_{p_i}^{p_f} \left(\acute{T}_v - T_v \right) \frac{dp}{p} \tag{12.10}$$

In the Midlatitudes, virtual temperature is usually within one Kelvin of simple temperature,[7] therefore a suitable approximation can be made by:

$$\boxed{E \cong -R_d \int_{p_i}^{p_f} \left(\acute{T} - T \right) \frac{dp}{p}} \tag{12.11}$$

The relation shown in (12.11) is difficult to solve analytically, because the temperature of both the lifted parcel and the environmental air are dependent on pressure. The integral is usually solved by adding up the sum of many algebraic approximations through the depth of the integrated region (Riemann Sums). Using this method, and the fact that the integral of dp/p is $ln(p)$, we obtain the approximate solution given by:

$$E \cong -R_d \sum_{n=1}^{N} (\acute{T}_n - T_n) ln \left(\frac{p_{tn}}{p_{bn}} \right)$$ (12.12)

where R_d is the individual gas constant for dry air (286.8 J/(kg K)), $\acute{T}_n$ and T_n are the mean temperatures of the lifted parcel and the environmental air in the small vertical region n, respectively [K], p_{tn} and p_{bn} are the pressures at the top and bottom of the small vertical region n, respectively [Pa],[8] and N is the total number of small regions over which the Riemann Sum is computed. The smaller the vertical sections are, the closer the numerical solutions to (12.12) will be to an analytical solution for (12.11).

Because p_{bn} is greater than p_{tn}, the negative sign out front on the RHS of (12.16) makes the result positive if the temperature of the parcel is greater than the temperature of the environmental air (indicating that the latter is denser). This is true because $ln(p_{tn}/p_{bn})$ evaluates to a negative number, because $p_{tn} < p_{bn}$. For a somewhat cleaner result, the leading negative sign can be eliminated by reversing the order of integration in (12.11), or by rewriting the approximation in (12.12) as:

$$E \cong R_d \sum_{n=1}^{N} (\acute{T}_n - T_n) ln \left(\frac{p_{bn}}{p_{tn}} \right)$$ (12.13)

where we have inverted the pressure ratio.

Vertical energy density is the vertical distribution of the potential energy in the profile and is obtained by dividing (12.11) or (12.13) by the linear depth over which the integral (or Riemann Sum) is computed. Mathematically:

$$\dot{E} \equiv \frac{E}{z}$$ (12.14)

where $\dot{E}$ is the vertical energy density per unit mass (units discussed in the following text), and z is the depth of the region of integration in meters.

Relationship to vertical acceleration. In Chapter 11, the relationship in (11.24) stated that:

$$\frac{dw}{dt} = g \left(\frac{\acute{\alpha} - \alpha}{\alpha} \right)$$

where the dw/dt is the vertical acceleration imparted to a lifted parcel [m/s] with specific volume $\acute{\alpha}$ [m³/kg] in an environment with specific volume α [m³/kg]. g is the usual effective gravitational acceleration ($\cong 9.81$ m/s^2 in the Midlatitudes).[9] The relationship in (11.18) showed that:

$$\frac{dw}{dt} = g\left(\frac{\acute{T}_v - T_v}{T_v}\right)$$

which can be approximated (to within about 5 percent) by:

$$\frac{dw}{dt} \cong g\left(\frac{\acute{T} - T}{T}\right)$$

Both sides of all three relations have units of acceleration [m/s^2]. The unit analysis shown in (12.5) concludes that the potential energy computed by (12.11), or estimated by (12.13), has units of m^2/s^2. In other words, a unit analysis shows that:

$$\left[E\right] = \left[\frac{dw}{dt}\right]\left[dz\right] \tag{12.15}$$

where the LHS is the potential energy derived in this chapter. The RHS consists of a vertical acceleration (dw/dt) applied through some shallow region of the atmosphere (dz). Completing the unit analysis shows the relationship explicitly:

$$\left(\frac{m^2}{s^2}\right) = \left(\frac{m}{s^2}\right)(m) \tag{12.16}$$

Physically, this means that the specific energy per unit mass (E) imparted to a lifted parcel is equal to the vertical acceleration acting on the parcel (computed by one of the three relations shown previously) multiplied by the vertical distance over which the acceleration is applied (RHS). It also means that vertical energy density per unit mass ($\dot{E}$) has units of acceleration, which can be shown by dividing both sides of (12.15) through by dz, or conducting a unit analysis of (12.14):

$$\left[\dot{E}\right] = \left[\frac{E}{z}\right] = \left(\frac{m^2}{s^2}\right)\left(\frac{1}{m}\right) = \frac{m}{s^2} \tag{12.17}$$

Stability, CAPE, and CINH. If E is positive, then the vertical region (defined by the limits of integration) provides a positive (upward) acceleration to the lifted parcel. In this case, we define the quantity as Convectively Available Potential Energy (CAPE). If it is negative, then the vertical region provides a negative (downward) acceleration to the lifted parcel, and we define it as Convective Inhibition (CINH).

Conventional stability indexes, such as the SI and the LI, only estimate the potential energy at a single level, that is, 500 hPa. (They can also be thought of as estimates of the vertical energy density at 500 hPa.) If one of these indexes is *negative* (the lifted parcel is warmer than the environment), it has determined that CAPE exists for the lifted parcel at 500 hPa. If the index is *positive* (the environment is warmer than the lifted parcel), it has determined that CINH exists for the lifted parcel at 500 hPa. (Figure 12.1 shows the former case.) A key point is that the indexes only estimate the potential energy of the parcel at a single level (a narrow range of dz): *For this reason, there is no 1:1 correspondence between stability, as estimated from an index, and potential energy computed by* (12.11) *or* (12.13). This is especially true for the simpler indexes, such as the SI or LI.

There is often a negative energy region (CINH) near the surface, especially in the early morning hours, when a radiation inversion may be present.[10] Other types of inversions may also represent negatively buoyant regions.[11] There are two ways a rising parcel can escape these low-level negative energy regions: (1) By applying an external, mechanical lifting force (such as a front or a topographic barrier), or (2) by destroying the negative energy region with diabatic heating (usually by the Sun). Once the parcel reaches the positive energy region, it receives an upward acceleration because of its own buoyancy, with no further application of an outside source required. This means it will continue rising until reaching another region of CINH, which is usually in the stratosphere. Figure 12.1 shows an example where the break-through point for a mechanically lifted parcel originating at 850 hPa is near 745 hPa.

Mechanical lifting, LFC, and EL. Obviously, the choice of the limits of integration p_i and p_f play a dominant role in determining the numerical answer obtained in (12.11) or (12.13), and we would like to avoid making arbitrary choices. Most of the many variations in different versions of computed CAPE arise from the choice of these limits. In the most basic form, the initial pressure (p_i) should be the lowest point at which a lifted parcel taps into the positive energy region. For a parcel lifted by mechanical means, this point is called the Level of Free Convection (LFC). The LFC is defined as the elevation at which the temperature of a lifted parcel (following a dry adiabat until saturated, then a moist adiabat above the LCL) first exceeds the environmental temperature profile at a higher level, making the parcel warmer than the environment.[12] Below the LFC, the parcel temperature is colder than the environmental temperature, and this is therefore a region of CINH. Above the LFC, the parcel enters a region of CAPE.

The final pressure (p_f) for the integration is the highest point in the positive energy area. This is called the Equilibrium Level (EL), which is defined as the point at which the lifted parcel's temperature crosses the environmental temperature profile, causing it to become colder than the environment.[13] The EL is usually in the lowest part of the stratosphere, where the environmental temperature profile either becomes isothermal, or actually begins *increasing* with height. Figure 12.2 illustrates the relationship between the LCL, LFC, EL, CAPE, and CINH for a surface parcel lifted by

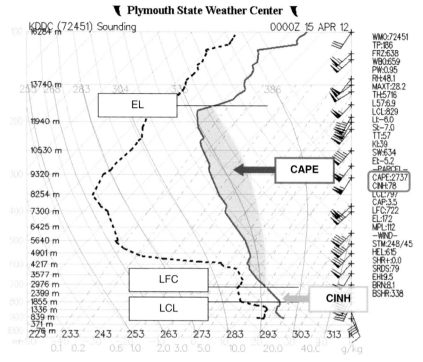

Figure 12.2. Parameters for a surface parcel lifted by mechanical forcing. Relationships between LCL, LFC, EL, CAPE, and CINH for a surface parcel lifted by mechanical means are shown. The yellow line indicates the vertical path of the parcel originating at the surface. Automatically computed values of CAPE and CINH [J/kg] are shown in the analysis column on the right. (Background image and profile data courtesy of Plymouth State University meteorology program, 2012.)

mechanical means. The sample sounding used is the same profile from Dodge City used in Figure 12.1. Computed values of CAPE and CINH are shown in the automated analysis column on the right side of the figure.

Convective lift. Solar heating of Earth's surface can also result in the positive vertical acceleration of a parcel, provided the diabatic heating is strong enough. If the environmental lapse rate (γ_e) near the surface becomes absolutely unstable ($\gamma_e > \Gamma_D$),[14] then any parcel originating in the superadiabatic region will be positively buoyant and rise given the slightest disturbance. In this case, visible cloud bases form at the Convective Condensation Level (CCL), which is usually slightly higher than the LCL[15]. The CCL is computed on a Skew-T by following the surface (2-m) dew point (T_d) vertically along a mixing ratio line, until it intersects the environmental temperature profile.[16] Cloud bases will begin forming at the CCL if the surface temperature reaches the Convective Temperature (T_c), which is computed by identifying the temperature at the CCL, and taking it dry adiabatically back to the surface. In the case of convective lift, the CCL (instead of the LFC) is the lower limit of integration in

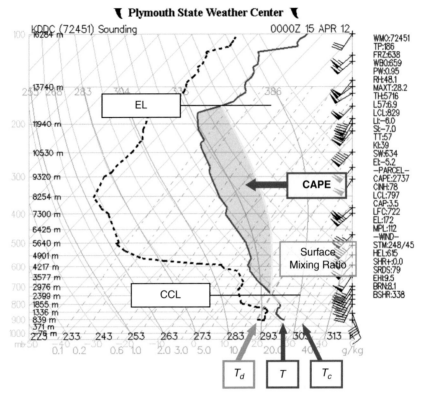

Figure 12.3. Parameters for a surface parcel lifted by convective forcing. Relationships between CCL, EL, CAPE and T_c for a parcel lifted by surface heating are shown. T_d is the surface (2-m) dew point. T is the surface (2-m) temperature. T_c is the convective temperature. The yellow line indicates the vertical path of parcel originating at the surface. (Background image and profile data courtesy of Plymouth State University meteorology program, 2012.)

(12.11), while the EL is still the upper limit. The relationship of the CCL and the T_c to CAPE and the EL are shown in Figure 12.3.

A final word of caution about using the methods described. CAPE and CINH represent the vertical accelerations acting on an isolated parcel. Putting aside for the moment that an "isolated parcel" is a mathematical fiction, you must keep in mind that a strong downward (upward) *acceleration* does not necessarily imply strong downward (upward) *motion*. For example, a lifted parcel tapping into a region of CAPE in the middle troposphere will inevitably encounter a strong region of CINH farther up, usually in the stratosphere above the EL. This does not mean that the rising parcel stops at the EL, but may in fact overshoot by a few thousand meters. This means that the tops of the cumuliform clouds developing in a strong updraft may exceed the EL and "punch" into the stratosphere, forming an overshooting dome.

12.3. Wind Shear and Helicity

Provided there is a sufficient amount of low-level moisture (water vapor, fog, or stratus clouds) to provide the raw material for building large clouds, a triggering mechanism (such as a front, orographic lift, or intense Solar heating), and plenty of static instability (quantified by CAPE), there's a very good chance a thunderstorm will develop. The thunderstorm begins as a series of intense updrafts, visible as building cumuliform clouds, which lift moist parcels from near the surface to the top of the troposphere. These are followed by the development of downdrafts, initially triggered by falling precipitation particles. If the precipitation evaporates on the way down, heat is removed from the surrounding air in the thunderstorm downdrafts. This cools the air, increases its density, and increases the speed and size of the downdrafts. Eventually, the downdrafts build a dome of cool, dense air directly below the cloud base. The cold dome (sometimes called the *mesohigh*) cuts off the supply of warm, unstable air, and chokes off the updrafts. After this, the storm rapidly breaks down into stratiform and debris clouds.

These "pulse" type thunderstorms ordinarily last between thirty and sixty minutes, which is not usually enough time for them to develop into *supercells*.[17] But, if the downdraft becomes physically separated from the updraft, so that the low-level cold pool does *not* interfere with the updraft's supply of warm, moist, unstable air, the thunderstorm can continue for a longer period of time and grow to a much larger size. These long-lived thunderstorms often become severe[18] and may develop into supercells. Providing a comprehensive review of supercell thunderstorm development is beyond the scope of this book, but we do have enough space to discuss it a little.

In the atmosphere, *wind shear* provides the mechanism to separate the updraft and downdraft. By wind shear, we mean variations in wind speed or direction (or both) with height. Variation in wind direction with height can also impart rotation to the storm, which, when conserved at progressively smaller and smaller horizontal scales, can cause the storm's central updraft to rotate rapidly, creating a *mesocyclone*. Mesocyclones are the sheath within which tornadoes usually develop, provided there are additional factors present.[19] Let's discuss two methods of evaluating wind shear.

Hodographs[20] are a method of graphically evaluating vertical wind shear in the atmosphere, and represent an alternative way of displaying the wind data recorded by radiosonde balloons. While thermodynamic diagrams focus on the pressure, temperature, and water vapor data recorded by the radiosonde, hodographs focus on the wind (kinematic) data.

To draw a hodograph, the observed wind at different altitudes is first plotted as a series of vectors, with the tail of each beginning at the origin of a rectangular coordinate system. Figure 12.4 shows a hypothetical example. Depending on how the wind direction changes with height, and *whether* it changes, the hodograph may rotate

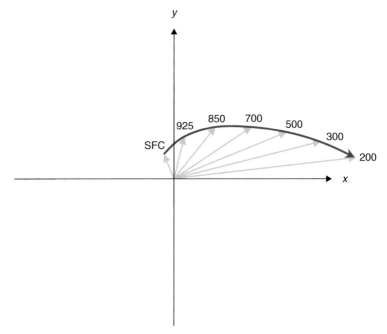

Figure 12.4. Constructing a hodograph using observed winds at different elevations. Wind vectors from the surface (SFC) through 200 hPa are plotted as light gray arrows with their tails at the origin. The lengths of the wind vectors are proportional to the associated wind speeds. The vectors point in the direction that the wind is blowing *toward*. The *x*-axis corresponds to the east-west dimension; the *y*-axis to the north-south dimension. The hodograph (red arrow) is drawn by connecting the heads of the wind vectors, going from lowest to highest. In this case, the wind speed increases and rotates clockwise (veers) with elevation, indicating the presence of warm-air advection through the vertical region represented.

clockwise, counterclockwise, or not at all. Clockwise hodographs (veering) are associated with warm-air advection (WAA), and counterclockwise hodographs (backing) are associated with cold-air advection (CAA).[21] Cartoon versions of three possibilities are shown in Figure 12.5. *Generally speaking, the longer the hodograph, the more shear that is present, and the better the chance thunderstorms have of developing into rotating supercells.* Supercell thunderstorms may occur with all three hodograph types.

• Straight-line hodographs (indicating a consistent direction, but with speed increasing with height; center panel of Figure 12.5) may produce both cyclonically and anticyclonically rotating supercells, but they do not favor either. A single, nonrotating cell may develop and split into two counterrotating cells. The *cyclonically* rotating supercell then moves to the *right* of the mean flow direction, and the *anticyclonically* rotating supercell moves to the *left* of the mean flow. These are usually called *right-movers* and *left-movers*.

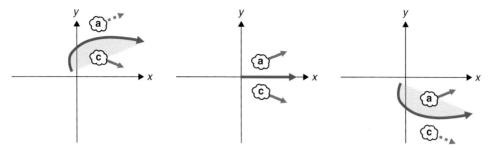

Figure 12.5. Hypothetical hodographs indicating clockwise rotation of the wind (veering), no rotation, and counterclockwise rotation of the wind (backing) with height. The *x*-axis corresponds to the east-west dimension; the *y*-axis to the north-south dimension. Storm clouds are labeled "c" (for cyclonically rotating storms) and "a" (for anticyclonically rotating storms). The hodographs (red arrows) were drawn by connecting the heads of the ambient wind vectors, going from lowest to highest. Wind vectors were then removed from the plots. The left panel shows directional *veering* (favoring *right-movers*); the center panel shows speed shear only; the right panel shows directional *backing* (favoring *left-movers*). Veering is associated with warm-air advection; backing with cold-air advection. All three panels indicate the presence of a significant amount of shear. Helicity (discussed in the following text) is shaded in red.

- Clockwise hodographs (left panel in Figure 12.5) may produce both left- and right-movers, but the right-movers will be more severe. (The reason for this is explained in the following text.) A real clockwise hodograph, based on a RAOB recorded above Dodge City, Kansas, on April 15, 2012, 0000 UTC, is shown in Figure 12.6. The latter was associated with a severe storm outbreak that produced more than 120 tornado reports across the Great Plains region of the United States. Severe weather (especially tornadoes) associated with clockwise hodographs and cyclonically rotating right-movers is more common than severe weather associated with counterclockwise hodographs in the United States.[22]
- Counterclockwise hodographs (right panel in Figure 12.5) may also produce both left- and right-movers, but the left-movers will be more severe. In the United States, this is less likely to produce tornadic thunderstorms that the reverse (mentioned previously), but it still occurs.

Hodographs are a useful tool for *qualitatively* diagnosing the amount of vertical shear present, and for making provisional forecasts of whether right-movers or left-movers with be favored by the large-scale flow. These diagrams can be constructed from radiosonde data, but there are also other sources of wind data, such as the VAD Wind Profiles (VWPs),[23] which are produced by the WSR-88D weather RADAR installations in the United States. VWPs are updated as often as once every four minutes, and can extend as high as 70,000 feet (21,300 m, which is usually well into the stratosphere) AGL. Hodographs can also be constructed from model data, using applications such as AWIPS[24] and FX-Net.[25]

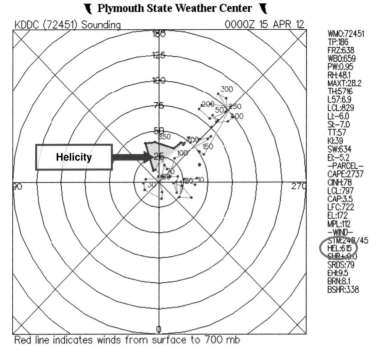

Red line indicates winds from surface to 700 mb

Figure 12.6. Hodograph for sounding from KDDC on April 15, 2012, 0000 UTC. The heavy red line indicates the hodograph from the surface through 700 hPa. The lightly shaded red area beneath the hodograph represents the surface-700 hPa helicity. Computed helicity (discussed in the following text) is highlighted in the automated analysis column on the right. The value circled has units of m^2/s^2. (Image courtesy of Plymouth State University meteorology program, 2012.)

Helicity provides a method for *quantitatively* computing the amount of vertical wind shear in a given profile. It is defined as the "property of a moving fluid, which represents the potential for helical flow (i.e., flow which follows the pattern of a corkscrew) to evolve."[26] We can imagine it rotating in all dimensions, but the helical flow of interest in the development of rotating supercell thunderstorms and *tornadogenesis* is in the horizontal plane. If the wind speed *increases* with height,[27] then the flow aloft is faster than the flow directly adjacent to the surface. This creates a secondary circulation, with its axis pointing 90 degrees to the *left* of the initial flow.[28] Figure 12.7 illustrates this using a purely *westerly* wind. If the speed *decreases* with height, the axis of the secondary circulation is 90 degrees to the *right* of the initial flow. Figure 12.8 illustrates this using a purely *southerly* wind.

There are two reasons these helical circulations are of interest:

1. If a horizontally rotating vortex tube moves into a region of strong static instability (i.e., a region of high CAPE) and upward vertical motion, it may be rotated so that its rotational

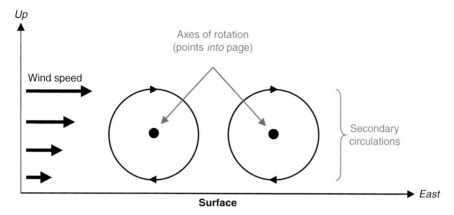

Figure 12.7. Creation of helical flow when wind speed *increases* with height. Example shown has wind from the west, with speed *increasing* with height, creating secondary overturning circulations. The axes of the new rotating cells point 90 degrees to the *left* of the original flow (i.e., toward the north, in this example), following the right-hand rule.

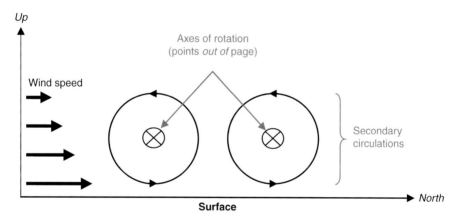

Figure 12.8. Creation of helical flow when wind speed *decreases* with height. Example shown has wind from the south, with speed *decreasing* with height, creating secondary overturning circulations. The axes of the new rotating cells point 90 degrees to the *right* of the original flow (i.e., toward the east, in this example), following the right-hand rule.

axis is moved into the vertical dimension. This can cause a preexisting, overlying thunderstorm updraft to begin rotating, or give rise to a new, rotating thunderstorm updraft.

2. If another horizontally rolling vortex tube is advected beneath an established mesocyclone, its axis of rotation may also be shifted into the vertical dimension. The connection between the new upended vortex tube and the base of the mesocyclone provides the final piece of kinematic architecture needed for the creation of a tornado.

Mathematically, helicity is similar to *vorticity*, which is defined as the curl of the velocity vector.[29] This involves the use of the "del" operator, defined by:

$$\vec{\nabla} \equiv \left(\frac{\partial}{\partial x} \hat{i} + \frac{\partial}{\partial y} \hat{j} + \frac{\partial}{\partial z} \hat{k} \right) \tag{12.18}$$

where $\hat{i}$, $\hat{j}$ and $\hat{k}$ are normal vectors in the x, y and z directions, respectively. Taking the curl of a vector means taking the cross-product of the del operator with the vector, so that the total *relative*[30] vorticity is given by:

$$\vec{\nabla} \times \vec{V} \equiv \begin{vmatrix} \hat{i} & \hat{j} & \hat{k} \\ \dfrac{\partial}{\partial x} & \dfrac{\partial}{\partial y} & \dfrac{\partial}{\partial z} \\ u & v & w \end{vmatrix} \tag{12.19}$$

where the brackets on the RHS indicate that you should compute the *determinant*, and u, v, and w are the x, y, and z components of the velocity vector $\vec{V}$. Carrying out the determinant results in:

$$\begin{vmatrix} \hat{i} & \hat{j} & \hat{k} \\ \dfrac{\partial}{\partial x} & \dfrac{\partial}{\partial y} & \dfrac{\partial}{\partial z} \\ u & v & w \end{vmatrix} = \left(\frac{\partial w}{\partial y} - \frac{\partial v}{\partial z} \right) \hat{i} - \left(\frac{\partial w}{\partial x} - \frac{\partial u}{\partial z} \right) \hat{j} + \left(\frac{\partial v}{\partial x} - \frac{\partial u}{\partial y} \right) \hat{k} \tag{12.20}$$

where the first term on the RHS is $\vec{\xi}$ (the component of vorticity in the $\hat{i}$ or eastward direction), the second term on the RHS is $\vec{\eta}$ (the component of vorticity in the $\hat{j}$ or northward direction), and the third term the RHS is $\vec{\zeta}$ (the component of vorticity in the $\hat{k}$ or upward direction).[31]

If we assume a hydrostatic condition, where vertical velocities are (by definition) identically zero, or simply note that, for most of the atmosphere, the vertical component of velocity (w) is at least two orders of magnitude smaller than the horizontal components (u and v), then the first two terms on the RHS of (12.20) can be approximated by:

$$\vec{\xi} \cong \left(-\frac{\partial v}{\partial z} \right) \hat{i} \tag{12.21a}$$

and:

$$\vec{\eta} \cong -\left(-\frac{\partial u}{\partial z} \right) \hat{j} \tag{12.21b}$$

If we further assume that u and v are constant with respect to the horizontal dimensions x and y, the third term ($\vec{\zeta}$) on the RHS goes to zero, and the relative vorticity becomes:

$$\vec{\nabla} \times \vec{V} \cong -\frac{\partial v}{\partial z}\hat{i} - \left(-\frac{\partial u}{\partial z}\right)\hat{j} = -\frac{\partial v}{\partial z}\hat{i} + \frac{\partial u}{\partial z}\hat{j} \tag{12.22}$$

Storm-relative helicity (SRH) combines the approximations of the horizontal vorticity vectors ($\vec{\xi}$ and $\vec{\eta}$) shown in the preceding text, with information about convergence. Mathematically, SRH is defined by:

$$SRH \equiv \left| \int_0^H \left[\underbrace{(\vec{V}_h - \vec{C})} \cdot \underbrace{\left(\hat{k} \times \frac{\partial \vec{V}}{\partial z} \right)} \right] dz \right| \tag{12.23}$$

$$\quad\quad\quad\quad\quad \text{Term A} \quad\quad\quad \text{Term B}$$

where $\vec{V}_h$ is the ground-relative horizontal wind vector [m/s], $\vec{C}$ is the ground-relative horizontal storm motion vector [m/s], $\hat{k}$ is the directional unit vector pointing upward [unitless], $\partial \vec{V}/\partial z$ is the vertical gradient in the ground-relative horizontal wind vector [s⁻¹], and dz is the variable of integration [m]. The integral is evaluated between the surface (0 meters AGL) and a selected height H, which is usually 3000 meters AGL (~700 hPa for stations near sea level).[32]

Convergence Term. Term A in (12.23) represents the effect of the ground-relative wind on a given storm cell. If $\vec{V}_h$ (the wind) is greater than $\vec{C}$ (the storm's motion), the term is positive. In this case, *low-level convergence* is indicated (because the integral is evaluated in the lowest 3000 m of the atmosphere), which is a factor that tends to make the storm stronger. If the term is negative, a *low-level divergence* is indicated, which is generally detrimental to storm development.

Term A also serves to scale the importance of Term B. If the storm's ground-relative *motion* is about the same as the ambient ground-relative *wind*, the storm is relatively unaffected by the vertical variations in the vertical variation of the wind's horizontal components (discussed discussed in the following text). But if the ambient wind is much *faster* than the storm's motion, then the wind's influence on the storm is much greater.

Vertical Shear Term. Term B in (12.23) represents the vertical shear. This can be illustrated by performing the cross-product; that is:

$$\hat{k} \times \frac{\partial \vec{V}}{\partial z} \equiv \begin{vmatrix} \hat{i} & \hat{j} & \hat{k} \\ 0 & 0 & 1 \\ \dfrac{\partial u}{\partial z} & \dfrac{\partial v}{\partial z} & \dfrac{\partial w}{\partial z} \end{vmatrix} = \left(0 - \frac{\partial v}{\partial z} \right)\hat{i} - \left(0 - \frac{\partial u}{\partial z} \right)\hat{j} + \left(0 - 0 \right)\hat{k} \tag{12.24}$$

or:

$$\hat{k} \, x \frac{\partial \vec{V}}{\partial z} = -\frac{\partial v}{\partial z} \hat{i} + \frac{\partial u}{\partial z} \hat{j} \tag{12.25}$$

which, by comparison to (12.22), is identical to our approximation of $\vec{\nabla} \times \vec{V}$. Let's take a close look at the two terms in (12.25).

- $\hat{i}$ direction: If the northward component of the low-level wind vector (v) *decreases* with height, the vertical gradient $\partial v / \partial z$ is *negative*, and the overall term is *positive* because of the additional negative sign out front. Furthermore, this implies a horizontally oriented vortex tube with its axis pointing in the *positive x* direction (Figure 12.8).
- $\hat{j}$ direction: If the eastward component of the low-level wind vector (u) *increases* with height, the vertical gradient $\partial u / \partial z$ is *positive*, and the overall term is also *positive*. This also implies a horizontally oriented vortex tube with its axis pointing in the *positive y* direction (Figure 12.7).

The *large-scale* implications of the combination of these two conditions is shown graphically in Figure 12.9: clockwise rotation of the overall wind direction with increasing height (*veering*), associated with WAA, which is a well-known severe weather signature. Wind direction veering with height is also associated with a clockwise hodograph, such as the example shown in Figure 12.6. But the *small-scale* implications indicate a horizontally rotating vortex tube with its axis oriented toward the northeast (because the $\hat{i}$-dimensional vortex tube points toward positive x, and the $\hat{j}$-dimensional vortex tube points toward positive y). If rotated into the vertical dimension in a region of high CAPE and strong upward vertical motion, this will create a *cyclonically* rotating (counterclockwise) thunderstorm.

With this analysis of the terms, we can rewrite (12.23) by:

$$\text{SRH} = \int_0^H \underbrace{\left[\left(u_{\text{wind}} - u_{\text{storm}} \right) \hat{i} + \left(v_{\text{wind}} - v_{\text{storm}} \right) \hat{j} \right]}_{Term\ A} \bullet \underbrace{\left[-\frac{\partial v}{\partial z} \hat{i} + \frac{\partial u}{\partial z} \hat{j} \right]}_{Term\ B} dz \tag{12.26}$$

where u_{wind} and v_{wind} represent the x and y components of the ground-relative wind vector, and u_{storm} and v_{storm} represent the x and y components of the ground-relative motion of the storm, respectively. Carrying out the dot product results in:

$$\text{SRH} = \int_0^H \left[\underbrace{\left(u_{\text{wind}} - u_{\text{storm}} \right) \left(-\frac{\partial v}{\partial z} \right)}_{\hat{i}\ Term} + \underbrace{\left(v_{\text{wind}} - v_{\text{storm}} \right) \left(\frac{\partial u}{\partial z} \right)}_{\hat{j}\ Term} \right] dz \tag{12.27}$$

To perform the integral in (12.27), the Riemann Sum method could be applied by beginning at Earth's surface and:

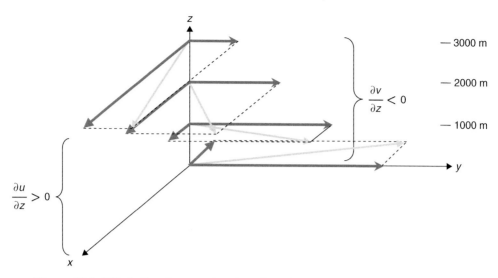

Figure 12.9. Wind direction veering (rotating clockwise) with increasing height. *x* represents east; *y* is north; *z* is up. Height scale is shown at right (AGL). Wind components in 1000-m increments are shown as thick, gray arrows. The total wind vectors are shown as lighter gray arrows. The surface wind is from the southeast, and therefore, has a negative *u*-component and positive *v*-component. Moving upward, the *u*-component first reverses sign, then grows increasingly positive, while the v-component steadily decreases. Seen from above, this corresponds to a clockwise-rotating hodograph.

- Taking the difference between the mean values of the components of the wind and the components of the storm's motion through a shallow vertical layer.
- Computing the vertical gradients in the two horizontal wind components through the same shallow vertical layer.
- Adding the $\hat{i}$ and $\hat{j}$ terms together.
- Multiplying the result by ΔZ (the depth of the shallow vertical layer).
- Moving upward to the next shallow vertical layer and repeating the first four steps. Add the new result to the previous total.
- Repeating until reaching H.

Correlations of SRH with tornado intensity. An observational study by Kerr and Darkow (1996), using 184 tornado proximity soundings, found that mean SRH for all events was 142 m²/s², and that SRH showed some utility at differentiating the destructive capacity of the tornado. Table 12.1 summarizes their results. While the mean SRH values showed differences, there was considerable overlap (i.e., large standard deviations) in the SRH associated with the different tornadic intensities.

SRH may be better at estimating the potential for supercell thunderstorms than it is at estimating tornado intensity, but it is still highly dependent on the depth of integration. Rasmussen and Blanchard (1998) computed 0–3 km SRH, and Rasmussen

Table 12.1. *Correlations of mean SRH based on observed proximity sounds with observed tornado intensity (Kerr and Darkow, 1996).*

The study involved 184 tornadoes, none of them reaching F5 intensity.

F-Scale[i]	F0	F1	F2	F3	F4
Mean helicity value $[m^2/s^2]$	66	140	196	226	249

[i] Fujita tornado intensity scale. Since 2007, the U.S. National Weather Service has used a revised version called the Enhanced Fujita (EF) scale.

Table 12.2. *Correlations of mean 0–3 km SRH and 0–1 km SRH to resulting thunderstorm types (Rasmussen and Blanchard, 1998; Rasmussen, 2001).*

ORD = Ordinary thunderstorms; SUP = Supercell thunderstorms; TOR = Tornadic supercells.

	ORD	SUP	TOR
0–3 km SRH	55	124	180
0–1 km SRH	15	33	89

(2001) computed 0–1 km SRH and compared the results to the thunderstorms that occurred. Both classified thunderstorms as ordinary (ORD), supercell (SUP), and tornadic (TOR). Table 12.2 summarizes their combined results. Once again, the mean results indicate some skill for separating storm types, with 0–1 km SRH outperforming 0–3 km SRH, but as before, the variation about the mean computed values causes considerable overlap between the SRH associated with storm types.

Figure 12.10 shows the 0–3 km (above sea level) *helicity* (as opposed to SRH, discussed further in the following text) contoured for the mainland United States, as well as parts of Mexico and Canada. The example shown is based on RAOBs recorded on April 15, 2012, at 0000 UTC, and is therefore valid for the same severe storm outbreak discussed in this and Chapter 11. Unlike the conventional stability indexes discussed in Chapter 11 (which usually identified the correct storm area, but also either underestimated the total area, or indicated the possibility of severe storms where none actually occurred), helicity rather accurately identifies the observed severe weather area, shown by Storm Prediction Center reports in Figure 12.11.

Unit analysis and final comments on helicity. Because the $\hat{i}$ and $\hat{j}$ terms in (12.27) are similar, we need only examine the units of one of them, plus the units of the variable of integration. Explicitly, we can write:

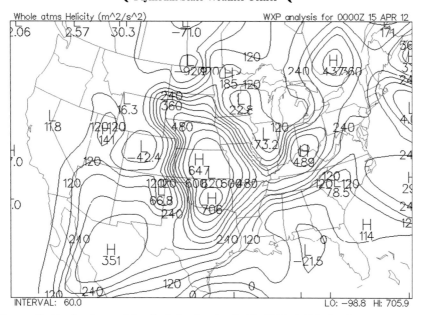

Figure 12.10. Surface – 3 km helicity for April 15, 2012, 0000 UTC. Severe storms occurred during the twelve-hour period prior to the valid time of this chart, over an area from west Texas through central Nebraska, and eastward across Iowa to western Illinois (see Figure 12.11). The helicity computed from RAOBs shows a prominent region of high values from Texas through Nebraska but much lower values in Illinois. (Plot courtesy of Plymouth State University meteorology program, 2012.)

$$\left[\left(u_{\text{wind}} - u_{\text{storm}}\right)\right]\left[\left(-\frac{\partial v}{\partial z}\right)\right][dz] = \left(\frac{m}{s}\right)\frac{\left(\dfrac{m}{s}\right)}{(m)}(m) = \frac{m^2}{s^2} \tag{12.28}$$

which, according to (12.3), is the same as J/kg. In other words, the units of SRH are equivalent to the units as CAPE and CINH. SRH and helicity therefore represent another form of specific energy – in this case, *kinetic* energy per unit mass.

SRH can only be computed if the individual storm motion is known, but simple helicity (using only the ground-relative wind velocity in Term A) can be computed directly from a RAOB, like any other scalar parameter. Both are proportional to the area beneath the relevant hodograph for the region selected and provide a method of quantitatively determining the amount of shear represented in the hodograph. Figure 12.6 highlights the area beneath the surface −700 hPa hodograph and shows the numerical value of the computed helicity in the automated analysis column on the right side of the figure.

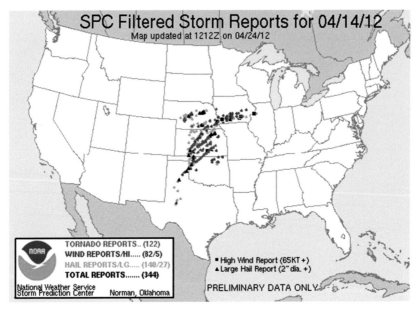

Figure 12.11. Severe storm reports for April 14, 2014. There were more than 120 reports of tornadoes in the region east of KDDC in one day. (Plot courtesy of Storm Prediction Center, 2012.)

12.4. Combining CAPE and SRH

Supercells need both static instability and shear-driven rotation to develop, but neither of these parameters need be extreme. Model studies have shown that mini-supercells[33] can develop with CAPE as low as 600 J/kg and that almost all supercells occur with CAPE values of less than 1500 J/kg. From this, we can conclude that once a certain level of instability is reached, additional CAPE does not necessarily imply more violent thunderstorms. Somewhat different arguments can be made about shear. Doswell (1991) states that, up to a certain point, adding more helicity tends to make storms more powerful. However, beyond some upper threshold, shear begins to be destructive to thunderstorms, breaking them apart into smaller storms more rapidly than they can develop into supercells, or preventing them from initially forming.[34] Doswell (1991) also points out that there are no "magic numbers," acting as lower and upper shear thresholds for supercell development. Each situation must be carefully evaluated in terms of instability, shear, and many other parameters indicative of severe storm development.[35]

The Severe Weather Threat (SW) index combines both thermodynamic information (using the TT index) and kinematic information (using the two wind speed terms and the shear term), and therefore is a more complete (and successful) method of estimation of the potential for severe convective storms than the simpler indexes that rely exclusively on thermodynamic parameters.[36] Similarly, we can combine CAPE (to capture thermodynamic information in the form of potential energy) and SRH (to

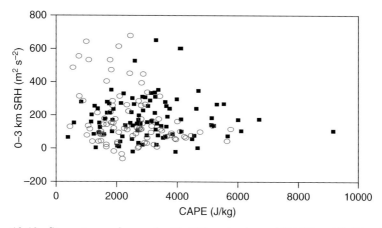

Figure 12.12. Storm types observed with different values of CAPE and 0–3 km SRH. Nontornadic supercells are shown as open circles; tornado-producing supercells are squares (Edwards and Thompson, 2000).

capture kinematic information in the form of kinetic energy) to create a method of estimating severe thunderstorm potential.

Johns et al. (1990) found a broad range of possible CAPE and 0–2 km SRH combinations in soundings associated with tornadoes, which they classified as either "strong" (F2 or F3) or "violent" (F4 or F5).[37] Strong tornadoes were associated with mean SRH values of about 360 m^2/s^2, and violent tornadoes were associated with 0–2 km SRH values of about 450 m^2/s^2. A more recent study used proximity soundings produced by the Rapid Update Cycle 2 model (RUC2), and showed some relationships between CAPE and 0–3 km SRH that may be used to help differentiate tornado-producing from nontornadic supercells (Edwards and Thompson 2000). Their results are shown in Figure 12.12.

Some of the conclusions that can be drawn from Figure 12.12 are:

- In all cases, the 0–3 km SRH was 700 m^2/s^2 or less, and the bulk of the tornadic supercells studied occurred with SRH of less than 400 m^2/s^2. That no cases of either type occurred with SRH values of more than 700 m^2/s^2 appears to confirm Doswell's (1991) point that excessive shear is destructive to severe thunderstorms.
- Both tornadic and nontornadic supercells occurred with CAPE values of less than 1500 J/kg and SRH values of less than 100 m^2/s^2. In some cases of both types, CAPE was less than 500 J/kg. Very high values of either or both parameters are not needed to produce severe weather.
- There were very few cases of either type occurring with negative SRH. (When they did occur, negative SRH values were small.) This appears to support the idea that some minimal amount of shear is necessary to separate the updraft from the downdraft and permit the development of larger, longer-lived storms.
- At high values of SRH (above about 400 m^2/s^2), and low to moderate values of CAPE (less than 4000 J/kg), nontornadic supercells dominate. Holding SRH steady while increasing CAPE made tornadic-supercells more likely.

- With CAPE values above 6000 J/kg, in spite of SRH as low as 100 m^2/s^2, there were no nontornadic supercells, although this observation suffers from a small sample size (only four events). If accurate, this and the previous point imply that the incidence of tornadoes is more sensitive to CAPE than it is to SRH.

Finally, there are clearly uncontrolled variables at work here, such as the presence of low-level jets, convergence regions near the surface (from dry lines, outflow boundaries, etc.), divergence regions in the upper atmosphere (such as the poleward exit region or equatorward entrance region of a jet streak), and high amounts of water vapor in the lowest 1500 m of the atmosphere (originating, in the United States, from the Gulf of Mexico). Just as with the conventional stability indexes, any forecasts generated from the method discussed previously should be considered provisional. Elliott (1988) discusses a more complete analysis routine for the diagnosis of severe weather potential.

12.5. Summary of Severe Weather Applications

In this chapter, we defined two important severe weather parameters than can be computed from radiosonde data. One is a measure of thermodynamic potential energy, and the other is a measure of kinematic energy.

CAPE and CINH are both measures of the thermodynamic potential energy that either promotes (the former) or deters (the latter) development of updrafts. Together with these, we defined the LFC and the EL. The former is the level at which a *mechanically* lifted parcel taps into the positive energy region aloft (CAPE), and the latter is the level at which a positively buoyant parcel reaches an area of negative energy (usually in the stratosphere). Along the way, we also defined the CCL, which is the altitude at which the water vapor carried aloft in surface parcels lifted by *convective* means (driven by diabatic heating of Earth's surface by the Sun) condenses into liquid droplets, creating visible cloud bases. We also defined the Convective Temperature (T_c), which is the temperature at which surface-based "free convection" will begin.

CAPE and CINH are computed by applying the general potential energy equation, which states:

$$E \equiv -\int_{p_i}^{p_f} (\acute{\alpha} - \alpha)\, dp$$

where,

E = potential energy in the vertical region of the profile [J/kg]
$\acute{\alpha}$ = specific volume of lifted parcel [m^3/kg]
α = specific volume of environmental air [m^3/kg]
dp = variable of integration [Pa]
p_i, p_f = pressure at base and top of integration region [Pa]

If the result of this integral is positive, a positive (upward) vertical acceleration is implied, and the energy is defined as CAPE. If the result is negative, a negative (downward) acceleration is implied, and the energy is defined as CINH. Because specific volume is not a standard RAOB parameter, we used the Ideal Gas Law to rewrite energy in terms of virtual temperature. By also using the approximation that most of time $T \cong T_v$, we obtained:

$$E \cong -R_d \int_{p_i}^{p_f} (\acute{T} - T) \frac{dp}{p}$$

where,

    R_d = individual gas constant for dry air (286.8 J/(kg K))
    $\acute{T}$ = temperature of the lifted parcel [K]
    T = temperature of the environmental air [K]

which is approximated by the series:

$$E \cong R_d \sum_{n=1}^{N} (\acute{T}_n - T_n) ln\left(\frac{p_{bn}}{p_{tn}} \right)$$

where,

    $\acute{T}_n$ = mean temperature of the lifted parcel through a small vertical region n [K]
    T_n = mean temperature of the environmental air through the same vertical region n [K]
    p_{bn}, p_{tn} = pressures at the base and top of the vertical region n [Pa]
    N = number of vertical regions added up to estimate the area of the integral

In addition to the thermodynamic potential energy, we also discussed the importance of wind shear (providing kinetic energy) in the development of severe convective storms. Wind shear can be analyzed qualitatively by plotting hodographs of the wind data recorded by radiosondes, or quantitatively by computing helicity. Helicity is closely related to vorticity. A specialized form of helicity is SRH, which includes a convergence and scaling term that compares the ground-relative wind to the ground-relative motion of the storm. SRH is defined by:

$$SRH \equiv \int_0^H \left[(\vec{V}_h - \vec{C}) \bullet \left(\hat{k} x \frac{\partial \vec{V}}{\partial z} \right) \right] dz$$

where,

    SRH = Storm-Relative Helicity [m²/s²]
    $\vec{V}_h$ = ground-relative wind vector [m/s]
    $\vec{C}$ = ground-relative storm motion [m/s]
    $\hat{k}$ = vertical unit vector [unitless]
    $\partial \vec{V} / \partial z$ = vertical variation in the ambient wind [s⁻¹]
    dz = variable of integration [m]
    H = top of the layer evaluated [m]

and is usually evaluated for the region between the surface (0 m AGL) to 3 km AGL (or about 700 hPa). In some cases, 0–1 km and 0–2 km SRH have been evaluated. The vector form of the preceding equation reduces to the scalar form, stating that:

$$SRH = \int_0^H \left[\left(u_{\text{wind}} - u_{\text{storm}} \right) \left(-\frac{\partial v}{\partial z} \right) + \left(v_{\text{wind}} - v_{\text{storm}} \right) \left(\frac{\partial u}{\partial z} \right) \right] dz$$

where,

u_{wind} = eastward component of the ground-relative ambient wind [m/s]
u_{storm} = eastward component of the ground-relative storm motion [m/s]
$\partial v / \partial z$ = vertical variation in the northward component of the wind [s⁻¹]
v_{wind} = northward component of the ground-relative ambient wind [m/s]
v_{storm} = northward component of the ground-relative storm motion [m/s]
$\partial u / \partial z$ = vertical variation in the eastward component of the wind [s⁻¹]

Practice Problems

1. Calculate the CAPE in the layer between 700 and 300 hPa if the mean temperature difference between the parcel and the environment is 8 °C.
2. Estimate the vertical acceleration of the same parcel, using the CAPE equation. Some information from a blank Skew-T may be necessary.
3. Beginning with (12.6) and (12.23), perform a unit analysis to show that CAPE and SRH have the same units.

Severe Weather Exercise 1

On April 24, 2007, a tornado killed at least ten people in Del Rio, Texas. Please refer to the following archive:

http://vortex.plymouth.edu/~stmiller/stmiller_content/archive_contents.html

Follow the link to "04/25/2007: Del Rio Texas Tornadic Thunderstorm." Download and decompress the tar-gzipped file.

In the archive, you will find satellite and radar products, various types of surface analyses (including isobaric, streamline, and constant dew point isopleths), upper air analyses, plotted soundings from Corpus Christi (KCRP) and Midland (KMAF), and storm reports from the Storm Prediction Center.

There are several types of radar products in the archive, including 0.5 and 1.5° elevation angle reflectivity, vertically integrated liquid (VIL), one-hour precipitation (OHP), base radial velocity (Vr), and storm-relative mean radial velocity (SRM). Please review the sequence of VIL images and note the important thunderstorm cell that crossed the U.S./Mexico border at about 0000 UTC on April 25, 2007. (VIL values coded in white – the highest values on the scale.) Then review the Vr images, and note the inbound/outbound (blue/red) velocity couplet that occurred with the same cell. The latter is a mesocyclone, and is almost certainly the physical mechanism that produced the tornado.

Next, please review the rest of the data. Speculate on the triggering mechanisms present, as well as the thermodynamic instability (CAPE?) of the atmosphere, and write a few paragraphs describing your conclusions.

Severe Weather Exercise 2[38]

A thunderstorm is produced by a cumulonimbus cloud, which requires the following ingredients to form:

- Sufficient moisture (water vapor) in the atmosphere.
- A substantial lifting mechanism.
- Enough instability to permit air parcels to rise to the height where glaciation (formation of ice crystals) can occur near the top of the cloud (typically the $-20\,°C$ isotherm).

There are many possible lifting mechanisms in the atmosphere, including *convection*, *convergence*, *fronts*, and *topography*. *Wind shear* is necessary for a *severe* thunderstorm to occur. (A thunderstorm is classified as severe if it produces hail with diameter one inch or larger, wind gusts of fifty knots or greater, or tornadoes.) *Wind shear* is a change in wind speed and/or direction with height, and it allows clouds to grow larger and last longer by keeping updrafts and downdrafts within the cumulonimbus cloud separated.

This lab exercise examines thunderstorms associated with a storm system that affected a large part of the eastern United States in November 2013.

1. **Where is the moisture?** Find areas with high dew point values.
   Go to the following website: http://vortex.plymouth.edu/sfccalc-u.html
   Choose the following options to get a contour analysis of dew point temperature:
   Region: "Contiguous U.S."
   Variable: "Dew point (F)"
   Interval: "5"
   Contour Type: "line"
   For Year, Month, Day and Time, use "2013," "Nov," "17," and "12Z"
   Lightly shade the region with dew point temperatures above $50\,°F$ on Map 1 *using a green color pencil.*
2. **Where is surface-based convection possible?** Find areas with warm temperatures.
   Go to the following website:
   http://www.hpc.ncep.noaa.gov/dailywxmap/index_20131110.html
   Scroll down and find the map with "Maximum Temperatures" for Sunday, November 17, 2013. Lightly shade the region with maximum temperatures above $70\,°F$ on Map 2 *using a red color pencil.*
3. **Where are areas of meteorologically driven convergence?** Identify low pressure areas and frontal zones.

3.1. At the same website, scroll up and look at the surface analysis. Look for low pressure centers and converging isobars to find regions of surface convergence. Lightly shade convergence areas on Map 3 *using a yellow color pencil.*

3.2. Where are the fronts? Looking at the same surface analysis, examine any fronts present. Draw the fronts on Map 3 with the appropriate symbols *using a black marker.*

4. **Is there any significant topography in the likely thunderstorm areas?** Use an atlas if you need help. Lightly shade the relevant topography on Map 3 using *a brown color pencil.*

5. **Spot evaluation of instability.** Examine several Skew-T's in the Midwest and evaluate their relative stability.
   Go to the following website: http://vortex.plymouth.edu/myo/
   Scroll down until finding "Upper Air." Select "Diagrams/Data," then "Archived." Enter the appropriate date and time (November 17, 2013, 1200Z), plot the following seven Skew-T's, and enter the required data in the table below:

	LI	SI	TT	SW	CAPE	CINH
KDVN						
KILX						
KAPX						
KDTX						
KPIT						
KBNA						
KBUF						

   The map on the webpage shows the locations of the RAOB sites.

6. **Quantitative instability contours.** Evaluate several stability indices.
   Go to the following website: http://vortex.plymouth.edu/myo/upa/ctrmap-a.html
   Plot contour maps of three different stability indices: LI, SI, and SWEAT:
   Region: "Contiguous U.S."
   Level: "500 mb"
   Variable: "Lifted Index," "Showalter," and "SWEAT" (sequentially)
   Interval: Experiment until finding an appropriate contour interval for each index
   Contour Type: "line"
   For Year, Month, Day and Time, use "2013," "Nov," "17," and "12Z"
   Shade the regions with negative LI (Map 4), negative SI (Map 5), and SW values of 150 or higher (Map 6) *using a blue color pencil.*

7. **Is there the possibility of severe thunderstorms?** Examine areas of strong vertical wind shear.
   On the same website: http://vortex.plymouth.edu/myo/upa/ctrmap-a.html
   Use the following options:
   Region: "Contiguous U.S."
   Level: "all"
   Variable: "Pos. Vertical Shear"
   Interval: "1"

Contour Type: "line"

For Year, Month, Day and Time, use "2013," "Nov," "17," and "12Z"

On Map 7, lightly shade the region with wind shear above 5 ($\times 10^{-3} s^{-1}$) *using a red color pencil.*

8. **Where will the thunderstorms be?** Combine information from Maps 1–7 using a light table.

   Based on the information that you gathered, where would you expect thunderstorms to occur? Carefully *outline* the region with the possibility of thunderstorms on the Map 8 using a red pencil. Label your outlined region as "Thunderstorms." *Shade in* areas where the thunderstorms may be severe using the information from No. 7, and label as "Possibly Severe."

9. **How does your forecast compare with what happened?**

   Go to the Storm Reports section of the Storm Prediction Center website: http://www.spc.noaa.gov/climo/

   Enter "131117" for November 17, 2013

   Click "Get Data"

   Examine the map with hail, wind damage, and tornado reports from that day. Outline the area where the reports occurred on Map 8 in black, and label as "Reports."

10. **Comment on the effectiveness of the forecast.** Did you correctly forecast the area of thunderstorms? Severe thunderstorms (if any)? Which stability index was the most effective? Discuss what worked, what didn't work, and other types of information that would have been helpful for improving the forecast.

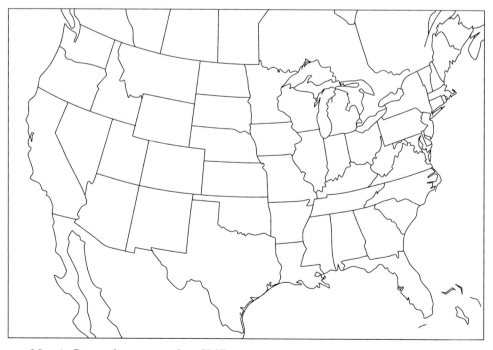

Map 1. Dew points greater than 50 °F.

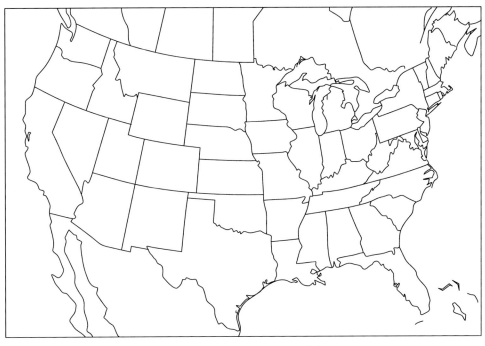

Map 2. Maximum surface temperatures greater than 70 °F.

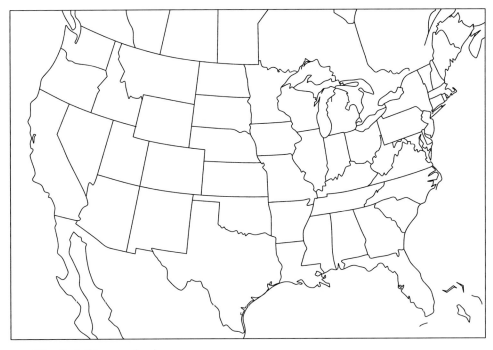

Map 3. Lifting mechanisms.

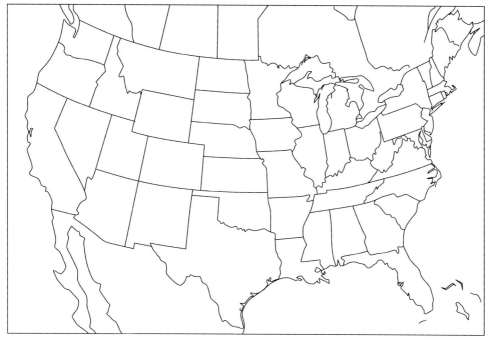

Map 4. Negative Lifted Index.

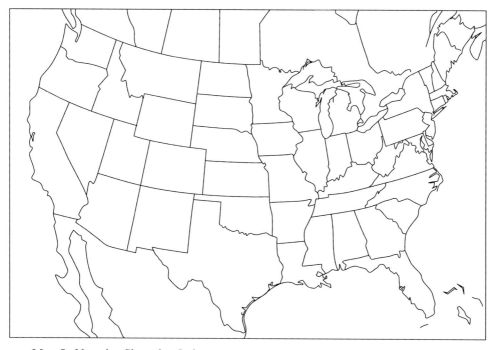

Map 5. Negative Showalter Index.

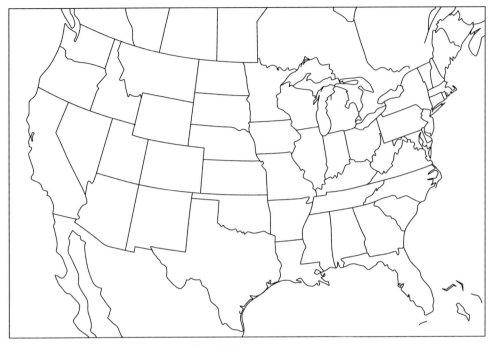

Map 6. SWEAT Index greater than 150.

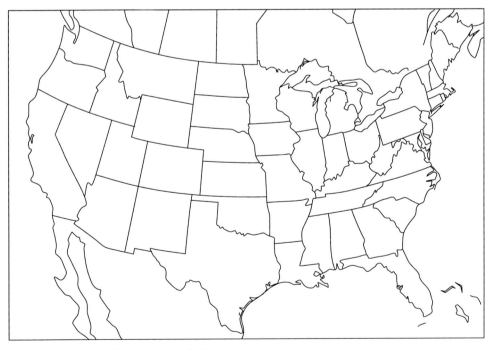

Map 7. Areas of strong vertical shear.

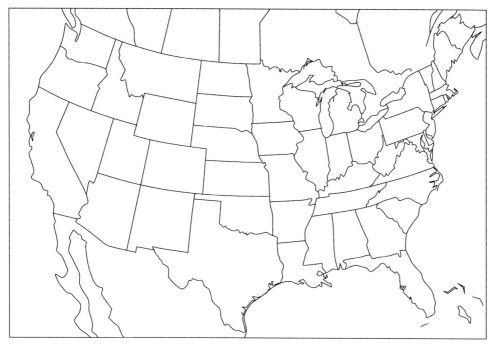

Map 8. Thunderstorm and severe thunderstorm forecast, and storm reports.

Appendix 1

List of Symbols

Symbol	Meaning
A	Area; pressure constant
$ALSTG$	Altimeter setting
a	1st Van der Waal's constant
$\vec{a}$	Acceleration
a_v	Vertical component of acceleration
B	Temperature constant
b	2nd Van der Waal's constant
C	Heat capacity
$CAPE$	Convectively available potential energy
CCL	Convective condensation level
$CINH$	Convective inhibition
CT	Cross totals
c	Specific heat coefficient
c_1, c_2, $c_3 \dots c_n$	Constants
c_p	Specific heat of a gas at constant pressure
c_p^{vapor}	Specific heat of water vapor at constant pressure
c_v	Specific heat of a gas at constant volume
c_v^{vapor}	Specific heat of water vapor at constant volume
E	Energy
$\dot{E}$	Vertical energy density
EL	Equilibrium level
e	Vapor pressure
e_0	Vapor pressure at reference temperature
e_s	Saturation vapor pressure
e_s^i	Saturation vapor pressure above ice
e_s^l	Saturation vapor pressure above liquid
$\vec{F}$	Force

(*continued*)

Symbol	Meaning
$\vec{F}_g$	Gravitational force
f	Degrees of freedom
G	Universal gravitational constant
g	Effective gravitational acceleration
$\vec{g}$	Gravitational acceleration (near Earth's surface)
g^*	Newtonian gravitational acceleration
g_0	Reference value of effective gravitational acceleration
H	Enthalpy; scaling height
h	Specific enthalpy
I	Luminous intensity
IPW	Integrated precipitable water
KI	K Index
k	Boltzmann's Constant
LCL	Lifted condensation level
LFC	Level of free convection
LI	Lifted index
l	Length; latent heat
l_f	Latent heat of fusion
l_s	Latent heat of sublimation
l_v	Latent heat of vaporization
l_{v0}	Latent heat of vaporization at $0\,°C$
M	Molar mass
$\bar{M}$	Mean molar mass of a gas ensemble
M_d	Mean molar mass of the dry air ensemble
M_v	Molar mass of water vapor
m	Mass
m_e	Mass of Earth
N	Total number of particles in an ensemble
N_A	Avogadro's Number
n	Number of moles
n_0	Number density
P	Power
PA	Pressure altitude
p	Pressure
p_0	Pressure at $0\,°C$
p_c	Critical pressure
p_d	Partial atmospheric pressure due to dry air
p_m	Melting pressure
p_{STN}	Station pressure

Symbol	Meaning
p_v	Partial atmospheric pressure due to water vapor
Q	Heat
q	Electric charge; specific heat; specific humidity
q_s	Saturation specific humidity
R^*	Universal gas constant
R_d	Individual gas constant for dry air ensemble
R_e	Equatorial radius of Earth
R_i	Individual gas constant
R_p	Polar radius of Earth
R_v	Individual gas constant for water vapor
R_φ	Radius of Earth at latitude φ
RH	Relative humidity
r	Radius; relative humidity
S	Entropy
SI	Showalter stability index
SLP	Sea-level pressure
SRH	Storm-relative helicity
SW	Severe weather threat index
$SWEAT$	Severe weather threat index
s	Specific entropy
T	Temperature
$\bar{T}$	Mean temperature through a vertical layer
T_0	Temperature at some reference point, such as sea level
T_c	Convective temperature; critical temperature
T_d	Dew point
T_e	Equivalent temperature
T_{ic}	Isentropic condensation temperature
T_M	Molecular temperature
T_v	Virtual temperature
$\bar{T}_v$	Mean virtual temperature in a vertical layer
T_w	Wet-bulb temperature
TT	Total totals
t	Time
U	Internal energy
u	Specific internal energy
V	Volume
VT	Vertical totals
v	Speed
$\vec{v}$	Velocity

(continued)

Symbol	Meaning
$\langle v \rangle$	RMS velocity
W	Work
w	Specific work; water vapor mixing ratio; vertical velocity
w_s	Saturation mixing ratio
Z	Compressibility factor
z	Distance above sea level (elevation)
z_{PA}	Pressure altitude
α	Specific volume; volume expansion coefficient
α_v	Specific volume of vapor
β	Pressure coefficient
β_0	Lapse rate in lowest layer of USSA
β_2	Lapse rate in the USSA middle stratosphere
β_3	Lapse rate in the USSA upper stratosphere
$\Delta\zeta$	Thickness
ε	Ratio of molar mass of water vapor to molar mass of dry air ensemble
Φ	Geopotential
Γ_D	Dry adiabatic lapse rate
Γ_M	Moist adiabatic lapse rate
γ	cp/cv
γ_e	Environmental lapse rate
γ_{as}	Environmental lapse rate: Absolutely stable
γ_{au}	Environmental lapse rate: Absolutely unstable
γ_c	Environmental lapse rate: Conditional
γ_{dn}	Environmental lapse rate: Dry neutral
γ_{mn}	Environmental lapse rate: Moist neutral
η	Efficiency of Carnot Cycle
κ	Ri/cp
θ	Angle; potential temperature
θ_e	Equivalent potential temperature
ρ	Density
ρ_0	Density at sea level
ρ_d	Density of dry air
ρ_v	Absolute humidity; density of water vapor
ζ	Geopotential height
Ω	Angular velocity of Earth

Appendix 2

Constants

Symbol	Name	Numerical Value	Unit
A	Pressure constant	2.5×10^{11}	Pa
B	Temperature constant	5.4×10^{3}	K
c_p	Specific heat of a dry air at constant pressure	1003.8	J/(kg K)
c_{pv}	Specific heat of water vapor at constant pressure	1844.8	J/(kg K)
c_v	Specific heat of dry air at constant volume	717.0	J/(kg K)
c_{vv}	Specific heat of water vapor at constant volume	1383.6	J/(kg K)
e_0	Vapor pressure at reference temperature	611.12	Pa
G	Universal gravitational constant	6.673×10^{-11}	Nm²/kg²
g_0	USSA sea-level effective gravitational acceleration	9.80665	m /s²
k	Boltzmann's Constant	1.38×10^{-23}	J/(Particle K)
l_f	Latent heat of fusion at 0 °C	0.3337×10^{6}	J/kg
l_s	Latent heat of sublimation at 0 °C	2.8345×10^{6}	J/kg
l_{v0}	Latent heat of vaporization at 0 °C	2.5008×10^{6}	J/kg
M_d	Mean molar mass of the dry air ensemble	28.97 (28.964)[i]	kg/kmol
M_v	Molar mass of water vapor	18.02	kg/kmol
m_e	Mass of Earth	5.988×10^{24}	kg
N_A	Avogadro's Number	6.02×10^{26}	Particles/kmol
p_c	USSA sea-level pressure	1013.25	hPa
p_c	Critical pressure of water vapor	2.22×10^{7}	Pa
R^*	Universal gas constant	8310 (8314.32)[ii]	J/(kmol K)
R_d	Individual gas constant for dry air ensemble	286.8 (287.1)[iii]	J/(kg K)

(*continued*)

Symbol	Name	Numerical Value	Unit
R_e	Equatorial radius of Earth	6378.39	km
R_p	Polar radius of Earth	6356.91	km
R_v	Individual gas constant for water vapor	461.2	J/(kg K)
T_0	USSA sea-level temperature	288.15	K
T_c	Critical temperature of water vapor	647	K
β_0	Lapse rate in lowest layer of USSA	6.5	K/km
β_2	Lapse rate in the USSA middle stratosphere	−1.0	K/km
β_3	Lapse rate in the USSA upper stratosphere	−2.8	K/km
ε	Ratio of molar mass of water vapor to molar mass of dry air ensemble	0.622	
Γ_D	Dry adiabatic lapse rate	9.77	K/km
$(\gamma-1)^{air}$	$2/f$	0.400	
$(\gamma-1)^{vapor}$	$2/f$	0.333	
$(-1/\gamma)^{air}$	$f/(f+2)$	−0.714	
$(-1/\gamma)^{vapor}$	$f/(f+2)$	−0.750	
κ^{air}	$2/(f+2)$	0.286	
κ^{vapor}	$2/(f+2)$	0.250	
ρ_0	USSA sea-level density of air	1.225	Kg/m^3
Ω	Angular velocity of Earth	7.292×10^{-5}	s^{-1}

[i–iii] USSA value in parenthesis.

Appendix 3

English Letters and Their Greek Equivalents

ENGLISH	GREEK	ENGLISH	GREEK
A	A	a	α
B	B	b	β
C	X	c	χ
D	Δ	d	δ
E	E	e	ε
F	Φ	f	φ
G	Γ	g	γ
H	H	h	η
I	I	i	ι
J	ϑ	j	ϕ
K	K	k	κ
L	Λ	l	λ
M	M	m	μ
N	N	n	ν
O	O	o	o
P	Π	p	π
Q	Θ	q	θ
R	P	r	ρ
S	Σ	s	σ
T	T	t	τ
U	Y	u	υ
V	ς	v	ϖ
W	Ω	w	ω
X	Ξ	x	ξ
Y	Ψ	y	ψ
Z	Z	z	ζ

Appendix 4

Additional Thermodynamic Functions: Helmholtz Free Energy and Gibbs Free Energy

In Chapter 4, we discussed two thermodynamic functions. The first was internal energy (u), which can be expressed as:

$$u = q - w \tag{A4.1}$$

where q is heat added and w is the work performed.[1] The second function was enthalpy (h), which is written as:

$$h = u + p\alpha \tag{A4.2}$$

where p is pressure, and α is specific volume.

In this appendix we introduce two new functions. The first of these is *Helmholtz Free Energy*[2] (f), which is an estimate of the work that can be performed by a closed thermodynamic system, such as an air parcel, at *constant temperature*. Helmholtz Free Energy is given by:

$$f = u - Ts \tag{A4.3}$$

where T is the temperature of the system, and s is its entropy.[3] The second new function is *Gibbs Free Energy*[4] (g), which is an estimate of the work obtainable from a closed thermodynamic system at *constant temperature and pressure*. Gibbs Free Energy[5] can be written as:

$$g = h - Ts \tag{A4.4}$$

Derivatives of these functions. The Ideal Gas Law (e.g., 2.33) expresses a relationship between the *state variables*, which (in this case) are pressure, volume, and temperature. Because the four thermodynamic functions shown in the preceding text are all functions of the state variables of an ideal gas, they should be mathematically related to each other as well. Beginning with (A4.1) and taking the derivative of both sides results in:

$$\boxed{du = dq - dw} \tag{A4.5}$$

The first term on the RHS can be rewritten using the Second Law of Thermodynamics (6.1), and assuming a reversible process (6.2). The second term on the RHS can be rewritten using the relation for work described by (3.12). Using these substitutions, (A4.5) becomes:

$$\boxed{du = T\,ds - p\,d\alpha} \tag{A4.6}$$

The relationship for enthalpy (A4.2) can also be written by first taking the derivative of both sides:

$$dh = du + d\left(p\alpha\right) \tag{A4.7}$$

or:

$$dh = du + p\,d\alpha + \alpha\,dp \tag{A4.8}$$

The first two terms on the RHS can be rewritten using the First Law of Thermodynamics (4.3), resulting in:

$$dh = dq + \alpha\,dp \tag{A4.9}$$

which can then be written in terms of Entropy using the Second Law of Thermodynamics (6.1), and assuming a reversible process:

$$\boxed{dh = T\,ds + \alpha\,dp} \tag{A4.9}$$

Next, we perform similar operations on the two new thermodynamic functions. Beginning with Helmholtz Free Energy (A4.2), we take the derivative of both sides:

$$df = du - d\left(Ts\right) \tag{A4.10}$$

or:

$$df = du - T\,ds - s\,dT \tag{A4.11a}$$

If the process is reversible, then the second term on the RHS can be rewritten using the Second Law:

$$df = du - dq - s\,dT \tag{A4.11b}$$

According to (A4.5),

$$du = dq - dw$$

therefore:

$$du - dq = -dw \tag{A4.12}$$

Substituting this identity into (A4.11b) allows us to write the first two terms on the RHS of (A4.11b), yielding:

$$df = -dw - s\,dT \tag{A4.13}$$

or:

$$df = -p\,d\alpha - s\,dT \tag{A4.14}$$

In the next step, we begin with expression for Gibbs Free Energy shown in (A4.4), and take the derivative of both sides:

$$dg = dh - d(Ts) \tag{A4.15}$$

Applying the chain rule results in:

$$dg = dh - s\,dT - T\,ds \tag{A4.16}$$

The identity in (A4.8) states that:

$$dh = du + p\,d\alpha + \alpha\,dp$$

Substituting this expression into (A4.16) yields:

$$dg = du + p\,d\alpha + \alpha\,dp - s\,dT - T\,ds \tag{A4.17}$$

Rearranging this somewhat results in:

$$dg = du - T\,ds + p\,d\alpha + \alpha\,dp - s\,dT \tag{A4.18}$$

Applying the Second Law to the second term on the RHS, and assuming a reversible process, allows us to rewrite (A4.18) as:

$$dg = du - dq + p\,d\alpha + \alpha\,dp - s\,dT \tag{A4.19}$$

But the relation in (A4.12) states that:

$$du - dq = -dw$$

which we can apply to the second two terms on the RHS of (A4.19), resulting in:

$$dg = -dw + p\,d\alpha + \alpha\,dp - s\,dT \tag{A4.20}$$

Table A4.1. *Summary of thermodynamic functions.*

Thermodynamic Function	Mathematical Expression	Function of Variables	Constants
Internal Energy	$du = T\,ds - p\,d\alpha$	s,v	T,p
Enthalpy	$dh = T\,ds + \alpha\,dp$	s,p	T,α
Helmholtz Free Energy	$df = -p\,d\alpha - s\,dT$	α,T	P,s
Gibbs Free Energy	$dg = \alpha\,dp - s\,dT$	p,T	α,s

or:

$$dg = -p\,d\alpha + p\,d\alpha + \alpha\,dp - s\,dT \tag{A4.21}$$

which allows some further cancellation, resulting in:

$$\boxed{dg = \alpha\,dp - s\,dT} \tag{A4.22}$$

Relationships between thermodynamic functions. Table A4.1 summarizes the derivatives of the functions shown in the preceding text. Each of the four functions is a unique combination of variables and constants, describing several directions along which the state of an ideal gas can vary as some of its state variables remain constant. Note that entropy has been added as one direction along which the ideal gas can vary.

Each of the thermodynamic functions is a function of two variables, allowing us to take two partial derivatives of each. These are:

$$\left(\frac{\partial u}{\partial \alpha}\right)_s = -p \tag{A4.23}$$

where the subscript s on the LHS indicates that s is held constant, so $ds = 0$. Similarly:

$$\left(\frac{\partial u}{\partial s}\right)_\alpha = T \tag{A4.24}$$

$$\left(\frac{\partial h}{\partial s}\right)_p = T \tag{A4.25}$$

$$\left(\frac{\partial h}{\partial p}\right)_s = \alpha \tag{A4.26}$$

$$\left(\frac{\partial f}{\partial \alpha}\right)_T = -p \tag{A4.27}$$

$$\left(\frac{\partial f}{\partial T}\right)_{\alpha} = -s \tag{A4.28}$$

$$\left(\frac{\partial g}{\partial p}\right)_{T} = \alpha \tag{A4.29}$$

$$\left(\frac{\partial g}{\partial T}\right)_{p} = -s \tag{A4.30}$$

From these we see that:

$$\left(\frac{\partial u}{\partial s}\right)_{\alpha} = \left(\frac{\partial h}{\partial s}\right)_{p} \tag{A4.31}$$

$$\left(\frac{\partial u}{\partial \alpha}\right)_{s} = \left(\frac{\partial f}{\partial \alpha}\right)_{T} \tag{A4.32}$$

$$\left(\frac{\partial h}{\partial p}\right)_{s} = \left(\frac{\partial g}{\partial p}\right)_{T} \tag{A4.33}$$

and,

$$\left(\frac{\partial f}{\partial T}\right)_{\alpha} = \left(\frac{\partial g}{\partial T}\right)_{p} \tag{A4.34}$$

Appendix 5

Computed Boiling Temperatures at Different Pressures

In Chapter 7 we derived an expression for vapor pressure above liquid water. The closed form of the Classius-Clapeyron Equation (7.46) stated that:

$$e = e_0 \exp\left[\frac{l_v}{R_v}\left(\frac{1}{T_0} - \frac{1}{T_d}\right)\right]$$

where e_0 is the vapor pressure at the triple point of water (611.12 Pa), l_v is the latent heat of vaporization (equal to 2.5008×10^6 J/kg) at $0\,°C$, and decreasing by about 10 percent as temperature increases from 0 to $100\,°C$), R_v is the individual gas constant for vapor (461.2 J/(kg K)), T_0 is the temperature at the triple point (273.16 K), and T_d is the dew point [K]. For the *saturation* vapor pressure, one simply substitutes the temperature for the dew point, resulting in:

$$e_s = e_0 \exp\left[\frac{l_v}{R_v}\left(\frac{1}{T_0} - \frac{1}{T}\right)\right] \qquad (A5.1)$$

Boiling has been described as a violent form of evaporation that occurs when the vapor pressure is equal to the total atmospheric pressure. With this in mind, equation (A5.1) can be recast to show the relationship between atmospheric pressure and the boiling temperature of water as follows:

$$p_B = e_0 \exp\left[\frac{l_v}{R_v}\left(\frac{1}{T_0} - \frac{1}{T_B}\right)\right] \qquad (A5.2)$$

where p_B is the pressure at which the water is boiled, and T_B is the boiling temperature of water at p_B. If we invert this equation, we can then compute the boiling temperature at any given pressure. First, we divide through by e_0, resulting in:

$$\frac{p_B}{e_0} = \exp\left[\frac{l_v}{R_v}\left(\frac{1}{T_0} - \frac{1}{T_B}\right)\right] \qquad (A5.3)$$

Taking the natural log of both sides yields:

$$ln\left(\frac{p_B}{e_0}\right) = \frac{l_v}{R_v}\left(\frac{1}{T_0} - \frac{1}{T_B}\right) \tag{A5.4}$$

Multiplying through by R_v / l_v results in:

$$\frac{R_v}{l_v} ln\left(\frac{p_B}{e_0}\right) = \frac{1}{T_0} - \frac{1}{T_B} \tag{A5.5}$$

Next, we subtract $1/T_0$ from both sides and multiply both sides by -1:

$$\frac{1}{T_B} = \frac{1}{T_0} - \frac{R_v}{l_v} ln\left(\frac{p_B}{e_0}\right) \tag{A5.6}$$

Finally, inverting both sides yields an expression for boiling temperature:

$$T_B = \frac{1}{\dfrac{1}{T_0} - \dfrac{R_v}{l_v} ln\left(\dfrac{p_B}{e_0}\right)} \tag{A5.7}$$

There is a slight problem in using this expression to compute all boiling temperatures, namely, that l_v is not really a constant. In fact, it varies with temperature (in this case T_B), which implies that we would need to compute l_v before computing T_B. To do that, we need to know T_B first. There are numerical methods for closing this loop, but another method that yields serviceable results in this case would be to simply use an average value of l_v in the known range of temperatures and pressures. With this in mind, we refer to Table 4.2 and find the value of l_v for 50 °C (which is a good approximation for the mean value between 0 and 100 °C): 2.3893×10^6 J/kg. Note that this is not the only simplifying assumption present in these calculations – the derivation of the Classius-Clapeyron equation involved several others (including the assumptions present in the Ideal Gas Law). These are described in detail in Chapter 7.

The results shown in Table A5.1 use equation (A5.7) and the mean value of l_v noted. Column 2 of the table shows the elevations corresponding to the pressures in column 2, representing the conditions valid for the U.S. Standard Atmosphere. (See Table 10.2 for the source of these elevations, and Chapter 10, generally, for a discussion of the U.S. Standard Atmosphere and the International Standard Atmosphere.)

Table A5.2 shows boiling-point data taken from the *CRC Handbook of Physics and Chemistry*,[1] and Table A5.3 summarizes the differences between the boiling points computed from (A5.7) and the boiling points taken from the *CRC Handbook*.

Table A5.1. *Boiling temperatures as a function of atmospheric pressure, computed from Equation (A5.7).*

Elevations correspond to the U.S. Standard Atmosphere.

Pressure [hPa]	Elevation [m]	Boiling Temperature [°C]	Boiling Temperature [K]
1013.25	0	100.76	373.92
1000	111	100.41	373.57
900	988	97.59	370.75
800	1947	94.49	367.65
700	3010	91.04	364.20
600	4203	87.14	360.30
500	5570	82.63	355.79
400	7180	77.26	350.42
300	9157	70.57	343.73
200	11766	61.56	334.72
100	15787	47.22	320.38

Table A5.2. *Boiling temperatures as a function of atmospheric pressure, from the CRC Handbook of Physics and Chemistry.*

Elevations correspond to the U.S. Standard Atmosphere.

Pressure [hPa]	Elevation [m]	Boiling Temperature [°C]	Boiling Temperature [K]
1013.25	0	100.00	373.16
1000	111	99.63	372.79
900	988	96.71	369.87
800	1947	93.51	366.67
700	3010	89.96	363.12
600	4203	85.95	359.11
500	5570	81.34	354.50
400	7180	75.88	349.04
300	9157	69.11	342.27
200	11766	60.07	333.23
100	15787	45.82	318.98

Table A5.3. *Comparison of boiling temperatures values computed from Equation (A5.7) to values from the CRC Handbook of Physics and Chemistry.*

Pressure [hPa]	Table A5.1 Value [K]	Table A5.2 Value [K]	Abs (A5.1–A5.2) [K]	Error [%]
1013.25	373.92	373.16	0.76	0.20
1000	373.57	372.79	0.78	0.21
900	370.75	369.87	0.88	0.24
800	367.65	366.67	0.98	0.27
700	364.20	363.12	1.08	0.30
600	360.30	359.11	1.19	0.33
500	355.79	354.50	1.29	0.36
400	350.42	349.04	1.38	0.40
300	343.73	342.27	1.46	0.43
200	334.72	333.23	1.49	0.45
100	320.38	318.98	1.40	0.44

Accepting the values taken from the *CRC Handbook* as correct, the error values in column 5 of Table A5.3 were then computed by:

$$\text{Error} = \text{abs}\left[\frac{(\text{Table A5.1 value}) - (\text{Table A5.2 value})}{(\text{Table A5.2 Value})} \right] \times 100\ \% \tag{A5.8}$$

These results indicate that, to within less than half a percentage point (and about one Kelvin), the Classius-Clapeyron equation can be used to estimate the boiling point temperature of water in pressures typical of Earth's troposphere and lower stratosphere, even when several simplifying assumptions are applied. This in turn appears to verify the notion that boiling is a form of evaporation that occurs when the vapor pressure is equal to the total atmospheric pressure.

Appendix 6

Variation of Saturation Mixing Ratio with Temperature and Pressure

In Chapter 7 we derived an expression for water vapor mixing ratio (w_s). Equation (7.67) stated that:

$$w = \frac{e\varepsilon}{p - e}$$

where e is the *in situ* vapor pressure, ε is the ratio of the molar mass of water vapor to the molar mass of dry air (0.622), and p is the total atmospheric pressure. The vapor pressure and the total atmospheric pressure can be in any units, as long as they are the same units. Overall, the *in situ* mixing ratio (w) has units of $kg_{vapor}/kg_{dry\ air}$, which is to say that it is unitless. By substituting the saturation vapor pressure (e_s) into this expression, we obtain an expression for the *saturation* mixing ratio:

$$w_s = \frac{e_s \varepsilon}{p - e} \tag{A6.1}$$

where w_s is the saturation mixing ratio. Using (7.68), we can rewrite this using the approximation:

$$w_s \cong \frac{e_s \varepsilon}{p} \tag{A6.2}$$

because the total atmospheric pressure is much greater than the saturation vapor pressure.

The present objective is to derive expressions for the variation of w_s with respect to pressure and temperature – in other words, its *partial derivatives*.[1] Beginning with (A6.2), the partial derivative with respect to pressure is approximated by:

$$\frac{\partial w_s}{\partial p} \cong \frac{\partial}{\partial p}\left(\frac{e_s \varepsilon}{p}\right) = \frac{\partial}{\partial p}\left(e_s \varepsilon p^{-1}\right) \tag{A6.3}$$

Epsilon is a constant, and the closed form of the Classius-Clapeyron Equation (7.46) shows that e_s is *not* a function of pressure, so that it too is a constant in this context. In other words, we can write:

$$\frac{\partial w_s}{\partial p} \cong e_s \varepsilon \frac{\partial}{\partial p}\left(p^{-1}\right) \tag{A6.4}$$

which is equal to:

$$\boxed{\frac{\partial w_s}{\partial p} \cong -\frac{e_s \varepsilon}{p^2}} \tag{A6.5}$$

The partial derivative with respect to temperature is a little more involved. Beginning with (A6.2) and taking the partial derivative with respect to T, we can write:

$$\frac{\partial w_s}{\partial T} \cong \frac{\partial}{\partial T}\left(\frac{e_s \varepsilon}{p}\right) \tag{A6.6}$$

In this context, ε and p are constants, so (A6.6) becomes:

$$\frac{\partial w_s}{\partial T} \cong \frac{\varepsilon}{p}\frac{\partial e_s}{\partial T} = \frac{\varepsilon}{p}\frac{de_s}{dT} \tag{A6.7}$$

because e_s is a function of T only. The differential form of the Classius-Clapeyron Equation (7.36b) states that:

$$\frac{de_s}{dT} = \frac{l_v e_s}{R_v T^2}$$

Inverting the approximation shown in (A6.2), we can write:

$$e_s \cong \frac{w_s p}{\varepsilon} \tag{A6.8}$$

So that (7.36b) becomes:

$$\frac{de_s}{dT} = \frac{l_v}{R_v T^2}\left(\frac{w_s p}{\varepsilon}\right) \tag{A6.9}$$

The definition of ε provided in (2.82) states that:

$$\varepsilon \equiv \frac{R_d}{R_v}$$

so that:

$$\frac{1}{\varepsilon} = \frac{R_v}{R_d} \tag{A6.10}$$

Substituting this in (A6.9), we obtain:

$$\frac{de_s}{dT} = \frac{l_v w_s p}{R_v T^2}\left(\frac{R_v}{R_d}\right) = \frac{l_v w_s p}{R_d T^2} \tag{A6.11}$$

If we now substitute the RHS of (A6.11) into (A6.7) we see that:

$$\frac{\partial w_s}{\partial T} \cong \frac{\varepsilon}{p}\left(\frac{l_v w_s p}{R_d T^2}\right) \tag{A6.12}$$

or:

$$\boxed{\frac{\partial w_s}{\partial T} \cong \frac{\varepsilon l_v w_s}{R_d T^2}} \tag{A6.13}$$

Summary. We have derived expressions for the variation in the saturation mixing ratio (w_s) as functions of pressure (p) and temperature (T). These are given by:

$$\frac{\partial w_s}{\partial p} \cong -\frac{e_s \varepsilon}{p^2}$$

where:

   e_s = Saturation vapor pressure
   p = total atmospheric pressure
   ε = ratio of molar masses of water and dry air (0.622)

and,

$$\frac{\partial w_s}{\partial T} \cong \frac{\varepsilon l_v w_s}{R_d T^2}$$

where:

   l_v = Latent heat of vaporization ($\cong 2.5008 \times 10^6$ J/(kg))
   R_d = Individual gas constant for dry air (286.8 J/(kg K))

Appendix 7

Derivation of the Moist Adiabatic Lapse Rate

Figure A7.1 illustrates the vertical ascent of a hypothetical parcel originating near the Earth's surface. Beginning with a temperature of 20 °C and dew point of −10 °C, the parcel initially ascends dry adiabatically. The temperature of the unsaturated parcel falls at the Dry Adiabatic Lapse Rate (Γ_D), and dew point tracks upward parallel to a mixing ratio line (w). At a pressure of about 640 hPa (the LCL), the temperature and dew point intersect at the isentropic condensation temperature. This indicates the parcel is saturated. If additional lift is supplied, the parcel ascends moist adiabatically (Γ_M), which, while initially slower than the dry rate, becomes parallel to the dry rate below temperatures of about −40 °C.

The current task is to derive an expression for the Moist Adiabatic Lapse Rate. The scenario under examination is as follows: after reaching the LCL, lifting continues, carrying the parcel upward and allowing it to expand and cool, which forces condensation to begin inside the saturated parcel. As water vapor condenses into liquid droplets, *latent* heat is released, which adds to the parcel's *sensible* heat. The internal heating of the parcel slows the rate of cooling resulting from the parcel's vertical motion. We assume that the water droplets produced by the condensation of the parcel's vapor remain inside, so that the process can be reversed, because the droplets can be converted back into vapor, reabsorbing the heat they surrendered when they condensed. This implies an *isolated, compound* parcel, consisting of two halves that are in equilibrium and open to each other. The first half consists of liquid water droplets, and the second half consists of dry air and water vapor. Water substance moves back and forth between the two halves of the compound parcel by evaporating and condensing.

Heat conservation. By assuming an isolated compound parcel, we are also stating that total amount of heat in the parcel is conserved, so the relation given by (3.15) applies:

$$\Delta Q_1 + \Delta Q_2 = 0$$

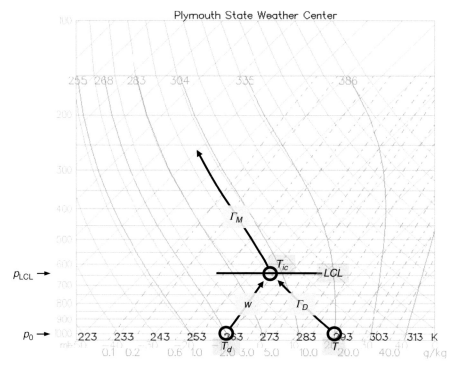

Figure A7.1. An unsaturated parcel begins near the Earth's surface, and is lifted aloft until reaching the LCL. Below the LCL, the temperature cools at the Dry Adiabatic Lapse Rate (Γ_D); above the LCL, the temperature cools at the Moist Adiabatic Lapse Rate (Γ_M). (Background image courtesy of Plymouth State University meteorology program, 2012.)

where, in this context, ΔQ_1 is the amount of latent heat surrendered when some of the parcel's vapor condenses, and ΔQ_2 is the amount of this heat subsequently taken up by the dry air in the parcel. We can rewrite this as:

$$-dQ_1 = dQ_2 \qquad\qquad (A7.1)$$

where the LHS is the latent heat *released*, and the RHS is the heat *absorbed* by the air in the parcel. Our next objective is to write both sides of (A7.1) more specifically.

To do this, we'll assume we have a parcel of moist air, with a total mass of 1 kg (i.e., a *unit mass*), and water vapor in an equilibrium (saturation) state. Because mixing ratio has units of $kg_{vapor}/kg_{dry\ air}$, we can estimate the mass of the vapor in the parcel by multiplying w by the total mass of the parcel. If the parcel is saturated, then $w = w_s$, and the mass of the vapor is $(1\ kg) \times w_s = w_s$.

Lifting the parcel. The initial condition of the parcel is p, T, w_s. After it is lifted a short distance, expanding adiabatically, its condition is given by $(p - dp)$, $(T - dT)$, $(w_s - dw_s)$. The pressure fell because the parcel moved upward, and the temperature

fell because it expanded. The saturation mixing ratio fell because it is directly dependent on temperature: if temperature falls, saturation mixing ratio does too.[1] This, in
turn, means that some of the water vapor in the parcel condensed into liquid water,
releasing latent heat. The amount of latent heat released during the condensation of
the fractional amount of vapor is given by:

$$dQ_1 = l_v dw_s \tag{A7.2}$$

which is then absorbed by the air in the parcel. Because the mass of the water vapor
in the parcel is much less than the mass of the *dry* air, most of the heat goes into the
dry air. Our third assumption is that *all* of this heat is absorbed by the dry air, which
is dQ_2, on the RHS of (A7.1). To understand what happens to the dry air in the parcel
when this heat is added, we appeal to (4.45), which states (in this context):

$$dQ_2 = c_p dT - \alpha dp$$

and which takes advantage of the fact that we are assuming a unit mass of air. Next
we use the identity for dQ_1 given in (A7.2), and the identity for dQ_2 described in the
preceding text, along with the assumption that all the heat released during condensation goes into the dry air component of the parcel, and plug them both into (A7.1),
resulting in:

$$\boxed{-l_v dw_s = c_p dT - \alpha_d d\left(p - e_s\right)} \tag{A7.3}$$

where e_s is the saturation vapor pressure, p is the total pressure, and the parenthetical
term on the RHS is the partial pressure resulting from the parcel's dry air content.
We'll rewrite the specific volume of the dry air (α_d) in the second term on the RHS by
beginning with the Ideal Gas Law in the form shown in (7.62):

$$p - e = \rho_d R_d T$$

which, using the definition of specific volume, and the fact that $e = e_s$ in this context,
can be rewritten as:

$$\left(p - e_s\right)\alpha_d = R_d T \tag{A7.4}$$

or:

$$\alpha_d = \frac{R_d T}{\left(p - e_s\right)} \tag{A7.5}$$

Substituting this identity into (A7.3), we obtain:

$$-l_v dw_s = c_p dT - R_d T \frac{d\left(p - e_s\right)}{\left(p - e_s\right)} \tag{A7.6}$$

Our fourth simplifying assumption is that $p \gg e_s$, so that $p - e_s \cong p$. Applying this to (A7.6) yields:

$$-l_v dw_s = c_p dT - R_d T \frac{dp}{p} \qquad (A7.7)$$

Next we go to the approximation for w_s, given by (7.68):

$$w_s \cong \frac{e_s \varepsilon}{p}$$

Taking the natural log of both sides, and assuming an equality (the fifth simplifying assumption), results in:

$$ln(w_s) = ln\left(\frac{e_s \varepsilon}{p}\right) = ln(e_s) + ln(\varepsilon) - ln(p) \qquad (A7.8)$$

Taking the derivative for both sides:

$$d[ln(w_s)] = d[ln(e_s) + ln(\varepsilon) - ln(p)] \qquad (A7.9)$$

results in:

$$\frac{dw_s}{w_s} = \frac{de_s}{e_s} + \frac{d\varepsilon}{\varepsilon} - \frac{dp}{p} \qquad (A7.10)$$

but the derivative of the constant ε is equal to zero, so we are left with:

$$\frac{dw_s}{w_s} = \frac{de_s}{e_s} - \frac{dp}{p} \qquad (A7.11)$$

or:

$$dw_s = w_s\left(\frac{de_s}{e_s} - \frac{dp}{p}\right) \qquad (A7.12)$$

We also have the hydrostatic relation (5.77), which states that:

$$\frac{dp}{dz} = -\rho g$$

But the Ideal Gas Law (2.37) states that:

$$p = \rho R_i T$$

which, in this context, can be written:

$$p = \rho R_d T$$

or:

$$\rho = \frac{p}{R_d T} \tag{A7.13}$$

Using this identity, we can rewrite the hydrostatic relation as:

$$\frac{dp}{dz} = -\frac{pg}{R_d T} \tag{A7.14}$$

Isolating all the pressure terms on the LHS results in:

$$\boxed{\frac{dp}{p} = -\frac{g}{R_d T} dz} \tag{A7.15}$$

Now let's return to (A7.7), which said:

$$-l_v dw_s = c_p dT - R_d T \frac{dp}{p}$$

and substitute the identity for dw_s shown in (A7.12)

$$-l_v w_s \underbrace{\left(\frac{de_s}{e_s} - \frac{dp}{p} \right)}_{dW_s} = c_p dT - R_d T \frac{dp}{p} \tag{A7.16}$$

Next, let's substitute the identity in (A7.15) for dp/p into the second term on the RHS of (A7.15):

$$-l_v w_s \left(\frac{de_s}{e_s} - \frac{dp}{p} \right) = c_p dT - R_d T \underbrace{\left(-\frac{g}{R_d T} dz \right)}_{dp/p} \tag{A7.17}$$

Canceling redundant terms and distributing the negative sign outside the parenthesis on the RHS yields:

$$-l_v w_s \left(\frac{de_s}{e_s} - \frac{dp}{p} \right) = c_p dT + g dz \tag{A7.18}$$

But now we'll also use the identity in (A7.15) on the LHS of (A7.18):

$$-l_v w_s \left[\frac{de_s}{e_s} - \underbrace{\left(-\frac{g}{R_d T} dz \right)}_{dp/p} \right] = c_p dT + g dz \tag{A7.19}$$

Simplifying the negative signs inside the square brackets on the LHS results in:

$$-l_v w_s \left[\frac{de_s}{e_s} + \frac{g}{R_d T} dz \right] = c_p dT + g\, dz \qquad (A7.20)$$

and dividing through by dz yields:

$$\boxed{-l_v w_s \left[\frac{1}{e_s} \frac{de_s}{dz} + \frac{g}{R_d T} \right] = c_p \frac{dT}{dz} + g} \qquad (A7.21)$$

which is an encouraging result, because we finally have a quantitative description of the vertical variation in temperature – the first term on the RHS. Our efforts from here forward will be focused on isolating this term and obtaining a fully explicit expression for it in terms of the other variables.

Isolating the vertical temperature gradient term. We'll begin by focusing on the term on the LHS of (A7.21) that describes the vertical variation in saturation vapor pressure (de_s/dz). The integrated form of the Classius-Clapeyron Equation states that:

$$e_s = e_s(T) \qquad (A7.22)$$

in functional notation, that is, e_s is a function of T only. If T is a function of height only (note assumption number six), the chain rule says that:

$$\frac{de_s}{dz} = \frac{de_s}{dT} \frac{dT}{dz} \qquad (A7.23)$$

and (A7.21) can be rewritten as:

$$-l_v w_s \left[\frac{1}{e_s} \frac{de_s}{dT} \frac{dT}{dz} + \frac{g}{R_d T} \right] = c_p \frac{dT}{dz} + g \qquad (A7.24)$$

Next, we'll isolate dT/dz on one side of the equation by

(a) distributing the $-l_v w_s$ term inside the square brackets on the LHS:

$$\frac{-l_v w_s}{e_s} \frac{de_s}{dT} \frac{dT}{dz} - \frac{l_v w_s g}{R_d T} = c_p \frac{dT}{dz} + g \qquad (A7.25a)$$

(b) moving the dT/dz term on the RHS to the LHS, and the term involving gravity on the LHS to the RHS:

$$\frac{-l_v w_s}{e_s} \frac{de_s}{dT} \frac{dT}{dz} - c_p \frac{dT}{dz} = \frac{l_v w_s g}{R_d T} + g \qquad (A7.25b)$$

(c) pulling the $-dT/dz$ term out of both terms on the LHS, and g out of both terms on the RHS:

$$-\frac{dT}{dz}\left(\frac{l_v w_s}{e_s}\frac{de_s}{dT}+c_p\right)=g\left(\frac{l_v w_s}{R_d T}+1\right) \qquad \text{(A7.25c)}$$

(d) pulling c_p out of the parenthesis on the LHS:

$$-\frac{dT}{dz}c_p\left(\frac{l_v w_s}{c_p e_s}\frac{de_s}{dT}+1\right)=g\left(\frac{l_v w_s}{R_d T}+1\right) \qquad \text{(A7.25d)}$$

(e) dividing through by c_p:

$$-\frac{dT}{dz}\left(\frac{l_v w_s}{c_p e_s}\frac{de_s}{dT}+1\right)=\frac{g}{c_p}\left(\frac{l_v w_s}{R_d T}+1\right) \qquad \text{(A7.25e)}$$

(f) and finally, dividing through by the parenthetical term on the LHS:

$$-\frac{dT}{dz}=\frac{g}{c_p}\frac{\left(\dfrac{l_v w_s}{R_d T}+1\right)}{\left(\dfrac{l_v w_s}{c_p e_s}\dfrac{de_s}{dT}+1\right)} \qquad \text{(A7.25f)}$$

In Chapter 5, we defined the Dry Adiabatic Lapse Rate by (5.83):

$$\Gamma_D \equiv -\frac{dT}{dz}=\frac{g}{c_p}$$

so let's substitute this definition into (A7.25f) and rewrite it as:

$$\boxed{\Gamma_M = \Gamma_D \frac{\left(\dfrac{l_v w_s}{R_d T}+1\right)}{\left(\dfrac{l_v w_s}{c_p e_s}\dfrac{de_s}{dT}+1\right)}} \qquad \text{(A7.26)}$$

where Γ_M indicates the Moist Adiabatic Lapse Rate.

Remaining details. There are still a few details to clean up. Some of the expressions in the denominator on the RHS of (A7.26) are a bit ambiguous, and we could use more explicit methods for writing them. That denominator is:

$$\left(\frac{l_v w_s}{c_p e_s}\frac{de_s}{dT}+1\right) \qquad \text{(A7.27a)}$$

Let's look at w_s first. We can rewrite this using (7.68), which states (in this context):

$$w_s \cong \frac{e_s \varepsilon}{p}$$

Assuming an equality (the second time we've done this, making this the seventh simplifying assumption), and substituting this identity into (A7.27a) results in:

$$\left(\frac{l_v e_s \varepsilon}{c_p e_s p} \frac{de_s}{dT} + 1 \right) \tag{A7.27b}$$

or:

$$\left(\frac{l_v \varepsilon}{c_p p} \frac{de_s}{dT} + 1 \right) \tag{A7.27c}$$

Next, we'll take a look at the de_s/dT term. In (7.36b), we saw that:

$$\frac{de_s}{dT} \cong \frac{l_v e_s}{R_v T^2}$$

But if we apply (7.68), which states:

$$e \cong \frac{pw}{\varepsilon}$$

or, in this context:

$$e_s \cong \frac{pw_s}{\varepsilon} \tag{A7.27c1}$$

and assume equalities (assumption number eight), then the de_s/dT term can be rewritten as:

$$\frac{de_s}{dT} = \frac{l_v p w_s}{R_v T^2 \varepsilon} = \left(\frac{1}{\varepsilon} \right) \frac{l_v p w_s}{R_v T^2} \tag{A7.27c2}$$

but, because $(1/\varepsilon) = (R_v/R_d)$ (inversion of 2.69), we also have:

$$\frac{de_s}{dT} = \left(\frac{1}{\varepsilon} \right) \frac{l_v p w_s}{R_v T^2} = \left(\frac{R_v}{R_d} \right) \frac{l_v p w_s}{R_v T^2} = \frac{l_v p w_s}{R_d T^2} \tag{A7.27c3}$$

Substituting this into (A7.27c) yields:

$$\left[\frac{l_v \varepsilon}{c_p p} \underbrace{\left(\frac{l_v p w_s}{R_d T^2} \right)}_{de_s / dT} + 1 \right] \tag{A7.27d}$$

or, after canceling p and combining l_v, the denominator of (A7.26) becomes:

$$\left(\frac{l_v^2 \varepsilon w_s}{c_p R_d T^2} + 1 \right) \tag{A7.27e}$$

We can now substitute this expression into the denominator of (A7.26) to obtain the final form of the Moist Adiabatic Lapse Rate:

$$\boxed{\Gamma_M = \Gamma_D \frac{\left(\dfrac{l_v w_s}{R_d T} + 1 \right)}{\left(\dfrac{l_v^2 \varepsilon w_s}{c_p R_d T^2} + 1 \right)}} \tag{A7.28}$$

Computed values of the Moist Adiabatic Lapse Rate are shown and discussed in Chapter 8.

Reminder of the assumptions. We made many simplifying assumptions to derive the result shown in (A7.28). Before you use this result, let's just go over what these assumptions are again.

- We assumed an isolated, compound parcel, with water vapor coexisting in equilibrium with liquid water.
- We assumed that, as vapor condensed, the resulting drops remained inside the parcel, so that when they evaporated again, they could take up exactly the same amount of latent heat that they released when they condensed. This made the processes reversible and allowed the heat to be conserved. That is: *the Moist Adiabatic Lapse Rate is a model of heat conservation.*
- We assumed that the combined mass of the parcel's vapor and dry air was equal to one kilogram.
- We assumed that all latent heat released during condensation was taken up by the parcel's *dry air*, and none of it went into the parcel's remaining load of water vapor.
- We assumed that the total atmospheric pressure (p) was much greater than the saturation vapor pressure (e_s), so that $p - e_s \cong p$.
- We assumed that $w_s = e_s \varepsilon / p$, and equivalently, that $e_s = p w_s / \varepsilon$.
- Finally, we assumed that temperature varies only as a function of elevation.

Appendix 8

Sea-Level Pressure Calculations Using the Smithsonian Meteorological Tables[1]

1. Calculate the *adjusted* mean virtual temperature, defined by:

$$\bar{t}_v \cong \bar{t} + L + c \qquad (A8.1)$$

where:

$$\bar{t} \equiv \frac{t_0 + t_{12}}{2}$$

t_0 is the current temperature [°C]

t_{12} is the temperature twelve hours earlier [°C]

L = Estimate of temperature [°C] half way through the fictitious air column, in an "average" atmosphere with lapse rate one half the dry rate (Γ_D), that is,

$$L \equiv \left(\frac{z_{STN}}{2} \right)\left(\frac{\Gamma_D}{2} \right) = \frac{z_{STN}}{2} \frac{1}{200} = \frac{z_{STN}}{400}$$

c = Correction for humidity in the column, calculated using the observed dew point on station [°C], the station elevation [m ASL], and Smithsonian Meteorological Table 48A (see following text).

2. Calculate the *adjusted* geopotential height difference between sea level and the station [gpm], given by:

$$\Delta\tilde{\zeta} = \frac{T_0}{T_0 + \bar{t}_v} \Delta\zeta \qquad (A8.2)$$

where:

$\Delta\zeta = \zeta_{STN} - \zeta_{SL} \cong \zeta_{STN}$ = The difference in *unadjusted* geopotential height between the station and sea level.

Geopotential height can be found using Smithsonian Meteorological Table 50 (see following text), given station elevation [m ASL] and latitude [°N/S].

T_0 is the reference temperature (273.16 K)

363

3. Determine the geopotential height above the 1100 hPa surface [gpm] at station elevation by:

$$\tilde{\zeta}_{STN} = 67.442\, T_0\, log_{10}\left(\frac{1100}{p_{STN}}\right) \tag{A8.3}$$

where p_{STN} is the observed station pressure [hPa], 1100 hPa is the reference pressure (chosen because the sea-level pressure (SLP) never exceeds this value), and the constant 67.442 has units of gpm K^{-1}.

4. Determine *adjusted* geopotential height [gpm] at sea level by:

$$\tilde{\zeta}_{SL} = \tilde{\zeta}_{STN} - \Delta\tilde{\zeta} \tag{A8.4}$$

5. Plug value for $\tilde{\zeta}_{SL}$ into the following, and obtain SLP in hPa by:

$$SLP = (1100)10^{\left(\frac{-\tilde{\zeta}_{SL}}{67.442\, T_0}\right)} \tag{A8.5}$$

SMT Table 48A. *Correction for humidity c, used in determining $\bar{t}_v$ when reducing pressure to sea level.*

Station Elevation [m]	Station Dew-Point Temperature [°C]									
	−28	−26	−24	−22	−20	−18	−16	−14	−12	−10
0	0.1	0.1	0.1	0.1	0.1	0.1	0.2	0.2	0.2	0.3
500	0.1	0.1	0.1	0.1	0.1	0.2	0.2	0.2	0.3	0.3
1000	0.1	0.1	0.1	0.1	0.1	0.2	0.2	0.2	0.3	0.4
1500	0.1	0.1	0.1	0.1	0.2	0.2	0.2	0.3	0.3	0.4
2000	0.1	0.1	0.1	0.1	0.2	0.2	0.3	0.3	0.4	0.4
2500	0.1	0.1	0.1	0.2	0.2	0.2	0.3	0.4	0.4	0.5

Station Elevation [m]	Station Dew-Point Temperature [°C]									
	−8	−6	−4	−2	0	2	4	6	8	10
0	0.3	0.4	0.5	0.6	0.7	0.8	0.9	1.0	1.2	1.3
500	0.4	0.4	0.5	0.6	0.7	0.8	1.0	1.1	1.3	1.5
1000	0.4	0.5	0.6	0.7	0.8	1.0	1.1	1.3	1.5	1.7
1500	0.5	0.6	0.7	0.8	0.9	1.1	1.2	1.4	1.6	1.9
2000	0.5	0.6	0.8	0.9	1.1	1.2	1.4	1.6	1.8	2.1
2500	0.6	0.7	0.9	1.0	1.2	1.4	1.6	1.8	2.1	2.4

Station Elevation [m]	Station Dew-Point Temperature [°C]									
	12	14	16	18	20	22	24	26	28	30
0	1.5	1.7	1.9	2.2	2.5	2.8	3.2	3.6	4.1	4.6
500	1.7	1.9	2.2	2.5	2.8	3.2	3.6	4.0	4.6	5.1
1000	1.9	2.2	2.5	2.8	3.2	3.6	4.0	4.6	5.1	5.8
1500	2.1	2.4	2.8	3.1	3.6	4.0	4.6	5.1	5.8	6.5
2000	2.4	2.7	3.1	3.5	4.0	4.5	5.1	5.8	6.5	7.3
2500	2.7	3.1	3.5	4.0	4.5	5.1	5.8	6.5	7.3	8.2

SMT Table 50. *(Partial) Geometric meters to geopotential meters.*

Geometric Meters [m]	Latitude										
	0°	10°	20°	30°	40°	45°	50°	60°	70°	80°	90°
	[gpm]	[gpm]	[gpm]	[gpm]	[gpm]	[gpm]	[gpm]	[gpm]	[gpm]	[gpm]	[gpm]
1000	998	998	998	999	1000	1000	1001	1002	1002	1003	1003
2000	1995	1996	1997	1998	2000	2001	2002	2003	2005	2006	2006
3000	2993	2993	2994	2997	2999	3000	3002	3004	3007	3008	3008
4000	3989	3990	3992	3995	3998	4000	4002	4005	4008	4010	4011
5000	4986	4987	4989	4993	4997	4999	5002	5006	5009	5012	5012
6000	5982	5983	5986	5990	5995	5998	6001	6006	6010	6013	6014
7000	6978	6979	6983	6987	6993	6997	7000	7006	7011	7014	7015
8000	7974	7975	7979	7984	7991	7995	7999	8006	8011	8015	8016
9000	8969	8971	8975	8981	8989	8993	8997	9005	9011	9015	9017
10,000	9964	9966	9970	9977	9986	9991	9995	10 004	10 011	10 015	10 017
11,000	10 959	10 961	10 966	10 973	10 983	10 988	10 993	11 002	11 010	11 015	11 017
12,000	11 953	11 955	11 961	11 969	11 979	11 985	11 990	12 001	12 009	12 015	12 017
13,000	12 947	12 949	12 955	12 964	12 976	12 982	12 988	12 999	13 008	13 014	13 016
14,000	13 941	13 943	13 950	13 960	13 972	13 978	13 984	13 997	14 006	14 013	14 015
15,000	14 935	14 937	14 944	14 954	14 967	14 974	14 981	14 994	15 004	15 011	15 014
16,000	15 928	15 930	15 938	15 949	15 963	15 970	15 977	15 991	16 002	16 010	16 012
17,000	16 921	16 923	16 931	16 943	16 958	16 965	16 973	16 988	17 000	17 008	17 010
18,000	17 913	17 916	17 924	17 937	17 952	17 960	17 969	17 984	17 997	18 005	18 008
19,000	18 905	18 908	18 917	18 930	18 947	18 955	18 964	18 980	18 994	19 003	19 006
20,000	19 897	19 900	19 909	19 923	19 941	19 950	19 959	19 976	19 990	20 000	20 003
21,000	20 889	20 892	20 902	20 916	20 934	20 944	20 954	20 972	20 987	20 996	21 000

22,000	21 880	21 883	21 893	21 909	21 928	21 938	21 948	21 967	21 982	21 993	21 996
23,000	22 871	22 875	22 885	22 901	22 921	22 931	22 942	22 962	22 978	22 989	22 992
24,000	23 862	23 865	23 876	23 893	23 914	23 925	23 936	23 956	23 973	23 984	23 988
25,000	24 852	24 856	24 867	24 885	24 896	24 918	24 929	24 951	24 968	24 980	24 984
26,000	25 842	25 846	25 858	25 876	25 898	25 910	25 922	25 945	25 963	25 975	25 979
27,000	26 832	26 836	26 848	26 867	26 890	26 903	26 915	26 938	26 957	26 970	26 974
28,000	27 821	27 825	27 838	27 858	27 882	27 895	27 908	27 932	27 952	27 964	27 969
29,000	28 810	28 815	28 828	28 848	28 873	28 886	28 900	28 925	28 945	28 959	28 963
30,000	29 799	29 804	29 817	29 838	29 864	29 878	29 892	29 918	29 939	29 952	29 957
35,000	34 738	34 743	34 759	34 784	34 814	34 830	34 846	34 877	34 901	34 917	34 923
40,000	39 669	39 676	39 694	39 722	39 756	39 775	39 793	39 828	39 856	39 874	39 881
45,000	44 593	44 600	44 621	44 652	44 691	44 712	44 732	44 771	44 803	44 824	44 831
50,000	49 509	49 517	49 540	49 575	49 618	49 641	49 664	49 707	49 742	49 765	49 773
55,000	54 417	54 426	54 451	54 490	54 537	54 562	54 588	54 635	54 674	54 699	54 708
60,000	59 318	59 327	59 355	59 397	59 449	59 476	59 504	59 556	59 598	59 626	59 635
65,000	64 211	64 221	64 251	64 297	64 353	64 383	64 412	64 469	64 514	64 544	64 555
70,000	69 096	69 107	69 139	69 188	69 249	69 281	69 313	69 374	69 423	69 455	69 467
75,000	73 974	73 986	74 020	74 073	74 137	74 172	74 206	74 271	74 324	74 359	74 371
80,000	78 844	78 857	78 893	78 949	79 018	79 055	79 092	79 161	79 218	79 255	79 268
85,000	83 707	83 720	83 759	83 819	83 892	83 931	83 970	84 044	84 104	84 143	84 156
90,000	88 561	88 576	88 617	88 680	88 758	88 799	88 841	88 919	88 982	89 024	89 038
95,000	93 409	93 424	93 467	93 534	93 616	93 660	93 704	93 786	93 853	93 897	93 912

Notes

1   Basic Concepts and Terminology

1 See the bibliography for the complete citation.
2 All real systems are open!
3 Quantum mechanics has found a way around this, through a process known as *quantum tunneling*. Alpha decay of heavy atomic nuclei is impossible from a classic potential energy point of view, but it does in fact occur.
4 "Ideal gas" is defined in Chapter 2.
5 After Swedish scientist Anders Celsius, who initially proposed the scale in 1742.
6 After Dutch scientist Daniel Fahrenheit, who originally proposed the scale in 1724. There are several additional temperature scales, such as the Rankine scale, but they are not usually used in meteorology.
7 As originally defined, although, it's obvious that the important parameter here is the *weight* of the substance, not its chemical composition.
8 Von Baeyer (1999).
9 In this text, we will place a box around equations that are either "key ideas" or referred to in later derivations.
10 This discussion is based closely on Tsonis (2007), with additional input from Boucher (2012).
11 E.g., on the Skew-T diagram published by the U.S. Air Force (WPC 9–16), one square centimeter is equal to 28 Joules per kilogram (J/kg) of air. See Air Weather Service (1990) for more.

2   Equations of State

1 This is overstating the case just a bit. Ice cores and other proxy records indicate that the two major gases, N_2 and O_2, have been at fairly stable concentrations for a very long time – a billion years or more.
2 North and Erukhimova (2009).
3 Tsonis (2007).
4 After Dutch scientist J. H. van der Waals.
5 Thall (2012); Tsonis (2007).
6 This discussion closely follows that of Tsonis (2007).
7 The same relation was independently discovered by John Dalton in 1801.
8 Although this has been called into question. According to Barnett (1941), French scientist Guillaume Amontons described the relationship in 1700.
9 Loosely based on Tsonis (2007).

10 See Chapter 1, Section 1.3.
11 After Austrian physicist Ludwig Boltzmann.
12 This discussion is loosely based on Espinola (1994).
13 After English meteorologist John Dalton.
14 This is the same result noted in Tsonis (2007), as well as Rogers and Yau (1989).
15 This is refined to three decimals places in Chapter 10.

3   Work, Heat, and Temperature

1 Eighteenth-century American-born physicist and inventor, who served with the loyalist army during the American Revolution.
2 Recall your basic calculus: The *dot* (•) product of two vectors produces a *scalar* result in a direction common to both original vectors; the *cross* product (x) of two vectors produces a *vector* result perpendicular to both original vectors.
3 A fluid with no internal resistance to movement. This is a pretty good approximation for air.
4 R. H. Fowler, twentieth-century British mathematician and physicist.
5 Espinola (1994).
6 RMS velocity defined by $v = \sqrt{\sum_{i=1}^{N}\left(v_i\right)^2}$, where v_i are the individual velocities of particles 1 – N.
7 Parker (1993).
8 This discussion is based in part on North and Erukhimova (2009).
9 There has been some refinement to this number since Fahrenheit's original work. The currently understood value for "normal" mean body temperature (measured orally) is 98.6°F, which corresponds to 37.0°C. In the former case, normal includes a variation of about 1°F about this mean. For the latter, normal includes a variation of about ½°C.
10 Espinola (1994).

4   The First Law of Thermodynamics

1 After James Prescott Joule.
2 North and Erukhimova (2009).
3 Internal energy and Enthalpy, as well as *Helmholtz Free Energy* and *Gibbs Free Energy* are discussed further in Appendix 4.
4 The chain rule states that $d(xy) = xdy + ydx$.
5 Specific heats for *liquid* water and ice (*solid* water) are discussed in Chapter 3.
6 We can actually add two more to this list: *Superconductor* (a solid state distinct from ordinary solids), which occurs with some substances, and *plasma* (a superheated, electrically charged gaseous state), which occurs with all substances. These don't occur at ordinary terrestrial temperatures. There is also an ambiguous state, simply called *fluid*, that occurs at extremely high temperatures and pressures, but this is also not a factor in the Earth system. We'll discuss fluids a bit more later.
7 Contrary to popular belief, liquid water is compressible. The density of sea water in the abyss is greater than at the surface, in part because it's cold and salty, and in part because it's under tremendous pressure.
8 You'll learn more about this when you take a course in cloud microphysics. Also see Rogers and Yau (1996).
9 Don't confuse the terms in this equation with internal energy and work. $c_p dt$ does *not* equal the change in internal energy (du), and αdp does *not* equal work (dw).

5 Adiabatic Processes

1 Glickman (2000).
2 See *sign conventions for work* in Chapter 3.
3 The specific heat at constant pressure (c_p) is *approximately* constant between 0°C and 100°C. See Chapter 4.
4 The logarithm with a base of e (the natural number), i.e., $log_e(x)$, or the inverse function of e^x, also written as exp(x). Performing exp[$ln(x)$] yields the value of x. (Now you know why we made you take that class.)
5 Use the inverse function of ln(x).
6 See Table 3.3 in Chapter 3.
7 After Siméon Poisson, French physicist and mathematician.
8 See Fujita (1986).

6 The Second Law of Thermodynamics

1 Glickman (2000); North and Erukhimova (2009).
2 One described using the algebraic approximation $\Delta T/\Delta x$, rather than the calculus-based infinitesimal value dT/dx.
3 In classical physics. Some "impossible" processes are not impossible from a quantum mechanical point of view. Instead, they are simply *extremely unlikely*. One important example is alpha decay; a form of radioactivity.
4 Ludwig Boltzmann, Austrian physicist.
5 More recently, entropy has been redefined as the *dispersal of energy*.
6 This doesn't violate our rule that entropy must always increase, because for adiabatic process to occur, we must assume an isolated parcel. There are no *real* isolated parcels – this is a mathematical construct.
7 See Chapter 5.
8 See Chapter 1.
9 Approximate mean value for specific heat of ice between −20°C and 0°C. See Table 3.1.
10 This is the latent heat of fusion. See Table 4.2.
11 See Chapter 4, discussion following equation (4.65).
12 Approximate mean value for specific heat of liquid water between 0°C and 100°C. See Table 3.1.
13 This is the latent heat of vaporization at 100°C. See discussion after Equation (4.65).
14 This also explains why it only takes about half as much heat to raise the temperature of ice as it does to raise the temperature of liquid water by the same amount. Notice the difference in the specific heats of ice and liquid water.
15 Additional contributions were made later by Benoit Clapeyron.
16 Actually, (4.50) states that $nR* = mR_i$, but because we know that the "individual gas" in this case is air, we can substitute R_d for R_i.
17 See Chapter 3, Section 3.2, and Figure 3.4.
18 Originally R_i, the generic individual gas constant, but is replaced here with the R_d, the gas constant for the dry air ensemble. See Chapter 2.
19 This is the environmental temperature at 250 hPa, if we begin with a temperature of 15°C, and assume that the atmosphere cools at the dry adiabatic rate between 1000 and 250 hPa. It's not usually this cold at the tropopause, so this example overstates the efficiency by several percent.

7 Water Vapor and Phase Transitions

1 Named for Rudolf Classius and Benoit Clapeyron.
2 See Chapter 1, Section 1.3.

3 This derivation is partially adapted from Hess (1959).

4 The vapor pressure above liquid water is known (from laboratory experiments) to increase as the temperature of the water increases. The volume of liquid water also increases as its temperature increases. Once the water is entirely in vapor form, any additional heat causes its volume to greatly expand, while the pressure from the widely dispersed water vapor molecules drops off. Compare to Figure 7.4

5 State variables are pressure, temperature, and volume. See Chapter 1, Section 1.4.

6 The Ideal Gas Law ($pV = mR_iT$) states that, if p decreases sufficiently while V increases, then T must also decrease. See Chapter 2.

7 See Chapter 1, Section 1.9, and Equation (1.21).

8 See Table 4.2.

9 This is anomalous. For most substances, the solid form is denser than the liquid form, i.e., $\alpha_2 > \alpha_1$. Water is different than most other substances in this respect.

10 See Appendix 5 for more.

11 The atmosphere of Jupiter, as much as 65,000 km deep, is sometimes described as an "ocean" somewhere below the tops of the visible clouds. This ocean must consist of compressed Jovian atmosphere (at very high temperatures and pressures), forcing it past the critical point. The same is true of Saturn. Superfluid water occurs on Earth at mid-ocean hydrothermal vents, where the heat of the subsurface molten rock heats the water past the critical temperature, and the pressure from the overlying ocean keeps the water at pressures above the critical point.

12 See Wexler (1976), Wexler (1977), and Buck (1981) for additional forms.

13 This equation can be derived analytically be referring to Rogers and Yau (1989), combining equations (2.10) and (2.14) in their text, and integrating the result. It is also listed in Brock and Richardson (2001), in a slightly different form.

14 See Chapter 4, Table 4.2.

15 See Chapter 2, Equation (2.68).

16 See Chapter 3, Table 3.1.

17 See Chapter 4, Equation (4.64).

18 Rogers and Yau (1989).

19 Brock and Richardson (2001).

20 See Chapter 2, Equation (2.56).

21 Brock and Richardson (2001) and Buck (1981).

22 American Meteorological Society (2012).

23 See Chapter 1, Table 1.4.

24 Ibid.

25 Or computed using (4.66).

8 Moisture Considerations: Effects on Temperature

1 For example, N_2, which has a molecular weight of 28.02 kg/kmol, or O_2, which has a molecular weight of 32.00 kg/kmol. See Table 2.1.

2 North and Erukhimova (2009).

3 Air Weather Service (1990).

4 National Aeronautics and Space Administration (1966). This U.S. Standard Atmosphere is discussed in greater detail in Chapter 10.

5 See Chapter 4, Table 4.2.

6 See Chapter 4, Table 4.1.

7 Rogers and Yau (1989).

8 Hess (1959).

9 List (1963).

10 Rogers and Yau (1989).

11 See Chapter 5, and McIlveen (1992).
12 They are *equal* if the atmosphere is saturated.
13 See Chapter 2.
14 Hess (1959).
15 Ibid.
16 Rogers and Yau (1989).
17 See Chapter 5.
18 Hess (1959).

9   Atmospheric Statics

1 Jupiter rotates on its axis in slightly less than ten hours, causing its equatorial bulge to be much more pronounced than Earth's. Jupiter's polar radius is 66,854 km, but its equatorial radius is 71,492 km, which is about 6.94 percent greater than its polar radius. This is more than twenty times greater than the percentage difference between Earth's equatorial and polar radii.
2 Purists may not like the term "centrifugal acceleration," but will insist that the outward-pointing "centrifugal force" is proportional to the inward-pointing "centripetal acceleration." In this case, centrifugal acceleration is used to indicate outward-pointing centrifugal force divided by the mass of the object it is acting on.
3 Holton (2004).
4 The *sidereal day* is 23 hr, 56 min, 4.091 s, which is the time it takes the Earth to make a complete rotation relative to the background stars.
5 There are still a few small corrections that can be made to improve this analysis even further, but this model is complete enough for our purposes here. See Holton (2004) for more.
6 Hess (1959).
7 Holton (2004).
8 Holton (2004).
9 See Table 4.2.
10 See Equation (2.68).
11 This reference point is often used as the dividing line between liquid and frozen precipitation.
12 See Section 9.2, this chapter.
13 See Section 9.2.
14 See Chapter 1, (1.9)–(1.11), for a unit analysis.
15 Precipitation processes are the subject of a more advanced course, usually called Atmospheric Physics or Cloud Physics. See Wallace and Hobbs (2006) for more.
16 See Chapter 1.
17 See Bourgouin (2000).
18 Assuming a Northern Hemisphere, Midlatitude station.
19 See Chapter 4.
20 Stability is discussed in detail in Chapter 11.
21 Well, not always. There's another form of freezing drizzle that can occur with much lower temperatures. We're putting that aside for now to keep this manageable.
22 Mathworks (2012).
23 One standard deviation ($\pm 1\sigma$) encompasses only about 68 percent of a "normally" (Gaussian, or "bell curve") distributed sample. We have to go to $\pm 3\sigma$ to encompass more than 99 percent of the sample.
24 A "sounder" is a specialized form of *radiometer*, or quantitative camera, recording upwelling radiance at many wavelengths, and connected to a telescope. An "imager" is another kind of specialized radiometer. See Kidder and Vonder Harr (1995) for more.
25 "Reduction" refers to the hypothetical reduction in altitude, which corresponds to an increase in pressure.

26 See Appendix 8.
27 See Appendix 8.
28 These numbers refer to the visible cloud tops.

10   Model and Standard Atmospheres

1 Stability is covered in detail in Chapter 11.
2 Probably the second semester of a four-semester sequence.
3 See Chapter 9, Section 9.5.
4 Approximately 1.03 hPa for each 10 meters of elevation above sea level. See (10.64).
5 See (1.4f) for conversion.
6 National Aeronautics and Space Administration, 1962; 1966; 1976.
7 The following discussion is not exhaustive, but it's enough for our purposes here. If you want to delve into these references in more detail, it's easy to find the 1962, 1966, and 1976 *U.S. Standard Atmosphere* publications as PDF files on the Internet.
8 Note that this is referenced to a point 0.01 K below the standard melting point of 273.16 K. The reason for this choice is explained in National Aeronautics and Space Administration (1976).
9 The turbopause is approximately 85 km above sea level, and separates the homosphere (below) from the heterosphere (above). In the homosphere, the major (long residence-time) constituents of the atmosphere are well mixed, that is, don't vary noticeably from place to place or over ordinary meteorological time scales. This isn't true in the heterosphere. (See American Meteorological Society, 2012, for more.)
10 This small difference also propagates into computed values of c_p and c_v, but, once again, the differences are on the order of a tenth of a percent, so we'll stick with the values obtained in Chapter 4.

11   Stability

1 In the atmosphere, a moving boundary separating two air masses of differing densities (i.e., a *front*) is one form of external lifting force. Another is a topographic barrier, such as a mountain, which forces air upward through mass conservation.
2 See Chapter 10, section 10.2.
3 This value was computed using a value of the Universal Gas Constant (R^*) equal to 8,310 J/kmol and a mean molecular mass of the dry-air ensemble of 28.97 kg/kmol. The latter was determined using the four most important (by mass) gases in the ensemble. See Chapter 2, Equations (2.62)–(2.64). The USSA and ISA report a value of about 287.1 kg/kmol, which differs from the former value by about a tenth of a percent. See Chapter 10, Equation (10.45).
4 Holton (2004).
5 Recall definitions of open, closed, and isolated parcels; see Chapter 1.
6 See Rogers and Yau (1989).
7 See Chapter 10, Tables 10.1 and 10.2.
8 See Chapter 9, Table 9.1.
9 See Chapter 4, relation (4.61), or Table 4.1.
10 See Chapter 9, Table 9.1.
11 See Chapter 4, relation (4.62), or Table 4.1.
12 Temperature-dependent variations occur in c_v and c_p because of the increasing importance of vibrational degrees of freedom (f) at progressively higher temperatures. See Chapter 4.
13 There is also a pressure dependency. See Table 8.3.
14 See Chapter 10, section 10.3.
15 See Chapter 10, section 10.2, describing the homogeneous atmosphere, i.e., one with density independent of altitude, and its corresponding lapse rate, computed in (10.17). Provided the

lapse rate is less than 34.2 k/km, the density decreases with height, and the atmosphere will not autoconvect.

16 Air Weather Service (1990).

17 See Chapter 6, (6.21).

18 See Chapter 9, Figure 9.3.

19 Air Weather Service (1990).

20 Showalter (1947); Air Weather Service (1990); Elliott (1988).

21 National Weather Service (2012a).

22 Reported wind "speed" is the 1-min average of instantaneous wind speeds recorded by an anemometer, and wind "gusts" are the highest of the instantaneous values over the previous ten minutes, provided there has been at least a 10 kt (5 m/s) difference between the minimum and maximum instantaneous values (Office of Federal Coordinator for Meteorology, 1982).

23 Storm Prediction Center (2012).

24 Galway (1956); Air Weather Service (1990); Elliott (1988).

25 See Elliott (1988) for more details and additional T_{max} techniques. Elliott also discusses techniques for forecasting *minimum* temperature (T_{min}).

26 In North America, "CLR" is used at automated stations (without a human weather observer), and indicates "clear below 12,000 FT (3700 m) AGL." Outside North America, "CLR" indicates the absence of clouds at any altitude. In North America, "SKC" is used at stations *with* human weather observers and indicates the absence of clouds at any altitude.

27 Approximately equal to 1,000 m; see Chapter 10, Equation (10.63) and Table 10.2.

28 Air Weather Service (1990).

29 PSU Weather Center (2012a).

30 Air Weather Service (1990); Elliott (1988).

31 Air Weather Service (1990); Elliott (1988).

32 *Storm-Relative Helicity* (*SRH*) – the influence of the horizontal component of vorticity – is a quantitative measure of this, and we will return to it in greater detail in Chapter 12.

33 *Veering* means the direction rotates clockwise; *backing* means it rotates counterclockwise. Veering with height implies warm-air advection (WAA); backing with height implies cold-air advection (CAA).

12 Severe Weather Applications

1 The Severe Weather Threat index, or SW.

2 The first stability indexes were developed in the 1940s and 1950s, before the advent of inexpensive, fast digital computers. The Skew-T was the "graphical computer" developed to fill this gap, with indexes acting as some of the "algorithms" you could run on the "computer."

3 Using the Ideal Gas Law. See (8.7).

4 See Chapter 1. On the USAF's Skew-T diagram (WPC 9–16), one square centimeter is equal to 28 Joules per kilogram (J/kg) of air. See Air Weather Service (1990) for more.

5 See Chapter 1, Section 1.2.

6 For kinetic energy, there's an additional factor of ½.

7 See Chapter 8, Section 8.2.

8 The pressure units actually cancel, so the choice of units is not as critical as in other places, as long as they are consistent. I recommend sticking with the correct physical units as a matter of good practice.

9 See Chapter 9, Table 9.1.

10 See Chapter 11, Figure 11.7.

11 See Chapter 11, Figures 11.9 and 11.10.

12 Air Weather Service (1990).

13 Air Weather Service (1990).

14 See Chapter 11, (11.26) and (11.28).

15 See Rogers and Yau (1989) for more. The physics of convection are usually discussed in greater detail in a more advanced course, sometimes called Cloud Physics or Atmospheric Physics.

16 Air Weather Service (1990). For a slightly improved method, the average dew point in the lowest 100 hPa (~1000 m) can be estimated using the equal area method shown in Figure 11.20. The average dew point is then lifted along the mean mixing ratio until it intersects the environmental temperature profile.

17 A thunderstorm with a persistent, rotating updraft, called a *mesocyclone*, and persisting for longer than it takes a parcel near Earth's surface to reach the top of the thunderstorm (American Meteorological Society, 2014). While these storms are a minority in terms of the total number of thunderstorms occurring in the world, supercells are responsible for the large majority of death, injury, and property damage caused by thunderstorms (Doswell, 1991).

18 See Chapter 11, discussion following Table 11.3.

19 Mesocyclones are usually between about 2 and 4 km in diameter, and are often visible as a *velocity couplet* in Doppler-enabled weather RADAR displays. Velocity couplets represent *inbound* and *outbound* velocities (implying cyclonic rotation), as measured with respect to the RADAR's position, in close proximity to each other. The tornado is an order of magnitude smaller in the horizontal dimension and is not usually visible in data collected by operational weather RADAR.

20 See Doswell (1991) for a more detailed discussion.

21 See Holton (2004).

22 Doswell (1991).

23 VAD: Velocity Azimuth Display. See Warning Decision Training Branch (2005).

24 AWIPS: Advanced Weather Interactive Processing System. This is the primary work-station software used by forecasters working in field offices of the U.S. National Weather Service. See National Weather Service (2012b).

25 FX-Net is an application that reproduces some of the data-processing and visualization capability of AWIPS. See National Oceanic and Atmospheric Administration (2012).

26 National Weather Service (2012a).

27 A common condition, because of the frictional drag imposed on the wind flow by the surface of the Earth. Simple conservation arguments imply that the wind speed must go to zero at an elevation of zero, but this is not always the case in the real atmosphere, at least at scales larger than microscale.

28 You can visualize the axis of rotation using your right hand. Wrap your fingers in the direction of the rotation, and your thumb will point in the direction of the rotational axis. This is often called the *Right-Hand Rule.*

29 Holton (2004).

30 To compute the *absolute* vorticity, the *planetary vorticity* (proportional to the *Coriolis parameter*) must be added to the relative vorticity computed here. See Holton (2004) for more.

31 Don't confuse this with geopotential height (ζ), discussed in Chapter 9.

32 Doswell (1991).

33 Similar to standard supercells, but smaller in both the horizontal and vertical dimensions.

34 Newton (1963).

35 See Elliott (1988), section 9.

36 See Chapter 11, (11.43).

37 Fujita (F) intensity scale.

38 This exercise is based on one originally developed by Aviles (2012).

Appendix 4 Additional Thermodynamic Functions: Helmholtz Free Energy and Gibbs Free Energy

1 In this appendix, we'll start with the "closed" forms of these functions for clarity.
2 Named for nineteenth-century German physicist Hermann von Helmholtz.
3 See Tsonis (2007) and Espinola (1994).
4 Named for American physicist Josiah W. Gibbs, who lived from the mid-nineteenth century to the early twentieth century.
5 See Tsonis (2007) and North and Erukhimova (2009).

Appendix 5 Computed Boiling Temperatures at Different Pressures

1 Lide (1997).

Appendix 6 Variation of Saturation Mixing Ratio with Temperature and Pressure

1 See Bluestein (1992), p. 383, for similar derivations.

Appendix 7 Derivation of the Moist Adiabatic Lapse Rate

1 See (7.66). The direct relationship holds because the saturation vapor pressure is directly dependent on temperature.

Appendix 8 Sea-Level Pressure Calculations Using the Smithsonian Meteorological Tables

1 List (1963).

Bibliography

Air Weather Service. 1990: *The Use of the Skew-T Diagram in Analysis and Forecasting* (AWS TR-79/006, rev.), Air Weather Service (MAC).

American Meteorological Society Glossary of Meteorology. Cited 2012: Precipitable Water. Available online at http://amsglossary.allenpress.com/glossary/search?id=precipitable-water1.

  Cited 2014: Supercell. Available online at http://glossary.ametsoc.org/wiki/Supercell.

Aviles, L. 2012: Personal communication.

Barnett, M. K. 1941: "A Brief History of Thermometry," *Journal of Chemical Education* 18 (8): 358–64.

Bluestein, H. W. 1992: *Synoptic-Dynamic Meteorology in Midlatitudes, Vol .1: Principles of Kinematics and Dynamics*, Oxford University Press.

Bolton, D. 1980: "The Computation of Equivalent Potential Temperature," *Mon. Wea. Rev.* 108: 1046–53.

Boucher, T. R. 2012: Personal communication.

Bourgouin, P. 2000: "A Method to Determine Precipitation Types," *Weather and Forecasting* 15: 583–92

Brock, F. V., and S. J. Richardson. 2001: *Meteorological Measurement Systems*, Oxford University Press.

Buck, A. L. 1981: "New Equations for Computing Vapor Pressure and Enhancement Factor," *Journal of Applied Meteorology* 20: 1527–32.

Cox, R. A., and N. D. Smith. 1959: "The Specific Heat of Sea Water," *Proceedings of the Royal Society of London, Series A, Mathematical and Physical Sciences* 252 (1268): 51–62.

Craven, J. P., R. E Jewell, and H. E. Brooks. 2002: "Comparison between Observed Convective Cloud-Base Heights and Lifting Condensation Level for Two Different Lifted Parcels," *Weather and Forecasting* 17: 885–90.

Doswell, C. A. 1991: "A Review for Forecasters on the Application of Hodographs to Forecasting Thunderstorms," *National Weather Digest* 16 (1): 2–16.

Edwards, R., and R. L. Thompson, 2000: RUC-2 Supercell Proximity Soundings, Part II: An Independent Assessment of Supercell Forecast Parameters. Adapted from Preprints, 20th AMS Conference on Severe Local Storms, Orlando, Sep 2000. Available on-line at http://www.spc.noaa.gov/publications/edwards/part2.htm.

Elliott, G. 1988: *Weather Forecasting – Rules, Techniques and Procedures*, American Press.

The Engineering Toolbox. Cited 2012: Air Composition. Available online at http://www.engineeringtoolbox.com/air-composition-d_212.html.

Cited 2012: Ice – Thermal Properties. Available online at http://www.engineeringtoolbox.com/ice-thermal-properties-d_576.html.

Cited 2012: Water – Thermal Properties. Available online at http://www.engineeringtoolbox.com/water-thermal-properties-d_162.html.

Espinola, T. P. 1994: *Introduction to Thermophysics*, Wm. C. Brown Publishers.

Fujita, T. T. 1986: "Mesoscale Classifications: Their History and Their Application to Forecasting," *Mesoscale Meteorology and Forecasting*, ed. P. S. Ray, American Meteorological Society, 18–35.

Galway, J. G. 1956: "The Lifted Index as a Predictor of Latent Instability," *Bull. Amer. Meteor. Soc.*: 528–9.

Glickman, T. S. 2000: *Glossary of Meteorology*, 2nd ed., American Meteorological Society.

Hess, S. L. 1959: *Introduction to Theoretical Meteorology*, Krieger Publishing Company.

Hilsenrath, J., C. W. Beckett, W. S. Benedict, L. Fano, H. J. Hoge, J. F. Masi, R. L. Nuttall, Y. S. Touloukian, and H. W. Woolley, 1955: *Tables of Thermal Properties of Gases (National Bureau of Standards Circular No. 564)*, U.S. Department of Commerce. Accessed through Ohio University, Russ College of Engineering and Technology. Cited 2012. Available online at http://www.ohio.edu/mechanical/thermo/property_tables/air/air_Cp_Cv.html.

Holton, J. R. 2004: *An Introduction to Dynamic Meteorology*, 4th ed., Elsevier Academic Press.

Hydrometeorological Prediction Center (National Oceanic and Atmospheric Administration). Cited 2012: Daily Weather Maps. Available online at http://www.hpc.ncep.noaa.gov/dailywxmap/index.html.

Johns, R. H., J. M. Davies, and P. W. Leftwich, 1990: An Examination of the Relationship of 0-2 km agl "Positive" Wind Shear to Potential Buoyant Energy in Strong and Violent Tornado Situations. 16th Conf on Severe Local Storms, Kananaskis Park, Alberta, 593–8.

Kerr, B. W., and G. L. Darkow. 1996: "Storm-Relative Winds and Helicity in Tornadic Thunderstorm Environments," *Weather and Forecasting* 11: 489–505.

Kidder, S. Q., and T. H. Vonder Harr. 1995: *Satellite Meteorology – An Introduction*, Academic Press.

Kluge, S. Cited 2012: cloudbasechart.pdf. Available online at http://stevekluge.com/geoscience/regentses/labs/cloudbasechart.pdf.

Lide, D. R. 1997: *CRC Handbook of Chemistry and Physics*, 78th ed., CRC Press.

List, R. J. 1963: *Smithsonian Meteorological Tables*, 6th rev. ed., 2nd repr., Smithsonian Institution Press.

Mathworks, cited 2012: Available on-line at www.mathworks.com.

McIlveen, J. F. R. 1992: *Fundamentals of Weather and Climate*, Stanley Thornes Publishers, Ltd.

Miller, R. C. 1975: *Notes on Analysis and Severe-Storm Forecasting Procedures of the Air Force Global Weather Central* (AWS TR-200, rev.), Air Weather Service (MAC).

National Aeronautics and Space Administration. 1962: *U.S. Standard Atmosphere 1962*, U.S. Government Printing Office. Available online at http://www.dtic.mil/cgi-bin/GetTRDoc?Location=U2&doc=GetTRDoc.pdf&AD=AD0659893.

1966: *U.S. Standard Atmosphere Supplement 1966*, U.S. Government Printing Office. Available online at http://www.dtic.mil/cgi-bin/GetTRDoc?Location=U2&doc=GetTRDoc.pdf&AD=AD0659543.

1976: *U.S. Standard Atmosphere 1976*, U.S. Government Printing Office. Available online at http://ntrs.nasa.gov/archive/nasa/casi.ntrs.nasa.gov/19770009539_1977009539.pdf.

National Oceanic and Atmospheric Administration. Cited 2012: Earth System Research Laboratory. Available online at http://fx-net.noaa.gov/.

National Weather Service. Cited 2012: Present Weather Symbols. Available online at http://www.srh.noaa.gov/jetstream/synoptic/ww_symbols.htm.

   Cited 2012 (a): Glossary. Available online at http://www.weather.gov/glossary/.

   Cited 2012 (b): Field Systems Operations Center Test and Evaluation Branch (OPS24). Available online at http://www.nws.noaa.gov/ops2/ops24/awips.htm.

Newton, C. W. 1963: "Dynamics of Severe Convective Storms," *Meteor. Monogr.* 5 (27): 33–58.

North, G. R., and T. L. Erukhimova, 2009: *Atmospheric Thermodynamics – Elementary Physics and Chemistry*, Cambridge University Press.

Office of Federal Coordinator for Meteorology, 1982: *Surface Weather Observations and Reports, Federal Meteorological Handbook No. 1*, Office of Federal Coordinator for Meteorology.

Parker, S. P., ed. 1993: *McGraw-Hill Encyclopedia of Physics*, McGraw-Hill Companies.

Plymouth State Weather Center. Cited 2012: Base Diagram for Skew-T Log P. Available online at http://vortex.plymouth.edu/chart.html.

   Cited 2012 (a): Sounding data for KDDC, April 15, 2012, 0000 UTC. Available online at http://vortex.plymouth.edu/cgi-bin/gen_uacalplt-u.cgi?id=KDDC&pl=out1b&yy=12&mm=04&dd=15&hh=00&pt=parcel&size=640x480.

Rasmussen, E. N. 2001: "Refined Supercell and Tornado Forecast Parameters from the 1992 Baseline Climatology, *noted in* Warning Decision Training Branch (WDTB) Professional Development Series – Severe Convective Forecasting and Warnings." Available online at http://www.wdtb.noaa.gove/resources/PDS/PCU5_References.html.

Rasmussen, E. N., and D. Blanchard. 1998: "A Baseline Climatology of Sounding-Derived Supercell and Tornado Forecast Parameters," *Weather and Forecasting* 13: 1148–64.

Rogers, R. R., and M. K. Yau. 1989: *A Short Course in Cloud Physics*, 3rd ed., Butterworth-Heinemann.

Showalter, A. K. 1947: "A Stability Index for Forecasting Thunderstorms." *Bull. Amer. Meteor. Soc.* 34: 250–2.

State University of New York. 1994: "Earth Science Reference Tables," State Education Department.

Storm Prediction Center. Cited 2012: Storm Reports. Available online at http://www.spc.noaa.gov/climo/reports/120414_rpts.html.

Thall, E. Cited 2012: Thall's History of Gas Laws. Available online at http://web.fccj.org/~ethall/gaslaw/gaslaw.htm.

Tsonis, A. A. 2007: *An Introduction to Atmospheric Thermodynamics*, 2nd ed., Cambridge University Press.

Von Baeyer, H. C., 1999: *Warmth Disperses and Time Passes – A History of Heat*, The Modern Library.

Wallace, J. H., and P. V. Hobbs, 2006: *Atmospheric Science – An Introductory Survey*, 2nd ed., Academic Press.

Warning Decision Training Branch. 2005: *Distance Learning Operations Course*, National Oceanic and Atmospheric Administration, Norman, OK, Version 0409.

Weast, R. C., ed. *CRC Handbook of Chemistry and Physics*, 53rd ed., Chemical Rubber Co.

Wexler, A. 1976: "Vapor Pressure Formulation for Water in Range from 0 °C to 100 °C. A Revision." *J. Res. Nat. Bur. Stand.* 80A: 775–85.

   1977: "Vapor Pressure Formulation for Ice." *J. Res. Nat. Bur. Stand.* 81A: 5–20.

Wikipedia. Cited 2012: Image of Sling Psychrometer. Available online at http://en.wikipedia.org/wiki/File:Sling_psychrometer.JPG.

   Cited 2012: List of Weather Records. Available online at http://en.wikipedia.org/wiki/Temperature_extremes.

Index

acceleration, vertical, as a function of density, 258
acceleration, vertical, as a function of specific volume, 261
acceleration, vertical, as a function of virtual temperature, 258
adiabatic process (defined), 86
altimeter setting, 235
atmosphere, absolutely stable, 272
atmosphere, absolutely unstable, 265
atmosphere, conditionally stable, 269
atmosphere, constant lapse-rate, 233
atmosphere, dry neutral, 267
atmosphere, homogeneous, 228
atmosphere, isothermal, 231
atmosphere, moist neutral, 271
atmospheres, model, 228
Avogadro's Number, 11

boiling (defined), 157
Boltzmann's Constant, 34
Boyle's Law, 27
Brunt–Väisälä Frequency, 100

caloric, 2
calorie, 52
CAPE and SRH, combining, 322
Carnot Cycle for an atmospheric parcel, 123
Carnot Cycle (defined), 122
Carnot Cycle, efficiency, 136
Carnot Cycle, heat transferred, 130
Carnot Cycle, work performed, 128
Classius-Clapeyron Equation (advanced form), 158
Classius-Clapeyron Equation (derivation), 145
Classius-Clapeyron Equation (integrated form), 151
composition of air, 36t.2.1.
compound system, 5
convective condensation level, 301
convective inhibition, 307
convectively available potential energy, 307
Cross Totals Index, 292

Dalton's Law, 37
degrees of freedom, 56
derived quantities and units, 8

diabatic process (defined), 86
Dry Adiabatic Lapse Rate (defined), 103

energy (defined), 4
energy density, 306
energy, potential, in air column, 303
enthalpy, 67
entropy (defined), 114
entropy, net changes with systems and environment, 121
entropy, variations with adiabatic processes, 115
entropy, variations with diabatic temperature increases, 118
entropy, variations with irreversible processes, 118
entropy, variations with phase changes, 119
entropy, variations with reversible processes, 115
equation of state, Van der Waals', 35
equations of state, 26
equations of state, power series, 36
equilibrium, 4
equilibrium level, 301
equipartition principle, 56
external energy, 4
external equilibrium, 4

First Law of Thermodynamics, 65
fundamental quantities and units, 6

Gay-Lussac's gas laws, 27
geopotential, 198
geopotential height, 200
geopotential surfaces, 200
Gibbs Free Energy, 342
gravity (defined), 12
gravity, effective, 193
gravity, Newtonian, 193

heat capacity, 52
heat (defined), 50
heat, extensive, 61
heat, latent, 77
heat, specific, 51
helicity, 314
Helmholtz Free Energy, 342

Printed in the United States
by Baker & Taylor Publisher Services